Lecture Notes in Control and Information Sciences

Edited by A. V. Balakrishnan and M. Thoma

Vol. 1: Distributed Parameter Systems: Modelling and Identification
Proceedings of the IFIP Working Conference, Rome, Italy, June 21–26, 1976
Edited by A. Ruberti
V, 458 pages. 1978

Vol. 2: New Trends in Systems Analysis
International Symposium, Versailles, December 13–17, 1976
Edited by A. Bensoussan and J. L. Lions
VII, 759 pages. 1977

Vol. 3: Differential Games and Applications
Proceedings of a Workshop, Enschede, Netherlands, March 16–25, 1977
Edited by P. Hagedorn, H. W. Knobloch, and G. J. Olsder
XII, 236 pages. 1977

Vol. 4: M. A. Crane, A. J. Lemoine
An Introduction to the Regenerative Method for Simulation Analysis
VII, 111 pages. 1977

Vol. 5: David J. Clements, Brian D. O. Anderson
Singular Optimal Control: The Linear Quadratic Problem
V, 93 pages. 1978

Vol. 6: Optimization Techniques
Proceedings of the 8th IFIP Conference on Optimization Techniques, Würzburg, September 5–9, 1977
Part 1
Edited by J. Stoer
XIII, 528 pages. 1978

Vol. 7: Optimization Techniques
Proceedings of the 8th IFIP Conference on Optimization Techniques, Würzburg, September 5–9, 1977
Part 2
Edited by J. Stoer
XIII, 512 pages. 1978

Vol. 8: R. F. Curtain, A. J. Pritchard
Infinite Dimensional Linear Systems Theory
VII, 298 pages. 1978

Vol. 9: Y. M. El-Fattah, C. Foulard
Learning Systems:
Decision, Simulation, and Control
VII, 119 pages. 1978

Vol. 10: J. M. Maciejowski
The Modelling of Systems with Small Observation Sets
VII, 241 pages. 1978

Vol. 11: Y. Sawaragi, T. Soeda, S. Omatu
Modelling, Estimation, and Their Applications for Distributed Parameter Systems
VI, 269 pages. 1978

Vol. 12: I. Postlethwaite, A. G. J. McFarlane
A Complex Variable Approach to the Analysis of Linear Multivariable Feedback Systems
IV, 177 pages. 1979

Vol. 13: E. D. Sontag
Polynomial Response Maps
VIII, 168 pages. 1979

Vol. 14: International Symposium on Systems Optimization and Analysis
Rocquentcourt, December 11–13, 1978;
IRIA LABORIA
Edited by A. Bensoussan and J. Lions
VIII, 332 pages. 1979

Vol. 15: Semi-Infinite Programming
Proceedings of a Workshop, Bad Honnef, August 30 – September 1, 1978
V, 180 pages. 1979

Vol. 16: Stochastic Control Theory and Stochastic Differential Systems
Proceedings of a Workshop of the „Sonderforschungsbereich 72 der Deutschen Forschungsgemeinschaft an der Universität Bonn"
which took place in January 1979 at Bad Honnef
VIII, 615 pages. 1979

Vol. 17: O. I. Franksen, P. Falster, F. J. Evans
Qualitative Aspects of Large Scale Systems
Developing Design Rules Using APL
XII, 119 pages. 1979

Vol. 18: Modelling and Optimization of Complex Systems
Proceedings of the IFIP-TC 7 Working Conference
Novosibirsk, USSR, 3–9 July, 1978
Edited by G. I. Marchuk
VI, 293 pages. 1979

Vol. 19: Global and Large Scale System Models
Proceedings of the Center for Advanced Studies (CAS)
International Summer Seminar
Dubrovnik, Yugoslavia, August 21–26, 1978
Edited by B. Lazarević
VIII, 232 pages. 1979

Vol. 20: B. Egardt
Stability of Adaptive Controllers
V, 158 pages. 1979

Vol. 21: Martin B. Zarrop
Optimal Experiment Design for Dynamic System Identification
X, 197 pages. 1979

For further listing of published volumes please turn over to inside of back cover.

Lecture Notes in
Control and
Information Sciences

Edited by A.V. Balakrishnan and M. Thoma

54

Control Theory for Distributed Parameter Systems and Applications

Edited by
F. Kappel, K. Kunisch, W. Schappacher

Springer-Verlag Berlin Heidelberg GmbH 1983

Editors
Franz Kappel
Institut für Mathematik
Universität Graz
Elisabethstraße 16
A-8010 Graz, Austria

Karl Kunisch
Institut für Mathematik
Technische Universität Graz
Kopernikusgasse 24
A-8010 Graz, Austria

Wilhelm Schappacher
Institut für Mathematik
Universität Graz
Elisabethstraße 16
A-8010 Graz, Austria

AMS Subject Classifications (1980): 93C20, 49B22

ISBN 978-3-540-12554-9 ISBN 978-3-540-38647-6 (eBook)
DOI 10.1007/978-3-540-38647-6

Library of Congress Cataloging in Publication Data
Main entry under title:
Control theory for distributed parameter systems and applications.
(Lecture notes in control and information sciences ; 54)
Proceedings of the Conference on Control Theory for Distributed Parameter Systems, held at the Chorherrenstift Vorau, Styria, July 11–17, 1982.
Bibliography: p.
1. Control theory–Congresses.
2. Distributed parameter systems–Congresses.
I. Kappel, F. II. Kunisch, K. (Karl), 1952- . III. Schappacher, Wilhelm.
IV. Conference on Control Theory for Distributed Parameter Systems
(1982 : Chorherrenstift Vorau) V. Series.
QA402.3.C644 1983 003 83-10597

Originally published by Springer-Verlag Berlin Heidelberg New York Tokyo in 1983

2061/3020-543210

PREFACE

This volume comprises the proceedings of the "Conference on Control Theory for Distributed Parameter Systems" held at the Chorherrenstift Vorau (Styria), July 11 - 17, 1982.

Control theory for distributed parameter systems presently is a very thriving part of applied mathematics with problems equally challenging for theoretically and practically minded researchers. The aim of the conference was to stimulate the exchange of ideas and to provide information on recent advances in various directions of research. It was a great pleasure for us to welcome 30 participants coming from 8 different countries. The program of the meeting included 19 lectures. Our thanks go to the lecturers, to all participants and especially to the authors of the contributions contained in this volume.

The conference was made possible by grants from the European Research Office of the US Army (under Grant No. DAJA 45-82-M-0282), from the Amt der Steiermärkischen Landesregierung and from the Bundesministerium für Wissenschaft und Forschung. We greatly appreciate the financial support rendered by these institutions.

In particular we want to thank the staff of the Bildungshaus Chorherrenstift Vorau, especially Direktor P. Riegler for all their efforts which made the stay at Vorau so pleasant. Finally, special thanks go to Missis G. Krois for her invaluable help in all administrational matters and for her excellent typing of the manuscript for these proceedings.

March 1983

F. Kappel. K. Kunisch, W. Schappacher

CONTENTS

LIST OF PARTICIPANTS

In the following list lecturers are indicated by an asterisk.

* A.V. BALAKRISHNAN	Los Angeles
* M.J. BALAS	Troy
* H.T. BANKS	Providence
* R.F. CURTAIN	Groningen
* G. DA PRATO	Pisa
W. DESCH	Graz
* A. FAVINI	Bologna
* L. GRANEY	Middlesex
P. JANSSEN	Delft
F. KAPPEL	Graz
S. KASPAR	Graz
* W. KRABS	Darmstadt
K. KUNISCH	Graz
* I. LASIECKA	Gainesville
N. MATZL	Graz
* S. NAKAGIRI	Kobe
* L. PANDOLFI	Torino
R. PEER	Graz
G. PEICHL	Graz
* A.J. PRITCHARD	Coventry
G. PROPST	Graz
* E. SACHS	Raleigh
* Y. SAKAWA	Osaka
* D. SALAMON	Bremen
W. SCHAPPACHER	Graz
* T.I. SEIDMAN	Catonsville
* M. SLEMROD	Troy
* R. TRIGGIANI	Gainesville
A. VENNI	Bologna
* C. ZALINESCU	Iaşi

THE MATHEMATICAL STRUCTURE OF THE FEEDBACK CONTROL PROBLEM FOR LINEAR DISTRIBUTED PARAMETER SYSTEMS WITH FINITE-DIMENSIONAL CONTROLLERS

M. J. Balas

Electrical, Computer and Systems
Engineering Department
Rensselaer Polytechnic Institute
Troy, NY 12181, USA

1. INTRODUCTION

In previous work (summarized in [1]), we have emphasized finite-dimensional feedback control of (usually) linear infinite-dimensional distributed parameter systems (DPS). This is the only situation of practical interest in engineering applications because the controllers must be implemented by on-line digital computers with finite wordlength and finite memory-access-time. Since our work on DPS control has been motivated by engineering systems, e.g. large aerospace structures [2], Tokomak fusion reactors, and other process control applications, we have been inclined to develop new DPS control theory with some practical constraints. This has been done in the hope that our results would help engineers to see the limitations of what can be accomplished with implementable DPS controllers and would make use of their experience and intuition in the design and operation of complex systems. In other words, we would like to understand the theoretical structure of the problem to see what can be accomplished with finite-dimensional control.

We do not mean to suggest that the above is the only important issue in DPS control; there are, of course, many mathematical problems of interest such as controllability, observability, and stabilizability of linear and nonlinear DPS by both interior and boundary control (e.g.[22,19,10]). However, not much attention has been paid to the finite-dimensional control of DPS; notable exceptions are [11,23,20].

In the past, we have concentrated on model reduction of DPS, i.e. obtaining finite-dimensional approximations of an infinite-dimensional system, and the synthesis of controllers based on these reduced-order models. This has meant that stability analysis must be an intrinsic part of the design because the stability of the closed-loop system, consisting of the actual DPS and a reduced-order controller, is not theoretically guaranteed. In finite-dimensions, when the controller and the plant have the same dimension, the (deterministic) separation principle saves the day (e.g.[17]), however, for DPS, the plant dimension must always be (substantially) larger than the controller dimension.

Of course, model reduction and reduced-order controller design are not new in the engineering community; they are the most natural approach to large-scale system control problems and have been used in various forms (and occasionally disguises) for DPS in mechanical, chemical, aerospace, and electrical engineering applications.

Often the stability analysis has been based entirely on computer simulation (i.e. a few initial situations appear stable; therefore, the system is stable) or has been entirely disregarded. Although the former is at least a step in the right direction, the latter is unconscionable. We have obtained various stability bounds for DPS via singular and regular perturbation techniques (e.g. [1],[3] - [6]).

The real problem is to apply stable and effective control to a complex DPS whose parameters and structure are usually not very well known. Put simply: controlling the heat equation in one space dimension is no big deal; in fact, engineers have been doing much more complicated things for a long time without the help of mathematical control theory (e.g. [21]). However, when the application is, for example, a large flexible structure which is to be constructed and operated in space (where no such things have been done before) where data like the damping and stiffness are poorly known and the vibration modes can only be approximated for a given configuration, then control theory may have something useful (and even comforting) to say. Perturbation methods seem to us to be especially well suited to this type of problem and may be able to give indications of stability and performance that can be used in the design (and redesign) of finite-dimensional controllers for DPS.

In this paper, we will take a somewhat different viewpoint: assuming that a finite-dimensional linear controller is available, what is the most we can expect to accomplish with it on a linear DPS? In [15], Gibson showed that compact perturbations can never produce exponential stability in a contractive, strongly stable system. Therefore, since most practical systems can only introduce feedback through a finite number of actuators, such finite-rank perturbations, being compact, can never produce a margin of stability (i.e. rate of exponential decay) in a DPS which does not already have such a margin initially. This type of result shows, for example, that a flexible structure without inherent damping can never be stabilized with an exponential rate of decay by feedback through a finite number of actuators. Luckily, real structures have some inherent damping; however, that is not the important point. The result of Gibson is exactly the sort of thing that is needed from DPS control theory, it tells us that we must be careful of the way we idealize (model) DPS for the purpose of control: no damping, no hope! Of course, the Gibson result assumes perfect state feedback into the actuators and this would never be available in practice. At best, observations can be made from a finite number of sensors and this data passed through a filter of finite-order to produce the control commands for a finite number of actuators. In the spirit (it not the same mathematical direction) of Gibson's result, we will present results that show what a given finite-dimensional controller is doing: it is asymptotically recreating the projection of the infinite-dimensional DPS state onto a finite-dimensional subspace and this finite-dimensional control is all that is available to modify the DPS by feedback. The finite-dimensional projection created by the controller is not necessarily the one the designer has chosen by model reduction. Hence, our results give a better insight to the structure of the control problem but do not necessarily indicate how to improve the design.

In Section 2, the preliminaries are presented for the class of linear DPS considered here. In Sections 3 and 4, our main results on the structure of the finite-dimensional feedback control problem for DPS are given. Some connections between the structural results of Sections 3 and 4 and our previous analysis of the controller design via model reduction are presented in Section 5. Although boundary control is usually treated as a separate problem from interior control of DPS, many boundary control problems can be converted to equivalent interior control problems; this is developed in Section 6 and it extends the results of the previous sections to a large class of practical boundary control problems for DPS. Our conclusions and recommendations form Section 7.

2. PRELIMINARIES FOR LINEAR DPS

The class of linear distributed parameter systems (DPS) considered here will have the following state space form:

$$\left.\begin{aligned} \frac{\partial v(t)}{\partial t} &= Av(t) + Bf(t); \quad v(0) = v_o \\ y(t) &= Cv(t) \end{aligned}\right\} \tag{2.1}$$

where the state $v(t)$ is in an infinite-dimensional Hilbert space H with inner product denoted by $(.,.)$ and corresponding norm $||.||$. The operator A is a closed, linear, unbounded differential operator with domain $D(A)$ dense in H, and A generates a C_o-semigroup of bounded operators $U(t)$ on H. The operators B & C have finite ranks M & P, respectively, and $f(t)$, $y(t)$ represent the inputs from M actuators and the outputs from P sensors, respectively. Thus,

$$Bf(t) = \sum_{i=1}^{M} b_i f_i(t) \tag{2.2}$$

and

$$y(t) = [y_1(t),\ldots,y_P(t)]^T \quad \text{where}$$

$$y_j(t) = (c_j, v(t)); \quad 1 \leq j \leq P \tag{2.3}$$

with b_i and c_j in H.

This is the form of most interior control problems and, as we shall point out in Section 6, it also represents many boundary control problems. When (2.1) - (2.3) is a model of an actual engineering system, the choice of Hilbert space H and the norm $||\cdot||$ are usually dictated by the practical problem (e.g. $||\cdot||$ is the energy norm). However, some care must be used in this choice because, unlike the finite-dimensional case, the state space forms for (2.1) need not be equivalent (even when (A,B,C) is controllable and observable).

From the Hille-Yosida Theorem [12] or [25], the operator A generates a C_0-semigroup U(t) satisfying:

$$||U(t)|| \leq Ke^{-\sigma t} \quad ; \quad t \geq 0 \tag{2.4}$$

where $K \geq 1$ and σ is real, when

$$||R(\lambda,A)^n|| \leq \frac{K}{(\lambda+\sigma)^n} \quad ; \qquad n = 1,2,... \tag{2.5}$$

for all real $\lambda > -\sigma$ in the resolvent set of A. The operator $R(\lambda,A) = (\lambda I-A)^{-1}$ is called the resolvent operator for A, and it is a bounded linear operator for each λ in the resolvent set $\rho(A)$; the spectrum $\sigma(A)$ of A is the set $\sigma(A) = \rho^c(A)$.

When $\sigma > 0$ in (2.4), the semigroup U(t) and the system (2.1) are exponentially stable with stability margin σ; for simplicity, we will say that the operator A is exponentially stable in (2.1), when $\sigma > 0$.

In some cases, A can be shown to satisfy dissipative conditions:

$$\left.\begin{aligned} (Av,v) &\leq -\sigma(v,v) \qquad \sigma > 0 \\ (A^*v,v) &\leq -\sigma(v,v) \end{aligned}\right\} \tag{2.6}$$

for all v in D(A) or D(A*) where A* is the adjoint operator for A. When (2.6) is true and A generates a C_0-semigroup U(t), then U(t) satisfies (2.4) with $K = 1$ and $\sigma > 0$ ([19]Theo. 2.4 or [25] Theo. 3.2). However, not every exponentially stable system operator A satisfies a dissipativity condition in the original norm; see [25] Theo. 3.2, p. 92.

The generation of a semigroup for (2.1) is the mathematical way of saying that the model (2.1) is well-posed and, hence, represents a physical system. The physical system modeled by (2.1) is the weak (or mild) formulation of the DPS:

$$\left.\begin{aligned} v(t) &= U(t)v_0 + \int_0^t U(t-\tau)Bf(\tau)d\tau \\ y(t) &= Cv(t) \end{aligned}\right\} \tag{2.7}$$

There are other types of stability besides exponential stability (in fact, these are all related to the types of convergence of solutions of (2.7) to zero); however, for engineering systems, a margin of stability is essential in order that the system be able to tolerate small parameter variations, noise, and nonlinearities which are ignored in the model (2.1). Of course, a more detailed model, including all these factors, could be developed, in theory, but in practice such detail is poorly known. Consequently, this is one of the trade-offs in controller design: either make a simplified model of the DPS and design a controller which yields exponential stability with as satisfactory a stability-margin as possible or make an extremely detailed DPS model containing all possible factors affecting performance and design a corresponding controller to deal with this system, e.g. make it strongly stable. The latter can lead ultimately to madness since the more closely you look at a system the more detail is revealed. Therefore, even a detailed model of the DPS may not incorporate all the possible factors; hence, such an approach is very likely to lead to an unstable closed-loop system if weaker stability than exponential stability is used in the design criterion. Furthermore, the level of detail of the model can quickly exhaust the available possibilities for controller design to handle such systems. Enough detail must be included so that the controller can be designed to yield a reasonable level of performance from the closed-loop system. Most control engineers would agree with this imprecise statement of what they do; however, it takes quite a bit of experience with specific engineering systems to decide what the words "enough" and "reasonable" mean (and it is not our intention to presume to do this here).

Feedback control for such a DPS as (2.1) should be accomplished with finite-dimensional, discrete - time controllers of the form:

$$\left.\begin{aligned} f(k) &= L_{11}\, y(k) + L_{12}\, z(k) \\ z(k+1) &= L_{21}\, y(k) + L_{22}\, z(k) \end{aligned}\right\} \qquad (2.8)$$

where $z(k)$ belongs to R^{α}. Such controllers can be implemented with on-line digital computers whose memory-access-time and memory capacity is related to the controller dimension α. Although the discrete-time acpect of the controller is not a trivial issue (e.g. [18]), for convenience here, we shall deal only with the continuous-time version of (2.8); therefore, the finite-dimensional linear controller will have the form:

$$f(t) = L_{11}\, y(t) + L_{12}\, z(t) \qquad (2.9a)$$

$$\dot{z}(t) = L_{21}\, y(t) + L_{22}\, z(t) = Fz(t) + Ky(t) + Ef(t) \qquad (2.9b)$$

where $z(t)$ belongs to R^{α}.

The matrices F, K, and E are related to L_{21} and L_{22} by:

$$\left.\begin{aligned} L_{21} &= K + EL_{11} \\ L_{22} &= F + EL_{12} \ . \end{aligned}\right\} \qquad \begin{aligned} &(2.10a) \\ &(2.10b) \end{aligned}$$

The controller dynamics (2.9b) provide a filtering effect on the sensor data; these dynamics can be very helpful but, as we shall point out in Secs. 3 and 4, they cannot perform miracles (such as reconstructing the full DPS state). Special cases of (2.9) are static (or output) feedback:

$$L_{12} = 0, \quad L_{21} = 0, \quad L_{22} = 0 \qquad (2.11)$$

where no dynamics are present in the controller, and full dynamic (or α-dimensional) feedback:

$$L_{11} = 0 \qquad (2.12)$$

where no direct feedthrough is present and all sensor measurements are passed through the controller dynamics.

3. FINITE-DIMENSIONAL OBSERVERS FOR DPS

In this section we will examine what can be accomplished with a finite-dimensional observer of the form:

$$\left.\begin{aligned} q(t) &= Q_{11}\, y(t) + Q_{12}\, z(t) \\ \dot{z}(t) &= Fz(t) + Ky(t) + Ef(t) \end{aligned}\right\} \qquad \begin{aligned} &(3.1a) \\ &(3.1b) \end{aligned}$$

where $z(t)$ belongs to R^{α} with $\alpha < \infty$. If this observer is used to estimate the state of the infinite-dimensional DPS (2.1), then at best it can asymptotically reconstruct only the projection of the DPS state onto a finite-dimensional subspace. This is made precise by the following result:

Theorem 1. Assume $f(t)$ in (2.1) is continuously differentiable.

If (a) F is stable (i.e. all eigenvalues of F are in the open left-half of the complex plane),

(b) there exists a bounded linear operator $T: H \rightarrow R^{\alpha}$ such that

$$(FT - TA + DC)v = 0 \qquad (3.2)$$

for all v in $D(A)$, and

(c) E is chosen so that $E = TB$ (3.3)

then $z(t)$ in (3.1b) is given by

$$z(t) = Tv(t) + e(t) \qquad (3.4)$$

where

$$\left.\begin{aligned} \dot{e}(t) &= Fe(t) \\ e(0) &= z_o - Tv_o \ . \end{aligned}\right\} \qquad (3.5)$$

Furthermore, there exists a pair of nontrivial subspaces $\tilde{H}_N$ and $\tilde{H}_R$ in H such that:

$$H = \tilde{H}_N \oplus \tilde{H}_R \qquad (3.6)$$

$$\dim \tilde{H}_N \equiv N \leq P + \alpha \qquad (3.7)$$

$$\lim_{t\to\infty} [q(t) - \tilde{P}_N v(t)] = 0 \qquad (3.8a)$$

$$\lim_{t\to\infty} [q(t) - v(t)] = - \lim_{t\to\infty} \tilde{P}_R v(t) \qquad (3.8b)$$

where $\tilde{P}_N$ and $\tilde{P}_R$ are the projections onto $\tilde{H}_N$ and $\tilde{H}_R$ defined by (3.6). In fact, these subspaces are given by

$$\tilde{H}_N = \tilde{N}(T)^{\perp}$$

$$\tilde{H}_R = \tilde{N}(T) \equiv \{v \in D(\tilde{T}) \mid \tilde{T}v = 0\}$$

where $\tilde{T} \equiv \begin{bmatrix} C \\ T \end{bmatrix} : H \to R^{P+\alpha}$.

In order to prove Theo. 1, we will need the following result about pseudo-inverses of operators:

<u>Theorem 2.</u> Given a bounded linear operator $T: H_1 \to H_2$ with H_i Hilbert spaces. If T is onto (surjective), then the pseudo-inverse $T^{\#}$ of T defined by

$$T^{\#}: H_2 \to H_1 \text{ with}$$

$$T^{\#}T = \tilde{P}_N \qquad (3.9)$$

where $\tilde{P}_N$ is orthogonal projection onto N(T) has the following properties:

(a) $T^{\#}$ is well defined and linear on H_2

(b) $T\,T^{\#}T = T$ (3.10)

(c) $T^{\#}$ is a bounded operator

(d) If $\dim H_2 < \infty$, then $\dim N(T)^{\perp} = \dim H_2$.

The proofs of Theos. 1 and 2 appear in Appendix I. Although properties (a) and (c) of Theo. 1 are easy to guarantee by the choice of the observer parameters F and E, property (b) may seem to be more formidable. However, the following result suggests otherwise:

Theorem 3. If the spectra of F and A are separated, i.e. there exists a simple closed curve Γ with positive direction in the complex plane such that Γ encloses the eigenvalues of F and excludes the spectrum $\sigma(A)$ of A, then a unique bounded linear operator T exists such that (3.2) is satisfied. In fact

$$Tv = \frac{1}{2\pi i} \int_{\Gamma} R(\lambda,F)KC\, R(\lambda,A)v d\lambda \tag{3.11}$$

for all v in D(A) where $R(\lambda,F)$ and $R(\lambda,A)$ are the resolvent operators for F and A, respectively. The proof of Theo. 3 is given in Appendix II.

Since A generates a C_o-semigroup with the growth property (2.4), it follows (from [19] Prop. 1.15, p. 485), that, for any λ in $\sigma(A) = \rho(A)^c$,

$$\mathrm{Re}\ \lambda \leq -\sigma \tag{3.12}$$

where σ is a real number. Note that if A is exponentially stable, then $\sigma > 0$ and it is easy to find locations for the α eigenvalues of F where (a) and (b) of Theo. 1 are both satisfied. Although in some applications A may be exponentially stable, in general this would not be true; however, it will be possible to find stable locations for the eigenvalues of F separated from $\sigma(A)$ unless A is so pathological that every open subset of the open left-half complex plane is contained in $\sigma(A)$.

Therefore, the assumptions (a) - (c) of Theo. 1 seem likely to be satisfied in most applications. Consequently, Theo. 1 indicates that at best, finite-dimensional observers will asymptotically reconstruct some finite-dimensional projection of the DPS state; the dimension of the finite-dimensional subspace $\tilde{H}_N$ upon which the projection is made is bounded in (3.7) by the sum of the observer dimension and the number of independent sensors available. The result of Theo. 1 seems quite natural; however, it does not provide easy access to the subspace $\tilde{H}_N$, i.e. one would need to construct the operators T and $\tilde{T}$ and then obtain the orthogonal complement of the null space of $\tilde{T}$. Nevertheless, Theo. 1 provides insight into the mathematical structure of finite-dimensional observers for infinite-dimensional systems.

It becomes clear that the infinite-dimensional state of (2.1) cannot be asymptotically reconstructed by (3.1) unless

$$\lim_{t\to\infty} \tilde{P}_R\, v(t) = 0. \tag{3.13}$$

This says that the full state of the DPS (2.1) must be attracted to the finite-dimensional subspace $\tilde{H}_N = N(T)^{\perp}$. This seems very unlikely especially if the input f(t) is not zero. Also, if we are lucky enough for (3.13) to hold for a particular DPS (2.1), then such a system will be very easy to stabilize.

4. STABILIZING SUBSPACES AND FINITE-DIMENSIONAL CONTROL OF DPS

In the previous section, finite-dimensional observer were shown to be capable of reconstructing the projection of the DPS state of (2.1) onto some finite-dimensional subspace $\tilde{H}_N$. Here we will show that stable finite-dimensional control of (2.1) is possible if and only if the subspace $\tilde{H}_N$ and its complement $\tilde{H}_R$ are stabilizing subspaces for (2.1).

The concept of stabilizing subspaces was introduced in [8] and used to establish links with discrete and continuous time DPS controllers [7] and time and frequency domain stability conditions [6]. We say that (A,B) in (2.1) has a pair of stabilizing subspaces (H_N, H_R) if the following conditions are satisfied:

(a) $H = H_N \oplus H_R$ (4.1)

(b) $\dim H_N \equiv N < \infty$

(c) $A_o \equiv A + BG$ is exponentially stable (with a desired stability margin σ_o) for some gain operator $G: H \to R^M$ such that

$$G = GP_N \equiv G_N \tag{4.2a}$$

or equivalently

$$GP_R = 0 \tag{4.2b}$$

where P_N and P_R are the projections on H_N and H_R defined by (4.1). This says that A_o can be stabilized by feedback of the projection of the infinite-dimensional state of (2.1) onto some finite-dimensional subspace H_N. Such feedback is not generally available from measurements, but this concept is still part of the structure of the control problem.

In the past, we have assumed the vectors in H_N are conforming elements, i.e. $H_N \subseteq D(A)$, as part of the definition; however, this is not essential and will not be assumed here. Thus, H_N may be a nonconforming subspace in the definition of stabilizing subspaces.

The exponential stability of the C_o-semigroup $U_o(t)$ generated by $A_o = A + BG$ above is given by

$$||U_o(t)|| \leq K_o e^{-\sigma_o t}, \quad t \geq 0 . \tag{4.3}$$

Conditions for the existence of stabilizing subspaces for (A,B) are given in the following:

<u>Theorem 4.</u> If the subspaces H_N and H_R satisfy (4.1) and H_N is finite-dimensional with

(a) either $H_N \subseteq D(A)$ or $H_R \subseteq D(A)$;

(b) (A_N, B_N) are stabilizable,

(c) A_R is exponentially stable;

(d) $||A_{NR}||$ is sufficiently small;

where $A_N \equiv P_N A P_N$, $B_N \equiv P_N B$, $A_R \equiv P_R A P_R$, and $A_{NR} \equiv P_N A P_R$, then (H_N, H_R) are stabilizing subspaces for (A,B) in (2.1).

The proof of Theo. 4 follows from the decomposition of A_o on H_N and H_R:

$$A_o = \begin{bmatrix} A_N + B_N G_N & A_{NR} \\ A_{RN} + B_R G_N & A_R \end{bmatrix} \tag{4.4}$$

which is true if and only if (4.2) is satisfied. From (b) in Theo. 4, we can choose a stabilizing $G = G_N$ on H_N. The result is obtained from (a) - (d) of Theo . 4 and use of the semigroup perturbation theorem (e.g. [12], Theo. 10.9, p. 210). It is clear that (a) is necessary so that AP_N and AP_R will make sense; if $H_N \subseteq D(A)$, then $P_R v$ is in D(A) when v is in D(A) and similarly when $H_R \subseteq D(A)$.

Note that, stabilizing subspaces may exist for (2.1) via Theo. 4, and yet these need not be the subspaces $(\tilde{H}_N, \tilde{H}_R)$ of Sec. 3. However, the following result shows that $(\tilde{H}_N, \tilde{H}_R)$ are indeed stabilizing subspaces; furthermore, there is a pair of stabilizing subspaces associated with every finite-dimensional controller capable of producing an exponentially stable closed-loop system:

Theorem 5. Assume the hypotheses (a) - (c) of Theo. 1 for the finite-dimensional controller (2.9). The closed-loop system consisting of the DPS (2.1) and the controller (2.9) is exponentially stable if and only if the subspaces $(\tilde{H}_N, \tilde{H}_R)$ of Theo. 1 are stabilizing subspaces for (A,B) in (2.1).

The proof of Theo. 5 is given in Appendix III.

This result separates stable finite-dimensional control of the DPS (2.1) into two parts:

(1) the asymptotic reconstruction of the projection of the infinite-dimensional state of (2.1) onto some finite-dimensional subspace (dictated by the dynamics of the controller), and

(2) the stabilization of (2.1) by feedback involving only the finite-dimensional projection of the state in (1), i.e. the existence of stabilizing subspaces generated by the controller (2.9).

The structure of finite-dimensional control of an infinite-dimensional DPS (2.1) is revealed by Theo. 5. If a desired result, such as achieving exponential stability of the closed-loop system, cannot be obtained by feedback of a finite-dimensional projection of the infinite-dimensional state of (2.1), then it cannot be accomplished by any reasonable finite-dimensional controller (2.9). The controller dynamics aid in the reconstruction of this projection, but they cannot produce more than a finite-dimensional projection of the full state of (2.1).

Two corollaries regarding special cases of the controller follow directly from Theo. 5:

Corollary 1. Under the hypotheses of Theo. 5, if the controller (2.9) is static feedback (2.11), then the closed-loop system is exponentially stable if and only if the subspaces $(\tilde{H}_N, \tilde{H}_R)$, where $\tilde{H}_N \equiv N(C)^{\perp}$ and $\tilde{H}_R \equiv N(C)$, are stabilizing subspaces for (A,B) in (2.1).

Corollary 2. Under the hypotheses of Theo. 5, if the controller (2.9) is full dynamic feedback (2.12), then the closed-loop system is exponentially stable if and only if the subspaces $(\tilde{H}_N, \tilde{H}_R)$, where $\tilde{H}_N \equiv N(T)^{\perp}$ and $\tilde{H}_R \equiv N(T)$ with T a solution of (3.2), are stabilizing subspaces for (A,B) in (2.1).

A different perspective can be obtained if we treat the closed-loop system (2.1) and (2.9) as an extended static feedback problem.

Let $w \equiv \begin{bmatrix} v \\ z \end{bmatrix}$ in the Hilbert space $\bar{H} \equiv H \times R^{\alpha}$.

From (2.1) and (2.9), we have

$$\frac{\partial w(t)}{\partial t} = (\bar{A} + \bar{B}\, L\, \bar{C})\, w(t) = \begin{bmatrix} A + BL_{11}C & BL_{12} \\ L_{21}C & L_{22} \end{bmatrix} w(t) \tag{4.5}$$

where the operators $\bar{A}$, $\bar{B}$, $\bar{C}$ are defined by

$$\bar{A} = \begin{bmatrix} A & 0 \\ 0 & 0 \end{bmatrix}, \quad \bar{B} = \begin{bmatrix} B & 0 \\ 0 & I_{\alpha} \end{bmatrix}, \quad \bar{C} = \begin{bmatrix} C & 0 \\ 0 & I_{\alpha} \end{bmatrix}$$

and the extended static feedback gain operator is

$$L = \begin{bmatrix} L_{11} & L_{12} \\ L_{21} & L_{22} \end{bmatrix}.$$

From this viewpoint even though the closed-loop system has dynamic feedback, it looks like a static (output) feedback problem on the extended space $\bar{H}$. This idea has often been used in finite-dimensional systems (e.g.[17]). Now, by Cor. 1, the closed-loop system (2.1) and (2.9), or equivalently (4.5), is exponentially stable if and only if the subspaces $\tilde{H}_N = \tilde{N}(\bar{C})^{\perp}$ and $\tilde{H}_R = \tilde{N}(\bar{C})$ are stabilizing subspaces for $(\bar{A},\bar{B})$ in (4.5).

5. MODEL REDUCTION AND REDUCED-ORDER CONTROL OF DPS

The results of Secs. 3 and 4 indicate that most finite-dimensional controllers have an associated pair of stabilizing subspaces. However, these subspaces are not easy to construct, and, more important, they are not necessarily the most natural subspaces to use for model reduction of the DPS, i.e. construction of finite-dimensional approximations of (2.1). The more natural candidates for these subspaces are related to numerical methods for approximating partial differential equations, such as finite difference or finite-element techniques.

In this section, we make use of our knowledge of the structure of the finite-dimensional control problem, but we start with some model reduction of the DPS (2.1) and synthesize a finite-dimensional controller from this reduced-order model. This conforms with the engineering approach to such problems. We will point out some relationships along the way.

5.1. Model Reduction

In order to produce finite-dimensional controllers for the DPS (2.1), we must make a lumped parameter approximation of it. This is done when numerical methods such as finite elements or finite differences are used to discretize the spatial variables. In general, such an approximation or reduced-order model (ROM) is a (not necessarily orthogonal) projection of (2.1) onto an appropriate finite-dimensional subspace H_N of H; usually, we will assume $H_N \subseteq D(A)$. The ROM subspace H_N has dimension N and its projection is denoted by P_N; the residual subspace H_R associated with H_N completes the decomposition $H = H_N \oplus H_R$, and its projection is denoted by P_R. The total DPS state v can be written:

$$v = v_N + v_R$$

where $v_N = P_N v$ and $v_R = P_R v$. The choice of the subspaces H_N and H_R is usually dictated by the physical application and/or the numerical procedures available for integrating the DPS. When feedback control is the ultimate purpose of the model reduction, certain choices of subspaces will yield advantages [3].

A modal subspace H_N consists of linear combinations of a finite number of modes or eigenfunctions of the operator A. Modal subspaces have very special properties in control applications, e.g., $A_{NR} = 0$ and $A_{RN} = 0$ in (5.2) later. However, since most engineering applications are too complex for the exact modes to be known, these subspaces are more conceptually, rather than practically, useful.

The projection of the DPS (2.1) onto the subspaces H_N and H_R decomposes the system into the following (where $v(0) = v_0$ in $D(A)$):

$$\frac{\partial v_N(t)}{\partial t} = A_N v_N(t) + A_{NR} v_R(t) + B_N f(t); \quad v_N(0) = P_N v_o \tag{5.2a}$$

$$\frac{\partial v_R(t)}{\partial t} = A_{RN} v_N(t) + A_R v_R(t) + B_R f(t); \quad v_R(0) = P_R v_o \tag{5.2b}$$

$$y(t) = C_N v_N(t) + C_R v_R(t) \tag{5.2c}$$

where $A_N = P_N A P_N$, $A_{NR} = P_N A P_R$, etc. The terms $A_{NR} v_R$ and $A_{RN} v_N$ are called modeling error and the terms $B_R f$ and $C_R v_R$ are called control and observation spillover, respectively. The reduced-order model is obtained from (5.2) by ignoring the residuals:

$$\left.\begin{aligned} \frac{\partial v_N(t)}{\partial t} &= A_N v_N(t) + B_N f(t) \\ y(t) &= C_N v_N(t) \quad . \end{aligned}\right\} \tag{5.3}$$

In any choice of model reduction scheme it makes no practical sense if the residuals are unstable; therefore, we will assume that A_R generates a C_o-semigroup $U_R(t)$ with the property:

$$||U_R(t)|| \leq K_R e^{-\sigma_R t}, \quad t \geq 0$$

with $K_R \geq 1$ and $\sigma_R > 0$. Such a condition is usually satisfied in practice, as long as one is careful in the selection of H_N and H_R.

To summarize the above, we will say that a pair of subspaces (H_N, H_R) are model-reducing subspaces for (2.1) if the following are satisfied:

(a) $H = H_N \oplus H_R$ (5.5)

(b) $H_N \subseteq D(A)$ (5.6)

(c) The ROM (A_N, B_N, C_N) is stabilizable and detectable [17]

(d) A_R generates a C_o-semigroup $U_R(t)$ satisfying (5.4).

Note that, from Theo. 4, if the modeling error $||A_{NR}||$ is sufficiently small, then the model reducing subspaces (H_N, H_R) are also stabilizing subspaces for (A,B) in (2.1). Furthermore, for modal subspaces, stabilizing or model reducing subspaces are the same thing, because the modal subspaces are always in D(A) and $A_{NR} = 0$ in (4.4).

There are many ways to produce a model reduction of (2.1), but the above definition includes the basic properties expected of any reasonable version. In the past, we have usually assumed the ROM (A_N, B_N, C_N) is controllable and observable, which can be easily checked by the standard rank conditions; however, the mathematical condition of (c) above is all that is really necessary. Also, (5.6) is not absolutely essential, but we will retain it for convenience in our development.

5.2. Reduced-Order Controller Synthesis

In order to control the DPS (2.1), a finite-dimensional controller is generated from the ROM (5.3):

$$\left.\begin{aligned} f(t) &= G_N\hat{v}_N(t) \\ \frac{\partial \hat{v}_N(t)}{\partial t} &= A_N\hat{v}_N(t) + B_N f(t) + K_N(y(t) - \hat{y}(t)) \\ \hat{y}(t) &= C_N\hat{v}_N(t), \quad \hat{v}_N(0) = 0 \, . \end{aligned}\right\} \qquad (5.7)$$

Note that (5.7) can be identified with a finite-dimensional controller by taking any basis for the subspace H_N and obtaining the matrices corresponding to all the finite-rank operators; this would be done for controller synthesis. This is the most obvious candidate for a feedback controller; however, there are many ways in which (5.7) can be modified and improved, as pointed out in [1]. Nonetheless, (5.7) is a good starting point for the controller synthesis; it is, in fact, what most control system designers do with both large-scale and distributed parameter systems. The controller gains G_N and K_N are designed so that $A_N + B_NG_N$ and $A_N - K_NC_N$ are stable. Such designs can be accomplished by pole placement or linear quadratic regulator techniques [17]; computer algorithms for these methods are readily available.

It is clear that the model reduction approach makes the most use of available engineering knowledge and experience gained from finite-dimensional systems; hence, its popularity for synthesizing controllers for DPS. What is too often forgotten is that (5.7) is designed to be stable in closed-loop with the ROM (5.3), but this does not guarantee its stability in closed-loop with the actual DPS (2.1). Because of the model reduction, we cannot appeal to a separation principle for stable linear control. The closed-loop stability analysis must be an intrinsic part of finite-dimensional controller design for DPS.

5.3. Closed-loop Stability Analysis

Although the above procedure for DPS controller synthesis is quite straightforward and heuristically motivated, it is not really so out-of-step with the theoretical structure of the DPS control problem as developed in Secs. 3 and 4. The controller (5.7) can be rewritten:

$$\left.\begin{aligned} f(t) &= G_N\hat{v}_N(t) \\ \frac{\partial \hat{v}_N(t)}{\partial t} &= L_N\hat{v}_N(t) + K_N y(t), \qquad \hat{v}(0) = 0 \end{aligned}\right\} \qquad (5.8)$$

where $L_N \equiv A_N - K_N C_N + B_N G_N$; therefore, (5.8) is equivalent to (2.9) with

$$L_{11} = 0, \quad L_{12} = G_N, \quad L_{21} = K_N, \quad L_{22} = L_N \tag{5.9a}$$

or

$$F = A_N - K_N C_N, \quad K = K_N, \quad E = B_N \tag{5.9b}$$

i.e., (5.8) is full dynamic feedback from a stable controller ($F = A_N - K_N C_N$ is stable by design). Theo. 5 (or Cor. 2) suggests that the best we can expect from the controller (5.8) is that it will asymptotically recreate the projection of the full DPS state onto the ROM subspace H_N, i.e. let

$$\hat{v}_N = v_N + e_N \tag{5.10}$$

where $v_N \equiv P_N v$, and, at best $\lim_{t\to\infty} e_N(t) = 0$. Then, if (H_N, H_R) are stabilizing subspaces for (2.1), exponential stability of the closed-loop will be achieved. As we shall show next, this is not quite what happens because $\lim_{t\to\infty} e_N(t)$ is not necessarily zero.

From (5.2a), (5.7) and (5.10), we have

$$\begin{aligned}\frac{\partial e_N(t)}{\partial t} &= A_N \hat{v}_N(t) + B_N f(t) + K_N(y(t) - \hat{y}(t)) - [A_N v_N(t) + A_{NR} v_R(t) + B_N f(t)] \\ &= A_N e_N(t) + K_N(C_N v_N(t) + C_R(t) v_R(t) - C_N \hat{v}_N(t)) - A_{NR} v_R(t) \\ &= (A_N - K_N C_N) e_N(t) + (K_N C_R - A_{NR}) v_R(t).\end{aligned}$$

Therefore, since $v_R = P_R v$, we obtain

$$\left.\begin{aligned}\frac{\partial e_N(t)}{\partial t} &= (A_N - K_N C_N) e_N(t) + \Delta_{NR} v(t) \\ e_N(0) &= -P_N v_o\end{aligned}\right\} \tag{5.11}$$

where $\Delta_{NR} \equiv K_N C_R - A_{NR}$ and

$$\Delta_{NR} = \Delta_{NR} P_R \,. \tag{5.12}$$

Consequently, even though $A_N - K_N C_N$ is stable, the term $\Delta_{NR} v(t)$, arising from modeling error and observation spillover, does not allow the error equation (5.11) to be decoupled from the controlled DPS (2.1); hence, the controller (5.7) does not necessarily asymptotically reconstruct the projection $v_N = P_N v$.

In addition, from (5.7) and (5.10),

$$f(t) = G_N \hat{v}_N(t) = G_N v_N(t) + G_N e_N(t).$$

Hence, from (2.1), we have, for any v_o in $D(A)$,

$$\left.\begin{aligned} \frac{\partial v(t)}{\partial t} \quad & Av(t) + Bf(t) = (A + BG_N P_N)v(t) + BG_N e_N(t) \\ v(0) &= v_o \ . \end{aligned}\right\} \qquad (5.13)$$

If (H_N, H_R) are stabilizing subspaces for (A,B), then, by (4.2), we can rewrite (5.13) as

$$\left.\begin{aligned} \frac{\partial v(t)}{\partial t} &= A_o v(t) + BG_N e_N(t) \\ v(0) &= v_o \end{aligned}\right\} \qquad (5.14)$$

where $A_o = A + BG = A + BG_N P_N$ is exponentially stable as in (4.3).

The closed-loop system consisting of the actual DPS (2.1) and the controller (5.7) can be rewritten as (5.11) and (5.14). The following result gives conditions under which it is stable:

Theorem 6. The closed-loop system consisting of the DPS (2.1) and the finite-dimensional controller (5.7) can be made exponentially stable by a choice of the controller gains G_N and K_N if

(a) a pair of subspaces (H_N, H_R) exist which are stabilizing subspaces for (A,B) in (2.1);

(b) $H_N \subseteq D(A)$ and $\dim H_N = N < \infty$:

(c) the reduced-order model (5.3) for (2.1) based on (H_N, H_R) is stabilizable and detectable in the finite-dimensional sense,

(d) $||\Delta_{NR}||$ is sufficiently small.

Proof: Let $w(t) = \begin{bmatrix} v(t) \\ e_N(t) \end{bmatrix}$ in $\tilde{H} = H \times H_N$.

The closed-loop system becomes

$$\left.\begin{aligned} \frac{\partial w(t)}{\partial t} &= \tilde{A}_c w(t) \\ w(0) &= \begin{bmatrix} v_o \\ -P_N v_o \end{bmatrix} \end{aligned}\right\} \qquad (5.15)$$

where $\tilde{A}_c \equiv \begin{bmatrix} A_o & BG_N \\ \Delta_{NR} & A_N - K_N C_N \end{bmatrix}$.

But $\tilde{A}_c = \tilde{A}_o + \Delta\tilde{A}$ where

$$\tilde{A}_o \equiv \begin{bmatrix} A_o & BG_N \\ 0 & A_N - K_N C_N \end{bmatrix} \quad \text{and } \tilde{\Delta A} \equiv \begin{bmatrix} 0 & 0 \\ \Delta_{NR} & 0 \end{bmatrix}.$$

Since A_o and $A_N - K_N C_N$ can be made exponentially stable due to (a) and (c), $\tilde{A}_o$ is exponentially stable also; in fact, $\tilde{A}_o$ generates the C_o-semigroup $\tilde{U}_o(t)$ with

$$||\tilde{U}_o(t)|| \leq \tilde{K}_o e^{-\tilde{\sigma}_o t}, \quad t \geq 0 \tag{5.16}$$

where $\tilde{K}_o \geq 1$ and $\tilde{\sigma}_o > 0$. From the semigroup perturbation theorem (e.g. [12], Theo. 10.9, p.210), $\tilde{A}_c$ generates a C_o-semigroup $\tilde{U}_c(t)$ with

$$||\tilde{U}_c(t)|| \leq \tilde{K}_c e^{-\tilde{\sigma}_c t}, \quad t \geq 0 \tag{5.17}$$

where

$$\left.\begin{aligned} \tilde{K}_c &= \tilde{K}_o \\ \tilde{\sigma}_c &= \tilde{\sigma}_o - \tilde{K}_o ||\tilde{\Delta A}|| . \end{aligned}\right\} \tag{5.18}$$

Note that $||\tilde{\Delta A}|| = ||\Delta_{NR}||$, and if

$$||\Delta_{NR}|| < \frac{\tilde{\sigma}_o}{\tilde{K}_o} \tag{5.19}$$

then $\tilde{\sigma}_c > 0$ and exponential stability follows. This completes the proof of Theo. 6.

Note that (5.4) was not required in Theo. 6; however, if (H_N, H_R) are model reducing subspaces for (2.1), then, as we have already pointed out in Sec. 5.1, this pair of subspaces satisfies hypotheses (a) - (c) of Theo. 6 when $||A_{NR}||$ is sufficiently small. Therefore, exponential closed-loop stability follows when both $||A_{NR}||$, and $||\Delta_{NR}|| = ||K_N C_R - A_{NR}||$ are sufficiently small for model reducing subspaces (H_N, H_R).

In the special case of modal subspaces, since $A_{NR} = 0$ and $A_{RN} = 0$ and stabilizing subspaces are the same as model reducing subspaces, Theo. 6 requires only that $||K_N C_R||$ be sufficiently small for some pair of model stabilizing subspaces for (2.1). Although the actual modes of the DPS (2.1) are rarely known in practice, the modal version of Theo. 6 provides a simple understanding of the controller design trade-off: one designs the gains K_N large so that $A_N - K_N C_N$ will have a large stability margin and the error $e_N(t)$ will have a rapid rate of exponential decay; however, this also emphasizes the effect of observation spillover in (5.11) via the term $K_N C_R$ which counteracts the large stability margin and reduces the decay rate of $e_N(t)$. In some cases, even modal controllers can make the closed-loop system unstable (e.g. [9]); however, Theo. 6 says that, as long as the DPS can be exponentially stabilized with a desired stability margin via modal feedback from a finite number of modes, a finite-dimensional controller can also achieve exponential stability (possibly with a smaller stability margin) using only sensor feedback if the choice of controller gain K_N can be made to satisfy hypothesis (d) in Theo. 6.

This leads us to the next important issue raised by Theo. 6: how small is "sufficiently small" for $||\Delta_{NR}||$ in (d)? The answer is given by the following result:

<u>Theorem 7.</u> Hypothesis (d) is satisfied in Theo. 6 if $||\Delta_{NR}||$ satisfies (5.19) with

$$\left.\begin{aligned}\tilde{\sigma}_o &= \min(\sigma_N, \tilde{\sigma}_o)\\ \tilde{K}_o &= K_o\tilde{K}_N(1+\gamma+\gamma^2)^{1/2} \leq K_o\tilde{K}_N(1+\gamma)\end{aligned}\right\} \tag{5.20}$$

where $\gamma \equiv \dfrac{||BG_N||}{|\sigma_o - \tilde{\sigma}_N|}$, (σ_o, K_o) are given by (4.3), and $(\tilde{\sigma}_N, \tilde{K}_N)$ are given

(from a choice of the controller gain K_N) by

$$||e^{(A_N - K_N C_N)t}|| \leq \tilde{K}_N e^{-\tilde{\sigma}_N t}, \quad t \geq 0 \tag{5.21}$$

where

$$\tilde{K}_N \geq 1, \quad \tilde{\sigma}_N > 0, \quad \tilde{\sigma}_N \neq \sigma_o.$$

The proof of Theo. 7 comes directly from the following stability lemma whose proof is given in [1]:

<u>Stability Lemma:</u> Let $\omega = \begin{bmatrix}\omega_1\\ \omega_2\end{bmatrix} \in H = H_1 \times H_2$ where H_i are Hilbert spaces.

Consider

$$\frac{\partial\omega}{\partial t} = \tilde{A}_c\omega = \begin{bmatrix} A_{11} & A_{12}\\ A_{21} & A_{22}\end{bmatrix}\omega \tag{5.22}$$

where A_{ij} are bounded for $i \neq j$ and A_{ii} generates the C_o-semigroup $U_i(t)$ with the growth property:

$$||\tilde{U}_i(t)|| \leq \tilde{K}_i e^{-\sigma_i t}, \quad t \geq 0 \tag{5.23}$$

for $i = 1,2$. Assume $\sigma_1 \neq \sigma_2$. Then $\tilde{A}_c$ generates the C_o-semigroup $\tilde{U}_c(t)$ with growth property

$$||U_c(t)|| \leq K_c e^{-\tilde{\sigma}_c t}, \quad t \geq 0 \tag{5.24}$$

where

$$\tilde{\sigma}_c = \tilde{\sigma}_o - \tilde{K}_c||A_{21}|| \tag{5.25}$$

with

$$\left.\begin{array}{ll}\tilde{\sigma}_o = \min(\sigma_1, \sigma_2) & (5.26a)\\ K_c = K_1K_2(1 + \psi + \psi^2)^{1/2} \leq K_1K_2(1 + \psi) & (5.26b)\end{array}\right\}$$

and

$$\psi = \frac{||A_{12}||}{|\sigma_1 - \sigma_2|} \quad \text{where} \quad ||\omega||^2 \equiv ||\omega_1||^2 + ||\omega_2||^2.$$

The dual result with A_{12} and A_{21} interchanged in (5.25) and (5.26) is true, also.

Taking $A_{11} = A_o$, $A_{12} = BG_N$, $A_{21} = \Delta_{NR}$, and $A_{22} = A_N - K_NC_N$ in this lemma yields Theo. 7. Of course, Theo. 7 is only one of several possible estimates of the required bound on Δ_{NR}; the sharpness of the estimates (5.19) - (5.21) is always a question when perturbation methods and norm bounds are used. Yet, in most practical problems, the designer will be lucky to know any more than the estimated norms of residual data; often, these must come from experiment and computer simulation. Nevertheless, Theos. 6 and 7 give some idea of what data will be necessary for stability analysis and how it should be used.

Although Theo. 5 and Cor. 2 indicate that the projection of the full state of (2.1) onto the finite-dimensional subspace $\tilde{H}_N$ will be asymptotically reconstructed by the controller (5.7) and that the pair $(\tilde{H}_N, \tilde{H}_R)$ will be stabilizing subspaces for (A,B) in (2.1), there is no reason to believe that $(\tilde{H}_N, \tilde{H}_R)$ will be the model reducing subspaces (H_N, H_R), i.e. that $\tilde{H}_N = H_N$ and $\tilde{H}_R = H_R$. In fact, in general they are not equal because $H_R = H_N^{\perp}$ but this is not necessarily true for $(\tilde{H}_N, \tilde{H}_R)$. Also, $H_N \subseteq D(A)$ by definition for model reducing subspaces, but $\tilde{H}_N = N(T)^{\perp}$ which is not necessarily in D(A).

In other words, subspaces $(\tilde{H}_N, \tilde{H}_R)$ which reveal the theoretical structure of the control problem may not be the ones (H_N, H_R) chosen for model reduction. As we have said earlier, the model reducing subspaces are chosen (in advance of the controller design) for their ability to approximate the open-loop DPS (2.1); consequently, it is unlikely that they would be the subspaces $(\tilde{H}_N, \tilde{H}_R)$ generated by the controller (5.7) in closed-loop with the DPS (2.1). Since $\tilde{H}_N = N(T)^{\perp}$ where T is a solution of (3.2):

$$(FT - TA + K_NC)v = 0; \quad v \in D(A) \tag{5.27a}$$

$$F = A_N - K_NC_N \quad \text{stable} \tag{5.27b}$$

it remains an open question as to whether, by choice of the controller gain K_N in (5.27), the solution T could be guided so that $N(T)^{\perp} = H_N$, i.e.

$$T^{\#}T = P_N \tag{5.28}$$

where (H_N, H_R) were any desirable orthogonal model reducing subspaces $(H_R = H_N^{\perp})$? Also, from (3.3), we would need:

$$TB = E = B_N = P_N B . \tag{5.29}$$

This seems unlikely to us unless (A,B) would have very special mathematical structure. Therefore, Theo. 5 indicates the underlying theoretical structure of the finite-dimensional feedback control problem for (2.1), but Theos. 6 and 7 yield the more practical stability analysis based on the designer's choice of the model reducing subspaces (H_N, H_R).

5.4. An Alternative Approach to Stable DPS Control

Since (2.1) and (2.9) can be rewritten as (4.5) which is an extended static feedback control problem, an alternative to the above separate design and stability analysis would be to adjust the extended gains L in (4.5) so that $\overline{A}_c \equiv \overline{A} + \overline{B}L\overline{C}$ generates an exponentially stable C_o-semigroup $\overline{U}_c(t)$ on $H = H \times R^{\alpha}$. It is not clear how one would proceed with such a search for L, except that either the Hille-Yosida condition (2.5) or the dissipative condition (2.6) would be sought for $\overline{A}_c$. The latter seems more tractable since the resolvent operator for $\overline{A}_c$ need not be calculated. Therefore, we would search for extended gains L such that $\overline{A}_c$ satisfied:

$$(\overline{A}_c w, w) \leq -\sigma_c (w,w); \quad w \in D(\overline{A}_c) = D(\overline{A}) \tag{5.30a}$$

$$(\overline{A}_c^* w, w) \leq -\sigma_c (w,w); \quad w \in D(\overline{A}_c^*) = D(\overline{A}^*) \tag{5.30b}$$

for some $\sigma_c > 0$.

It is clear that $\overline{A}_c$ generates a C_o-semigroup $\overline{U}_c(t)$ because $\overline{A}$ does (due to the fact that A does) and $\overline{B}L\overline{C}$ is a bounded perturbation of $\overline{A}$. Thus, it is only necessary to verify (5.30), so that $\overline{U}_c(t)$ will be exponentially stable. This is a straightforward approach but it is clearly not easy to do in general. It demands far more luck in the search for L than the previous approach of Secs. (5.1) - (5.3) which separates the stability analysis from the controller design.

The following special case is of some interest:

<u>Theorem 8.</u> If $\overline{A}_o$ generates a C_o-semigroup and is dissipative, i.e.

$$(\overline{A}_o w, w) \leq -\sigma_o (w,w); \quad w \in D(\overline{A}_o) \tag{5.31a}$$

$$(\overline{A}_o^* w, w) \leq -\sigma_o (w,w); \quad w \in D(\overline{A}_o^*) \tag{5.31b}$$

for some $\sigma_o > 0$ and if

$$\mathrm{Re}(\overline{\Delta A} w, w) \leq 0; \quad w \in D(\overline{A}_o) \tag{5.32}$$

then $\overline{A}_c \equiv \overline{A}_0 + \overline{\Delta A}$ is dissipative and generates an exponentially stable C_0-semigroup $\overline{U}_c(t)$ satisfying:

$$||\overline{U}_c(t)|| \leq e^{-\sigma_0(t)}; \quad t \geq 0 \ . \tag{5.33}$$

The proof of Theo. 8 follows directly from [17] Theo. 3.2 (p. 92).

This result is not as useful as it appears because, even though A may generate an exponentially stable semigroup, $\overline{A} \equiv \begin{bmatrix} A & 0 \\ 0 & 0 \end{bmatrix}$ will not, consequently, $\overline{A}$ cannot be dissipative, i.e. satisfy (5.31). Thus, if there are any dynamics present ($\alpha \neq 0$) in the controller (2.9), then taking $\overline{A}_0 = \overline{A}$ and $\overline{\Delta A} = \overline{B}L\overline{C}$ in Theo. 8 will lead nowhere even when (5.32) is satisfied. In the special case of a static feedback controller ($\alpha = 0$ and $L = L_{11}$), if the gain L_{11} can be chosen so that

$$\mathrm{Re}(BL_{11}Cw,w) \leq 0; \quad w \in D(A) \tag{5.34}$$

and if the open-loop system A is dissipative, then Theo. 8 will yield an exponentially stable closed-loop system. In particular, if $B = C^*$ (i.e. the actuators and sensors are "collocated") then choosing any gain $L_{11} = -Q$, where Q is a positive definite matrix, will satisfy (5.34); unfortunately, most control problems do not permit this collocation. When the devices are not collocated and any dynamics are present in the controller, we must go back to the beginning of this subsection and depend on our being able to discover an extended gain matrix L which would make $\overline{A}_c = \overline{A} + \overline{B}L\overline{C}$ dissipative.

The method of this subsection originated in a discussion with Prof. J. Walker, Dept. of Engr. Sciences and Appl. Math., Northwestern University, Evanston, Illinois.

6. BOUNDARY CONTROL OF DPS

In many DPS applications, control can only be achieved from the boundary of the process, i.e. control enters through the boundary conditions; interior control may be impossible. Such boundary control would be the case if one desired to control the temperature of a steel ingot, e.g., in steel tempering, control of the temperature of the surface of the ingot is the only practical possibility. A dual problem occurs when only boundary observation is available, for example, this would be the case in the steel tempering problem where thermocouples could be used to sense the surface temperature. In this section, we will concentrate on the boundary control issue, but boundary observation can be handled in a similar way.

Our DPS model (2.1) - (2.3) appears to handle only interior control because in (2.2) the influence functions b_i are in the state space H. One approach is to use a boundary space that is different from H and restrict the influence functions to this boundary space. However, this two space approach does not permit our results of Secs.

3 - 5 to be applied to the boundary control problem.

Instead, we will follow a different route: from the boundary control problem, we will create an equivalent interior control problem of form (2.1) - (2.3); then the results of Secs. 3 - 5 will be applied to this interior problem and interpreted in terms of the actual boundary problem. This method was originated in [14] where it was used to investigate boundary controllability of DPS; here we will use it to develop finite-dimensional boundary controllers for DPS. See also [13].

Consider the following linear boundary control problem:

$$\left.\begin{array}{l} \dfrac{\partial v(t)}{\partial t} = A_b v(t); \qquad v(0) = v_o \\ \\ y(t) = Cv(t) \end{array}\right\} \tag{6.1}$$

where the state v(t) is in a Hilbert space H as in (2.1) and y(t) is the same as in (2.3). Consider a linear differential operator A_o whose domain in H is yet to be defined. The control enters through the boundary conditions in the following way:

$$D(A_b) \equiv \{v \in H \mid v \text{ is sufficiently smooth and } \tau_b v = \hat{B}f\}$$

where

$$\hat{B}f = \sum_{i=1}^{M} \hat{b}_i \hat{f}_i \quad . \tag{6.2}$$

"Sufficiently smooth" means if v belongs to $D(A_b)$ then $A_o v$ belongs to H; we define the operator A_b as $A_b = A$ on the domain $D(A_b)$. This is a non-homogeneous boundary value problem with the (clearly linear) boundary operator $\tau_b: H \rightarrow H_b$ defined by (6.2) where $\hat{b}_i$ are in H_b the boundary space; this boundary space is a different Hilbert space composed of functions defined only on the boundary of the process. The control f(t) enters through the linear operator $\hat{B}: R^M \rightarrow H_b$.

Since (6.1) - (6.2) is not in the form (2.1) - (2.3), we must convert it into an equivalent homogeneous boundary value problem. Let $D(A_o) \equiv \{v \in H \mid v$ is sufficiently smooth$\}$ and $\tau_b v = 0$, where "sufficiently smooth" means if v belongs to $D(A_o)$, then Av_o belongs to H. Note that A_o and A_b are the same differential operator but they operate on different domains in H. Define

$$v(t) = w(t) + hf(t) \tag{6.3}$$

where w belongs to $D(A_o)$ and h is chosen in H, such that:

$$\tau_b hf = \hat{B}f \tag{6.4a}$$

$$\tau_b h = \hat{B} \quad . \tag{6.4b}$$

Consequently, h extends the boundary conditions into the interior. Therefore, $\tau_b v = \tau_b w + \tau_b hf = \hat{B}f$ and v belongs to $D(A_b)$. From (6.1) - (6.3), we obtain (formally):

$$\frac{\partial w(t)}{\partial t} = \frac{\partial v(t)}{\partial t} + hf(t) = A_b v(t) + hf(t)$$

$$= A_o w(t) + A_b hf(t) + hf(t)$$

and

$$y(t) = Cv(t) = Cw(t) + Chf(t).$$

Let

$$\tilde{f}(t) \equiv \dot{f}(t); \quad f(0) \equiv 0 \tag{6.5a}$$

$$q(t) \equiv \begin{bmatrix} w(t) \\ f(t) \end{bmatrix}; \quad q(0) \equiv \begin{bmatrix} v_o \\ 0 \end{bmatrix} \tag{6.5b}$$

$$\left.\begin{aligned} \frac{\partial q(t)}{\partial t} &= \tilde{A}q(t) + \tilde{B}\tilde{f}(t); \quad q(0) = q_o \\ y(t) &= \tilde{C}q(t) \end{aligned}\right\} \tag{6.6}$$

where $\tilde{A} = \begin{bmatrix} A_o & A_b h \\ 0 & 0 \end{bmatrix}$, $\tilde{B} = \begin{bmatrix} -h \\ I_M \end{bmatrix}$, $\tilde{C} = [C \quad Ch]$

with $D(\tilde{A}) \equiv D(A_o) \times R^M$ dense in the Hilbert space $\tilde{H} \equiv H \times R^M$. Both $\tilde{B}$ and $\tilde{C}$ are finite-rank linear operators with $\tilde{B}: R^M \to \tilde{H}$ and $\tilde{C}: \tilde{H} \to R^P$.

Therefore, the homogeneous boundary value problem (6.6) is an equivalent interior control problem for the original boundary control problem (6.1) - (6.2). Furthermore, the boundary control f(t) is related to the equivalent interior control $\tilde{f}(t)$ by (6.5a) or

$$f(t) = \int_0^t \tilde{f}(\tau)d\tau \quad . \tag{6.7}$$

The choice of h in H is usually done by fitting a sufficiently differentiable polynominal through the nonhomogeneous boundary conditions; note that the choice of h is not unique and, hence, there may be many equivalent interior control problems. One special case of interest is when hf belongs to $N(A_b)$, i.e.

$$A_b hf = 0 \; . \tag{6.8}$$

We illustrate the above with a simple example. Consider the following heat conduction problem:

$$\frac{\partial v(x,t)}{\partial t} = \frac{\partial^2 v(x,t)}{\partial x^2}; \quad 0 \le x \le 1 \quad \text{and} \quad t \ge 0 \tag{6.9a}$$

$$v(0,t) = 0 \tag{6.9b}$$

$$\frac{\partial v(1,t)}{\partial x} = f(t) \tag{6.9c}$$

$$y(t) = \int_0^1 c(x)v(x,t)dx \ . \tag{6.9d}$$

The temperature distribution $v(x,t)$ in (6.9) is maintained at zero on one end of a uniform bar of unit length and the heat flow is controlled by $f(t)$ on the other end of the bar. Let $H = L^2(0,1)$ and $D(A_b) \equiv \{v \in H \mid$ sufficiently smooth, and (6.9b) - (6.9c) are satisfied$\}$.

Also, $D(A_o) \equiv \{v \in H \mid$ sufficentily smooth and $v(0,t) = 0$ and $\frac{\partial v(1,t)}{\partial x} = 0\}$, where

$$A_o \equiv \frac{\partial^2}{\partial x^2} \quad .$$

Consider

$$h(x) = x \quad . \tag{6.10}$$

Therefore, $hf(t) = xf(t)$ is in $D(A_b)$ and $v(t) = w(t) + hf(t) = w(t) + xf(t)$ is in $D(A_o)$ when $w(t)$ is in $D(A_o)$. Higher order polynominals could have been used for h but the choice (6.10) has the advantage that hf is in $N(A_b)$. The equivalent interior control problem for (6.9) with h chosen by (6.10) is the following:

$$\frac{\partial q(t)}{\partial t} = \begin{bmatrix} \frac{\partial^2}{\partial x^2} & 0 \\ 0 & 0 \end{bmatrix} q(t) + \begin{bmatrix} x \\ 1 \end{bmatrix} \tilde{f}(t) \tag{6.11}$$

$$y(t) = [(c,.)(c,x)]q(t)$$

where $q(t) \equiv \begin{bmatrix} w(t) \\ f(t) \end{bmatrix}$ and $\tilde{f}(t) \equiv \dot{f}(t)$.

Note that the above h produces hf in $N(A_b)$; however, if the boundary condition (6.9b) is changed to $\left.\frac{\partial v(x,t)}{\partial x}\right|_{X=0} = 0$, then $h(x) = 1/2\ x^2$ must be used and hf is not in $N(A_b)$.

In general, we must assume that $\tilde{A}$ in (6.6) generates a C_o-semigroup, but this follows if A_o generates one. The results of Secs.3 - 5 can now be applied to obtain finite-dimensional boundary controllers of the form

$$
\left.
\begin{aligned}
f(t) &= \int_0^t \tilde{f}(\tau)d\tau && \text{(6.12a)}\\
\tilde{f}(t) &= L_{11}y(t) + L_{12}z(t) && \text{(6.12b)}\\
\dot{z}(t) &= L_{21}y(t) + L_{22}z(t) = Fz(t) + Ky(t) + E\tilde{f}(t) && \text{(6.12c)}
\end{aligned}
\right\}
$$

where $\dim z = \alpha < \infty$. Note that the actual control signal applied at the boundary is the integral of the control generated by the equivalent interior control problem.

There are many interesting theoretical issues generated by this approach to boundary control:

(1) Can the boundary conditions always be extended into the interior, i.e. does an h always exist which satisfies (6.4)?

(2) What are the connections between this approach and the usual two space approach, and what is the effect of the "non-uniqueness" of h on these connections?

(3) Does the integral feedback in (6.12a) have any special significance?

(4) What do the stabilizibility and detectability conditions look like for reduced-order models of (6.6)?

It seems unlikely that the answer to (1) would be "yes", in general; there must be situations where the boundary conditions do not extend into the interior. The answer to (2) would reveal a great deal about the structure of boundary control in DPS. As far as (3) is concerned, we can say the following:

Suppose (6.5a) is replaced with

$$\dot{f}(t) + \varepsilon f(t) = \tilde{f}(t), \quad f(0) \equiv 0 \tag{6.13}$$

where $\varepsilon > 0$ then (6.6) would be unchanged except for

$$A = \begin{bmatrix} A_o & (A_b+\varepsilon)h \\ & \\ 0 & -\varepsilon \end{bmatrix} . \tag{6.14}$$

Since for practical control systems an exact integrator law like (6.12a) may be difficult to implement (and can lead to an unstable implementation, sometimes), (6.13) is more likely to be the form of the boundary control law which can be implemented. However, the stabilizibility and detectability conditions for reduced-order models of (6.6) will be different if $\tilde{A}$ is replaced by (6.14); consequently, the answer to (4) will depend on how the boundary control law is implemented.

7. SUMMARY AND CONCLUSIONS

As even a cursory perusal of these Proceedings or [22] reveals, there are (it would only be poetic justice if it were: infinitely) many interesting problems in distributed parameter control. In this paper, we have concentrated on the mathematical structure of a linear distributed parameter system (DPS) of the form (2.1) in closed-loop with a finite-dimensional controller (2.9). This is motivated by the implementation problem for online controllers of engineering systems.

Our main results are Theo. 1 (Sec. 3) and Theo. 5 (Sec. 4). The first of these results shows that any finite-dimensional observer for (2.1) can only asymptotically reconstruct the projection of the infinite-dimensional state of (2.1) onto some finite-dimensional subspace. The second of these results reveals what is going on in any stable finite-dimensional controller (2.9) which produces an exponentially stable closed-loop system with the DPS (2.1); such a controller is possible if and only if there exists a pair of stabilizing subspaces $(\tilde{H}_N,\tilde{H}_R)$ for (2.1). This separates the control problem into two parts:

(1) the asymptotic reconstruction of the projection of the full DPS state of (2.1) onto the finite-dimensional subspace $\tilde{H}_N$, and

(2) the exponential stabilization of the DPS (2.1) with only projection feedback on $\tilde{H}_N$.

The technical restrictions are that the controller (2.9) must be stable and (3.2) - (3.3) must have a solution, i.e. the controller must have an intrinsic finite-dimensional observer. This seems like a fairly natural and readily satisfied (Theo. 3) set of restrictions.

Although the usual separation principle for linear systems cannot be invoked to determine the stability of (2.1) in closed-loop with (2.9) because the controller is finite-dimensional and the DPS is infinite-dimensional, the above provides a kind of separation principle for the structure of this problem. This result reveals that exponential stability of (2.1) can only be achieved by a finite-dimensional controller when it can be achieved by finite-dimensional projection feedback. Of course, such a projection of the state of (2.1) can rarely be measured directly by the sensors, but the dynamics of the controller (2.9) make it possible to asymptotically reconstruct this projection from most sensor outputs.

Obviously, finite-dimensional projection feedback on $\tilde{H}_N$ can at best relocate a finite number of eigenvalues namely $\sigma(\tilde{A}_N)$, i.e. that part of the spectrum of the operator A projected onto the subspace $\tilde{H}_N$. In general, $(\tilde{H}_N,\tilde{H}_R)$ are not A-invariant subspaces; so, the effect of the projection feedback is not confined to $\sigma(\tilde{A}_N)$ alone. Also, finite-dimensional projection feedback yields a finite rank (hence, compact) perturbation of A in (2.1). Thus, from [15], if A generates a contraction semigroup which is strongly, but not exponentially, stable, then it is impossible to obtain exponential stability with the projection feedback; consequently, Theo. 5 shows that it cannot be achieved with a finite-dimensional controller either; the controller dynamics cannot perform miracles.

The construction of the stabilizing subspaces $(\tilde{H}_N, \tilde{H}_R)$ associated with the controller (2.9) involves solving (3.2) for the operator T and calculating the pseudo-inverse of $\tilde{T} = \begin{bmatrix} C \\ T \end{bmatrix}$. Although Theo. 3 gives simple conditions under which this can be done, i.e. the stable controller and the DPS must share no part of their spectra, it is not straightforward to actually calculate $(\tilde{H}_N, \tilde{H}_R)$. Also, the more natural subspaces to use for controller design are model reducing subspaces (H_N, H_R), which are chosen in advance for their ability to approximate the open-loop DPS (2.1); the controller is designed from a model reduction of (2.1). The odds are against the subspaces $(\tilde{H}_N, \tilde{H}_R)$ (which are associated with the controller structure) being the reducing subspaces chosen a priori.

Model reduction as a means for finite-dimensional controller synthesis is discussed in Sec. 5. This yields controllers of the form (5.7). Although they try, such controllers do not, in general, asymptotically reconstruct the projection of the DPS state onto the finite-dimensional model reducing subspace H_N. Consequently, it is not immediately clear whether such a controller will produce an exponentially stable closed-loop system.

Heuristically, if the reduced-order model (5.3) is a good approximation of the open-loop DPS (2.1), then a controller which stabilizes the reduced-order model should also stabilize the actual DPS, as long as the residuals (unmodeled part of the DPS) are stable. The problem is that, even though the open-loop residuals are stable (a prerequisite of model reduction), they can be made unstable in closed-loop with the controller through modeling error and spillover terms, see [9]. Theos. 6 and 7 give bounds on the norms of the relevant modeling error and spillover terms which guarantee exponential stability of the closed loop (2.1) and (5.7). A direct relationship between the model reducing subspaces and $(\tilde{H}_N, \tilde{H}_R)$ is of interest; however, Theo. 5 gives some motivation for the idea that (H_N, H_R) should be stabilizing subspaces for (2.1), as used in Theos. 6 and 7. An alternative approach for sythesizing finite-dimensional stabilizing controllers for (2.1) is described in Subsection 5.4. The closed-loop system (2.1) and (2.9) is rewritten as an extended static feedback control problem; gains are sought for which the extended problem is dissipative.

Many DPS problems only admit control through the boundary conditions. Such boundary control problems do not have the form (2.1) - (2.3). These problems have traditionally been handled using the trace theory of [18], Chapt. 1, which establishes a boundary Hilbert space, in addition to the usual state space. In [14], a method was developed for extending the boundary conditions into the interior; this creates an equivalent interior control problem. In Sec. 6, the method of [14] is used to cast boundary control problems into the form (2.1) - (2.3); so, the results of this paper can be extended to boundary control problems (at least those for which the method of [14] works). The finite-dimensional boundary controllers, thus produced, have the form (6.12); their structure in producing exponential stability of the closed-loop is much the same as that of interior control problems, except for the integral feedback in (6.12a). The relationship between boundary controllers obtained this way and those obtained via the "two space approach", e.g. [24] remains to be investigated; see also [13].

As far as examples are concerned, it is easy to illustrate the ideas in this paper with the heat or wave equation in one-space dimension with a variety of boundary conditions. This is the simplest situation and, at least as far as the heat equation goes, almost anything works because the diffusion process is very stable by nature. For the wave equation, things are a little more complicated because no damping is present; from the results of [15] and our Theo. 5, it is clear that a finite-dimensional control cannot make this wave equation exponentially stable. However, if some damping is present, then a finite-dimensional modal controller can improve the stability of any finite number of modes (possibly at the cost of some stability margin in the other modes). The heat and wave equation in more than one-space dimension are a bit more difficult but not intractable. For real enginerring systems, where (2.1) is an approximation (and sometimes, a gross one at that), the situation is often quite complicated, and analysis of a particular one of these is really deserving of a much more detailed treatment than our remaining space can encompass. It is clear how to proceed conceptionally, based on the results developed here, but the details are simultaneously difficult and instructive. Next we present a brief "sermon".

In the past, physics provided the impetus for applied mathematics and many mathematicians were also excellent physicists. We feel that engineering systems play a similar role vis à vis mathematics, and any serious applied mathematician will want to become expert in at least one area of engineering systems. We suggest that this entails more than a brief study of the literature on the highly theoretical side of the area, e.g. IEEE Transactions on Automatic Control: it means getting "dirty", i.e. learning to speak the language of the engineers in a given area and developing some of their kind of experience and intuition about the applications via computer simulation and laboratory experiment. This is often difficult and time consuming, and it involves overcoming a certain snobbery which says that only the fanciest mathematics is interesting. However, it is rewarding - your mathematics will be much richer for the experience, and it is much more likely to be used in applications; isn't that the whole point of it? Okay, end of sermon.

ACKNOWLEDGEMENTS

This research was supported in part by the National Science Foundation under Grant No. ECS-80-16173 and the National Aeronautics and Space Administration under Grant No. NAG-1-171. Any options, findings, and conclusions or recommendations expressed in this publication are those of the author and do not necessarily reflect the views of NSF or NASA.

REFERENCES

[1] Balas, M.: Toward A More Practical Theory of DPS Control, Advances in Dynamics and Control: Theory and Appl., Vol. 18, C.T. Leondes, ed., Academic Press, NY, 1982.

[2] Balas, M.: Trends in Large Space Structure Control Theory: Fondest Hopes, Wildest Dreams, IEEE Trans. Autom. Control, Vol. AC-27 (1982), 522-535.

[3] Balas, M.: The Galerkin Method and Feedback Control of Linear DPS, J. Math. Analysis and Appl. (to appear).

[4] Balas, M.: Reduced-Order Feedback Control of DPS via Singular Perturbation Methods, J. Math. Analysis and Appl. 87 (1982), 281-294.

[5] Balas, M.: Stability of DPS with Finite-Dimensional Controller-Compensators Using Singular Perturbations, J. Math. Analysis and Appl. (to appear).

[6] Balas, M.: Stable Feedback Control of DPS: Time and Frequency Domain Conditions, Invited Lecture at Workshop on Applications of Distributed Systems Theory to the Control of Large Space Structures, Jet Propulsion Laboratory, Pasadena, CA, 1982.

[7] Balas, M.: Discrete-Time Control of DPS, Proc. of Int'l. Symp. on Engr. Sci. and Mechanics, National Cheng Kung Univ., Tainan, Taiwan, R.O.C., 1981.

[8] Balas, M.: Stabilizing Subspaces and Linear DPS: Discrete and Continuous-Time Control, Proc. of 15th Asilomar Conf. on Circuits, Systems and Computers, Pacific Grove, CA, 1981

[9] Balas, M.: Feedback Control of Flexible Systems, IEEE Trans. Autom. Control. Vol. AC-23 (1978), 673-679.

[10] Ball, J., J. Marsden, M. Slemrod: Controllability for Distributed Bilinear Systems, SIAM J. Control and Opt. 20 (1982), 575-597.

[11] Curtain, R.: Finite-Dimensional Compensation Design for Parabolic Distributed Systems with Point Sensors and Boundary Input, IEEE Trans. Autom. Control, Vol. AC-27 (1982), 98-104.

[12] Curtain, R., A. Pritchard: Functional Analysis and Modern Applied Mathematics, Academic Press, NY, 1977.

[13] Curtain, R.: Finite-Dimensional Compensators for some Hyperbolic Systems with Boundary Input, Invited Lecture at Conf. on Control Theory for DPS, Vorau, Austria, 1982.

[14] Fattorini, H.: Boundary Control Systems, SIAM J. Control 6 (1968), 349-385.

[15] Gibson, J.S.: A Note on Stabilization of Infinite-Dimensional Linear Oscillators by Compact Linear Feedback, SIAM J. Control and Opt. 18 (1980), 311-316.

[16] Kato,T.: Perturbation Theory for Linear Operators, Springer, NY, 1966.

[17] Kwakernaak, H., R. Sivan: Linear Optimal Control Systems, J. Wiley and Sons, NY, 1972.

[18] Lions, J.L., E. Mangenes: Nonhomogeneous Boundary Value Problems and Applications, Vol. 1, Springer, NY, 1972.

[19] Pritchard, A., J. Zabczyk: Stability and Stabilizability of Infinite-Dimensional Systems, SIAM Review 23 (1981), 25-52.

[20] Pritchard, A.: Finite dimensional compensators for nonlinear infinite dimensional systems, Invited Lecture at Conf. on Control Theory for DPS, Vorau, Austria, 1982.

[21] Ray, W.H.: Some Applications of DPS State Estimation in Control Theory of Systems Governed by PDE, A. Aziz, J. Wingate and M. Balas, eds., Academic Press, NY, 1977.

[22] Russell, D.: Controllability and Stabilizability Theory for Linear PDE: Recent Progress and Open Questions, SIAM Review 20 (1978), 371-388.

[23] Schumacher, J.: Dynamic Feedback in Finite and Infinite Dimensional Linear System Systems, PhD Thesis, Dept. of Mathematics, Vrije Universiteit, Amsterdam, The Netherlands, 1981.

[24] Triggiani, R., I. Lasiecka: Boundary Feedback Stabilization Problems for Hyperbolic Equations, Invited Lecture at Conf. on Control Theory for DPS, Vorau, Austria, 1982.

[25] Walker, J.: Dynamical Systems and Evolution Equations: Theory and Applications, Plenum Press, NY, 1980.

APPENDIX I: Proofs of Theos. 1 and 2

Proof of Theo. 1: Let T be a solution of (3.2); since D(A) is dense in H and $\tilde{T}$ is bounded (in fact, it is finite rank), $\tilde{T}$ can be extended as a bounded operator to all of H. Consider $z(t) = Tv(t) + e(t)$. Let $v_0 \in D(A)$; hence $v(t) \in D(A)$ for $t \geq 0$. ([12] Theo. 8.10, p. 157). From (2.1) and (2.9), since v(t) in D(A) is differentiable and satisfies (2.1), we have

$$\dot{e}(t) = \dot{z}(t) - T\frac{\partial v}{\partial t}(t)$$

$$= Fz(t) + KCv(t) + Ef(t) - T[Av(t) + Bf(t)]$$

$$= Fe(t) + (E-TB)\,f(t) + (FT - TA + KC)v(t)$$

$$\therefore \dot{e}(t) = Fe(t) \quad \text{because of (3.2) and (3.3).}$$

Also, $e(0) = z_0 - Tv_0$, and, since F is stable by assumption, $\lim_{t\to\infty} e(t) = 0$.

Consider

$$
\begin{aligned}
q(t) &= Q_{11}\, y(t) + Q_{12}\, z(t) \\
&= (Q_{11}\, C + Q_{12}\, T)\, v(t) + Q_{12}\, e(t) \\
&= Q\, \tilde{T}\, v(t) + Q_{12}\, e(t)
\end{aligned}
$$

where $Q \equiv [Q_{11}\ Q_{12}]$ and $\tilde{T} \equiv \begin{bmatrix} C \\ T \end{bmatrix}$. From Theo. 2, let $Q = \tilde{T}^{\#}$, i.e.

$$
\begin{aligned}
q(t) &= \tilde{T}^{\#}\tilde{T}\, v(t) + Q_{12}\, e(t) \\
&= \tilde{P}_N\, v(t) + Q_{12}\, e(t)
\end{aligned}
$$

where $\tilde{P}_N$ is orthogonal projection onto $\tilde{H}_N \equiv N(\tilde{T})^{\perp}$.

$$\therefore \lim_{t\to\infty} [q(t) - \tilde{P}_N\, v(t)] = \lim_{t\to\infty} Q_{12}\, e(t) = 0 .$$

Furthermore,

$$\lim_{t\to\infty} [q(t) - v(t)] = \lim_{t\to\infty} [q(t) - \tilde{P}_N\, v(t)] - \lim_{t\to\infty} \tilde{P}_R\, v(t) = - \lim_{t\to\infty} \tilde{P}_R\, v(t)$$

where $\tilde{P}_R$ is orthogonal projection onto $\tilde{H}_R \equiv N(\tilde{T})$. Clearly,

$$H = N(\tilde{T}) \oplus N(\tilde{T})^{\perp} = \tilde{H}_N \oplus \tilde{H}_R .$$

Also, from Theo. 2 part (d), since $\dim R^{P+\alpha} = P + \alpha$, $\dim \tilde{H}_N = \dim N(\tilde{T})^{\perp} = \dim R(\tilde{T}) \leq P + \alpha$. This gives (3.6). Finally, the subspaces $\tilde{H}_N$ and $\tilde{H}_R$ are nontrivial: Suppose not. If $N(\tilde{T}) = \{0\}$, then $\tilde{T}$ is 1-1 and $\dim H \leq P + \alpha$, which is generally not true. If $N(\tilde{T})^{\perp} = \{0\}$, then $H = N(\tilde{T})$. Therefore, for any v in H, $\tilde{T}v = 0$ or equivalently, $Cv = 0$ and $Tv = 0$. This would mean $H = N(C) \cap N(T)$; hence, the measurements $y(t) \equiv 0$ and this is not generally true. Consequently, trivial subspaces H_N and H_R lead to trivial DPS (2.1). This completes the proof of Theo. 1.

Proof of Theo. 2: $T^{\#}$ is well defined on R(T) by (3.9); hence, since T is onto, $T^{\#}$ is well defined on H_2. Consider $v \equiv \alpha v_1 + v_2 \in H_2$. Thus, $v = \alpha Tw_1 + Tw_2 = T(\alpha w_1 + w_2)$ because T is linear and onto. Therefore, $T^{\#}(v) = T^{\#}[T(\alpha w_1 + w_2)] = \tilde{P}_N(\alpha w_1 + w_2) = \alpha \tilde{P}_N w_1 + \tilde{P}_N w_2 = \alpha T^{\#} Tw_1 + T^{\#} Tw_2 = \alpha T^{\#} v_1 + T^{\#} v_2$ and $T^{\#}$ is linear on H_2, because of the linearity of the projection $\tilde{P}_N$. This proves (a).

Consider, from (3.9), for any v in H_1, $T\, T^{\#}\, Tv = T\, \tilde{P}_N v = Tv$ because $v = \tilde{P}_N v + \tilde{P}_R v$ and $\tilde{P}_R v \in N(T)$. This proves (b). Let $v_n \to v$ and $T^{\#} v_n \to z$ for $v_n \in H_2$. Since $H_2 = R(T)$ is closed by hypothesis, $v \in H_2$ and $v_n = Tw_n \to v = Tw$. Thus, $T^{\#} v_n = T^{\#} Tw_n = \tilde{P}_N w_n \to z$ by (3.9). This, together with (from part (b)) $T\tilde{P}_N w_n = Tw_n \to v$ implies $z \in H_1$, and $Tz = v$ because T is bounded on H_1 and, hence, is closed, Now, $T^{\#} v = T^{\#} Tz = \tilde{P}_N z$. But $z = \lim_n \tilde{P}_N w_n$ and $N(T)^{\perp}$ is closed; therefore, $z \in N(T)^{\perp}$ and $T^{\#} v = \tilde{P}_N z = z$.

This proves $T^{\#}$ is a closed operator on H_2 into H_1 and, by the Closed Graph Theorem (e.g. [12] Theo. 3.3, p. 45), $T^{\#}$ is bounded since both H_1 and H_2 are Hilbert spaces by hypothesis. This proves (c).

Finally, it is easy to see that $R(T^{\#}) = N(T)^{\perp}$. Furthermore, $N(T^{\#}) = \{0\}$ because $y \in N(T^{\#})$ implies $y = Tv$ and $0 = T^{\#} y = T^{\#} Tv = \tilde{P}_N v$ and, $v \in N(T)$ or $y = Tv = 0$. Consequently, $T^{\#}$ is 1-1 and onto (bijective) from $H_2 = R(T)$ to $R(T^{\#}) = N(T)^{\perp}$. If $\dim H_2 < \infty$, then $\dim N(T)^{\perp} = \dim H_2 = \dim R(T)$. This proves (d) and completes the proof of Theo. 2.

APPENDIX II: Proof of Theo. 3

Let T be defined on D(A) by (3.11). Note that

$$R(\lambda,A)A = AR(\lambda,A) = \lambda R(\lambda,A) - I \quad . \tag{A.II.1}$$

For any v in D(A), we have (using (A.II.1)):

$$2\pi i\ (FT - TA)v = \int_\Gamma [FR(\lambda,F)\ KC\ R(\lambda,A) - R(\lambda,F)\ KC\ R(\lambda,A)A]v\ d\lambda$$

$$= \int_\Gamma [FR(\lambda,F)\ KC\ R(\lambda,A) + R(\lambda,F)\ KC - \lambda R(\lambda,F)\ KC\ R(\lambda,A)]v\ d\lambda$$

$$= \int_\Gamma R(\lambda,F)\ KC\ v\ d\lambda - \int_\Gamma KC\ R(\lambda,A)v\ d\lambda$$

$$= -\ 2\pi i\ KC\ v - 0 \tag{A.II.2}$$

This last equality follows from ([16] pp. 39-40 and Theo. 6.17, p. 178):

$$\int_\Gamma R(\lambda,F)z\ d\lambda = -\ 2\pi i\ z$$
$$\int_\Gamma R(\lambda,A)v\ d\lambda = 0 \tag{A.II.3}$$

because the curve Γ encloses all the eigenvalues of F and excludes the spectrum of A.

From (A.II.2), we have T satisfying (3.2) as desired. Since $R(T) \subseteq R^{\alpha}$, T has finite-rank and, hence, is a bounded linear operator defined on D(A). However, D(A) is dense in H; therefore, T can be extended to a bounded linear operator on all of H.

To show that T is unique, assume the bounded linear operators T_1 and T_2 satisfy (3.2). Let $\Delta T = T_1 - T_2$ and, from (3.2), we obtain, for any v in D(A):

$$(F\Delta T - \Delta TA)v = 0$$

$$[(\lambda I - F)\Delta T - \Delta T(\lambda I - A)]v = 0\ .$$

Consequently, if λ is chosen in $\rho(F) \cap \rho(A)$, i.e. λ is not an eigenvalue of F and λ is in $\sigma(A)^c$, then

$$[\Delta TR(\lambda,A) - R(\lambda,F)\Delta T]v = 0 \tag{A.II.4}$$

using (A.II.3) on (A.II.4), yields $\Delta Tv = 0$ for all v in D(A) which is dense in H. Therefore, $\Delta T = 0$; hence, T is unique. This completes the proof of Theo. 3.

APPENDIX III: Proof of Theo. 5

By Theo. 1, there exists a bounded linear operator $T: H \to R^\alpha$ such that $z(t) = Tv(t) + e(t)$ where

$$\dot{e}(t) = Fe(t) \tag{A.III.1}$$

Consider the control law of (2.9) with L_{11} and L_{12} given:

$$\begin{aligned} f(t) &= L_{11}\, y(t) + L_{12}\, z(t) \\ &= (L_{11}\, C + L_{12}\, T)\, v(t) + L_{12}\, e(t) \\ &= L\tilde{T}\, v(t) + L_{12}\, e(t) \end{aligned} \tag{A.III.2}$$

where $L \equiv [L_{11}\; L_{12}]$ and $\tilde{T} \equiv \begin{bmatrix} C \\ T \end{bmatrix}$.

From Theo. 2 part (b),

$$\tilde{T}v = \tilde{T}\,\tilde{P}_N v \tag{A.III.3}$$

where $\tilde{P}_N$ is orthogonal projection onto $N(\tilde{T})^\perp$. Thus, using (A.III.3) in (A.III.2), we have

$$f(t) = L\tilde{T}\tilde{P}_N v(t) + L_{12}\, e(t)\,. \tag{A.III.4}$$

From (A.III.1), (A.III.4), and (2.1), we obtain

$$\left.\begin{aligned} \frac{\partial v(t)}{\partial t} &= A_o v(t) + BL_{12}\, e(t) \\ \dot{e}(t) &= Fe(t) \end{aligned}\right\} \tag{A.III.5}$$

where $A_o \equiv A + BG$ and $G \equiv L\tilde{T}\,\tilde{P}_N$.

Clearly, the pair of subspaces

$$\tilde{H}_N \equiv N(\tilde{T})^\perp \text{ and } \tilde{H}_R \equiv N(\tilde{T})$$

satisfy $H = \tilde{H}_N \oplus \tilde{H}_R$ and $\dim \tilde{H}_N \equiv N \leq P + \alpha < \infty$ from Theo. 1 (3.7). Consequently,

$(\tilde{H}_N, \tilde{H}_R)$ are stabilizing subspaces for (A,B) in (2.1) if and only if (4.2) is satisfied and A_o is exponentially stable. However, $G\tilde{P}_N = L\tilde{T}\tilde{P}_N^2 = L\tilde{T}\tilde{P}_N = G$, i.e. (4.2) is satisfied no matter what L is. Furthermore, the closed-loop system (A.III.5) is exponentially stable if and only if $A_o = A + BG$ is exp. stable (because F is assumed stable). Therefore, Theo. 5 is proved.

INVERSE PROBLEMS FOR HYPERBOLIC SYSTEMS WITH UNKNOWN BOUNDARY PARAMETERS *)

H.T. Banks and K.A. Murphy

Lefschetz Center for Dynamical Systems
Division of Applied Mathematics
Brown University
Providence, RI 02912, USA

and

Department of Mathematics
Southern Methodist University
Dallas, Texas 75275, USA

In this note we present a scheme for estimation of parameters that can be considered an extension of the techniques and ideas of [5], [9] to allow treatment of identification problems that are typical in the 1-D seismic inverse problem [1], [10],[11]. It is shown in [5] how one can use cubic spline approximation techniques in parameter identification problems for hyperbolic systems with simple Dirichlet or Neumann boundary conditions. Here we are again interested in hyperbolic systems but with special boundary conditions which depend on unknown parameters. One possible approach is to make a change of the variables so as to reduce the problem to one with simple known boundary conditions where the unknown parameters have been transformed to the partial differential equation itself. While such a technique can prove fruitful for certain classes of problems (e.g., see the beam examples with damping in [3]), it is not feasible for the problems under consideration here. Rather we shall treat the boundary conditions and unknown parameters contained therein directly.

The problem we consider concerns the acoustic or 1-D elastic wave equation [1], [10],[11] with elastic boundary conditions at one (the upper or left) boundary and absorbing boundary conditions at the other (lower or right) boundary. Specifically we consider

$$\rho(x) \frac{\partial^2 u}{\partial t^2} = \frac{\partial}{\partial x} (E(x)\frac{\partial u}{\partial x}) \qquad 0 \leq x \leq 1,\ t > 0, \qquad (1...)$$

---

*) Research supported in part by the Air Force Office of Scientific Research under contract AFOSR 81-0198, in part by the National Science Foundation under grant MCS-8205355, and in part by the U.S. Army Research Office under contract ARO-DAAG-29-79-C-0161.

$$\frac{\partial u}{\partial x}(t,0) + q_1 u(t,0) = s(t,\tilde{q}), \quad \frac{\partial u}{\partial t}(t,1) + q_2 \frac{\partial u}{\partial x}(t,1) = 0, \qquad (\ldots 1)$$

$$u(0,x) = u_t(0,x) = 0,$$

where q_1 is a parameter (an elastic modulus) for the restoring force in the medium at the surface ($x = 0$), s is an unknown source term (which we do not assume is necessarily an impulse) resulting from a perturbing shock to the medium at the surface. Here q_2 ($= \sqrt{E(1)/\rho(1)}$) in the absorbing boundary condition (no upgoing or reflected waves) at the "bottom" of the field results from factoring the wave operator at $x = 1$, ρ is the mass density of the medium, and E is an elastic modulus.

The fundamental problem consists of estimating ρ, E, q_1, q_2, $\tilde{q}$ from observations of displacement $u(t,0)$ (or velocity $u_t(t,0)$) at the surface. There is a large literature on 1-D seismic inverse problems of this nature and it is well-known that it is, in general, impossible to determine both field mass density and elastic modulus from surface observations alone. It is therefore standard practice to make some assumptions in order to simplify the problem and reduce ill-posedness. Thus the problem we discuss (ρ = constant) is a restriction to a special case of the actual 1-D seismic problem of interest. However, we hasten to add that all the 1-D problems themselves fall short of addressing the "real" problems which are unquestionably 3-dimensional in nature.

Our purpose here is to indicate that methods developed and used in other contexts ([2],[3],[4],[5],[7],[8],[9]) are, in principle, applicable to seismic problems. Even though we employ a simple 1-D model problem to illustrate the ideas, the techniques are readily applicable to higher dimensions and indeed we have already established that certain aspects and features of our schemes can be adapted with relative ease to treat 2-D and 3-D problems.

We observe that in (1), the assumption ρ = constant leads to a problem in which knowledge of E, q_1 and $q_2 = \sqrt{E(1)/\rho}$ along with that of the source parameters $\tilde{q}$ resolves the inverse problem. This is the problem for which we have developed both theory and software packages based on the cubic spline approximation techniques of [5]. For ease in exposition here, we discuss the special case were E is constant and also assume that we have transformed the system (by a standard change of variables) to one with homogeneous boundary conditions. Thus the problem we discuss is the following.

Consider the system

$$u_{tt} = q_0 u_{xx} + f(t,x,q) \qquad 0 \leq x \leq 1,\ t > 0$$

$$u_x(t,0) + q_1 u(t,0) = 0 \qquad (2\ldots)$$

$$u_t(t,1) + q_2 u_x(t,1) = 0 \qquad (...2)$$

$$u(0,x) = \phi(x;q)$$

$$u_t(0,x) = \psi(x;q),$$

where f, ϕ, ψ are continuous and the vector parameter $q = (q_0,q_1,q_2,\tilde{q})$ is to be chosen from some given compact set Q contained in the set $\{q|q_0 > 0,\ q_1 < 0,\ q_2 > 0\}$. Given data (observations) $\tilde{y}_i \sim u(t_i,0)$, $i = 1,2,\ldots,m$, we seek to minimize

$$J(q) \equiv \sum_{i=1}^{m} |\hat{y}_i - u(t_i,0;q)|^2 \qquad (3)$$

over $q \in Q$, where $(t,x) \to u(t,x;q)$ is the solution of (2) corresponding to q.

Following the general approach in [5],[9], we rewrite (2) as an abstract system

$$\begin{aligned} \dot{z}(t) &= A(q)z(t) + F(t,q) \\ z(0) &= z_0(q) \end{aligned} \qquad (4)$$

in a Hilbert space X. To this end, we define V(q) as the Sobolev space $H^1(0,1)$ with inner product

$$\langle v,w\rangle_q \equiv q_0 \int_0^1 v'w'dx - q_0 q_1 v(0)w(0) \qquad (5)$$

and then take $X(q) = V(q) \times H^0(0,1)$. We further define

$$V_B(q) = \{w \in V(q)|w \in H^2(0,1),\ w'(0) + q_1 w(0) = 0\}$$

and

$$A(q) = \begin{pmatrix} 0 & 1 \\ q_0 D^2 & 0 \end{pmatrix}$$

on dom $A(q) \equiv \{(u,v)^T \in V_B(q) \times H^1 | v(1) + q_2 u'(1) = 0\}$ in X(q). Here $D = \frac{\partial}{\partial x}$ is the usual spatial differentiation operator.

With these definitions, (2) can be written as (4) with $z = (u,u_t)^T$ and $z_0(q) = (\phi,\psi)^T$ - (where we assume $\phi \in H^1$). It is then not difficult to establish that A(q) is dissipative with $\mathcal{R}(\lambda - A(q)) = X(q)$ for some $\lambda > 0$. It follows that A(q) is the infinitesimal generator of a strongly continuous semigroup S(t;q) on X(q) and that mild solutions of (4) have the representation

$$z(t;q) = S(t;q)z_0(q) + \int_0^t S(t-\sigma;q)F(\sigma,q)d\sigma. \qquad (6)$$

The corresponding form of the least squares criterion (3) is

$$J(q) = \sum_{i=1}^{m} \left| \hat{y}_i - z_1(t_i)|_{x=0} \right|^2$$

where z_1 is the first component of $z(t) = (u(t,\cdot), u_t(t,\cdot))^T$.

We approximate equation (6) to define a sequence of approximating estimation problems. Given $q \in Q$, define $X^N(q)$ to be the subspace of $S^3(\Delta^N) \times S^3(\Delta^N)$ satisfying the boundary conditions corresponding to q (i.e., the boundary conditions in the definitions of $V_B(q)$ and dom $A(q)$). Here $S^3(\Delta^N)$ is the standard subspace of C^2 cubic splines corresponding to the partition $\Delta^N = \{x_i\}_{i=0}^N$, $x_i = i/N$ (see p. 208-209 of [13]). More precisely, $X^N(q)$ is the linear span of the following basis elements: Let $\tilde{B}_j^N$, $j = -1,\ldots,N+1$ denote the standard C^2 basis elements for $S^3(\Delta^N)$. Then define $\beta_1^N(q),\ldots,\beta_{2N+3}^N(q)$ by

$$\beta_1^N = \begin{pmatrix} \frac{4q_1}{N}\tilde{B}_{-1}^N + (3 - q_1/N)\tilde{B}_0^N \\ 0 \end{pmatrix}, \qquad \beta_2^N = \begin{pmatrix} -\frac{4q_1}{N}\tilde{B}_1^N + (3 + q_1/N)\tilde{B}_0^N \\ 0 \end{pmatrix}$$

$$\beta_3^N = \begin{pmatrix} \tilde{B}_2^N \\ 0 \end{pmatrix}, \quad \ldots, \quad \beta_{N-1}^N = \begin{pmatrix} \tilde{B}_{N-2}^N \\ 0 \end{pmatrix},$$

$$\beta_N^N = \begin{pmatrix} \tilde{B}_{N-1}^N \\ \frac{3Nq_2}{4}\tilde{B}_N^N \end{pmatrix}, \qquad \beta_{N+1}^N = \begin{pmatrix} \tilde{B}_N^N \\ 0 \end{pmatrix}, \qquad \beta_{N+2}^N = \begin{pmatrix} \tilde{B}_{N+1}^N \\ -\frac{3Nq_2}{4}\tilde{B}_N^N \end{pmatrix},$$

$$\beta_{N+3}^N = \begin{pmatrix} -1/(3Nq_2)\tilde{B}_{N+1}^N \\ \tilde{B}_{N+1}^N \end{pmatrix}, \qquad \beta_{N+4}^N = \begin{pmatrix} -1/(3Nq_2)\tilde{B}_{N+1}^N \\ \tilde{B}_{N-1}^N \end{pmatrix}$$

$$\beta_{N+5}^N = \begin{pmatrix} 0 \\ \tilde{B}_{N-2}^N \end{pmatrix}, \quad \ldots, \quad \beta_{2N+1}^N = \begin{pmatrix} 0 \\ \tilde{B}_2^N \end{pmatrix},$$

$$\beta_{2N+2}^N = \begin{pmatrix} 0 \\ \frac{-4q_1}{N}\tilde{B}_1^N + (3+q_1/N)\tilde{B}_0^N \end{pmatrix}, \qquad \beta_{2N+3}^N = \begin{pmatrix} 0 \\ \frac{4q_1}{N}\tilde{B}_{-1}^N + (3-q_1/N)\tilde{B}_0^N \end{pmatrix}.$$

It is straightforward to show that these basis elements satisfy the boundary conditions corresponding to q.

In discussing our approximation schemes, it will be necessary to consider projections of $X(q)$ onto $X^N(q)$ in norms corresponding to different parameters $\tilde{q}$ (see

(5)). Note that the spaces $X(q)$, $q \in Q$, are, as sets, equal, with only the norms depending on q (and these are all equivalent as q ranges over Q). We thus define $P^N_{\tilde q}(q)$ as the orthogonal projection of $X(q)$ onto $X^N(q)$, the projection being taken with respect to the $X(\tilde q)$ inner product. Whenever it happens that q and $\tilde q$ are the same, we adopt the notation $P^N(q) = P^N_q(q)$. We then define approximations for $A(q)$ by $A^N(q) \equiv P^N(q)A(q)P^N(q)$ and the corresponding approximating system equations by

$$z^N(t;q) = S^N(t;q)P^N(q)z_0(q) + \int_0^t S^N(t-\sigma;q)P^N(q)F(\sigma,q)d\sigma, \tag{7}$$

where $S^N(t;q)$ is the semigroup generated by $A^N(q)$. The approximating parameter identification problems can then be stated as: Minimize over Q the function

$$J^N(q) \equiv \sum_{i=1}^m \left|\hat y_i - z_1^N(t_i)|_{x=0}\right|^2 \tag{8}$$

where z_1^N is the first component of z^N given by (7).

Assuming that we have solved the Nth approximating problem for best parameters $\bar q^N$ (this is a finite-dimensional state space problem which lends itself to solution with standard computational packages), we may invoke the compactness assumption on Q to obtain a subsequence $\bar q^{N_k}$ converging to some $q^* \in Q$. This q^* is obviously a candidate for a solution to our original problem of minimizing J subject to (6) if only we have $z^N(t;q) \to z(t;q)$ in an appropriate sense. Indeed, for the problem at hand, it is sufficient (see the arguments in [5, p. 12], [9, p. 820-822]) to establish that "for any q^N, q^* in Q, $q^N \to q^*$ implies $z^N(t;q^N) \to z(t;q^*)$" where this latter convergence must be carefully interpreted (since $z^N(t;q^N)$ is in $X^N(q^N)$ while $z(t;q^*)$ is in $X(q^*)$). Indeed, elements in $X^N(q^N)$ satisfy the boundary conditions corresponding to q^N while $z(t;q^*) \in \text{dom } A(q^*)$ and hence satisfies the boundary conditions corresponding to q^*. Thus, we must have, in discussing the convergence $z^N(t;q^N) \to z(t;q^*)$, a means of comparing elements in $X^N(q^N)$ with those in dom $A(q^*)$. To establish the convergence statement itself, we use a version of the Trotter-Kato approximation theorem (e.g., see [5] or [12]).

Theorem. Let $(\mathcal{B},|\cdot|)$ and $(\mathcal{B}^N,|\cdot|_N)$, $N = 1,2,\ldots$, be Banach spaces and let $\Pi^N\colon \mathcal{B} \to \mathcal{B}^N$ be bounded linear operators. Further assume that $T(t)$ and $T^N(t)$ are C_0-semigroups on $\mathcal{B}$ and $\mathcal{B}^N$ with infinitesimal generators $\tilde A$ and $\tilde A^N$, respectively. If

(i) $\lim_{N\to\infty} |\Pi^N z|_N = |z|$ for all $z \in \mathcal{B}$,

(ii) there exist constants M, ω independent of N such that $|T^N(t)|_N \le Me^{\omega t}$, for $t \ge 0$,

(iii) there exists a set $\mathcal{D} \subset \mathcal{B}$, $\mathcal{D} \subset \text{dom }(\tilde A)$, with $\overline{(\lambda_0-\tilde A)\mathcal{D}} = \mathcal{B}$ for some $\lambda_0 > 0$, such that for all $z \in \mathcal{D}$ we have $|\tilde A^N\Pi^N z - \Pi^N\tilde A z|_N \to 0$ as $N \to \infty$, then $|\mathcal{T}^N(t)\Pi^N z - \Pi^N\mathcal{T}(t)z|_N \to 0$ as $N \to \infty$, for all $z \in \mathcal{B}$, uniformly in t on compact intervals in $[0,\infty)$.

Given a sequence $\{q^N\}$ in Q converging to q^*, we employ this theorem with $\mathcal{B}^N = X(q^N)$, $\mathcal{B} = X(q^*)$, $\tilde{A} = A(q^*)$, $\tilde{A}^N = A^N(q^N) = P^N(q^N)A(q^N)P^N(q^N)$, $\mathcal{T}(t) = S(t;q^*)$, $\mathcal{T}^N(t) = S^N(t,q^N)$, and Π^N is chosen so that elements satisfying the q^* boundary conditions are mapped (under Π^N) into elements satisfying the q^N boundary conditions. Once the convergence of the semigroups is obtained, one can use standard arguments along with the representations (7) and (6) to establish the desired convergence $z^N(t,q^N) \to z(t;q^*)$.

Returning to discuss the mapping Π^N, we observe that we need to associate elements in dom $A(q^*)$ (which satisfy the q^* boundary conditions) with elements in dom $A(q^N)$ (which satisfy the q^N boundary conditions). We therefore define the function g^N depending on q^N, q^* by

$$g^N(x) \equiv \exp[q_1^* - q_1^N]x - (x^2/2)(q_1^* - q_1^N)\exp[q_1^* - q_1^N],$$

and for $h \in$ dom $A(q^*)$, define $h^N = (g^N h_1, (q_2^N/q_2^*)g^N h_2)^T$. Letting $\mathcal{I}^N$ be the canonical isomorphism of $X(q^*)$ to $X(q^N)$ (recall as sets these are equal; only the norms differ), we then define Π^N: $X(q^*) \to X(q^N)$ by $\Pi^N h = \mathcal{I}^N(h^N)$. It is readily seen that Π^N maps elements of $X(q^*)$ satisfying the q^* boundary conditions to elements of $X(q^N)$ satisfying the q^N boundary conditions.

Whenever $q^N \to q^*$, it easily argued that $g^N \to 1$, $D^j g^N \to 0$, $j = 1,2,3,4$, with the convergence being uniform in each case. Using these properties of g^N, condition (i) of the Trotter-Kato theorem is easily verified. The stability criterion (ii) is established via a uniform dissipative estimate similar to those found in [5], [8], [9]. Finally, to argue condition (iii), we may choose the set $\mathcal{D} \equiv$ dom $A(q^*) \cap (H^4 \times H^4)$. Verification that $(\lambda_0 - A(q^*))\mathcal{D}$ is dense can be reduced to an existence argument for a two point boundary value problem; this turns out to be tedious but straightforward. To argue the convergence of $A^N(q^N)$ to $A(q^*)$ required in (iii), one first employs the triangle inequality:

$$\begin{aligned}
&|P^N(q^N)A(q^N)P^N(q^N)\Pi^N z - \Pi^N A(q^*)z|_N \\
&\le |P^N(q^N)[A(q^N)P^N(q^N)\Pi^N z - \Pi^N A(q^*)P^N_{q^N}(q^*)z]|_N \\
&\quad + |P^N(q^N)[\Pi^N A(q^*)P^N_{q^N}(q^*)z - \Pi^N A(q^*)z]|_N \\
&\quad + |P^N(q^N)\Pi^N A(q^*)z - \Pi^N A(q^*)z|_N \\
&\le |A(q^N)P^N(q^N)\Pi^N z - \Pi^N A(q^*)P^N_{q^N}(q^*)z|_N + |\Pi^N||A(q^*)[P^N_{q^N}(q^*)z - z]| \\
&\quad + |(P^N(q^N) - I)\Pi^N A(q^*)z|_N .
\end{aligned}$$

The last two terms are easily estimated using standard interpolating spline estimates (e.g., see [14, p. 54]) modified (see Lemmas 4.1, 4.2 of [5]) to take into account the fact that one is using essentially the H^1 norm on the first component of $z = (z_1, z_2) \in X(q)$. In addition to such estimates, convergence of the first term is facilitated by arguing that

$$|D[P^N(q^N)\Pi^N z - \Pi^N P^N_{q^N}(q^*)z]|_N \to 0$$

and

$$|D^2[(P^N(q^N)\Pi^N z)_1] - g^N D^2[(P^N_{q^N}(q^*)z)_1]|_{H^0} \to 0$$

for $z \in \mathcal{D}$. Here $(\ )_1$ denotes the first component h_1 of any element $h = (h_1, h_2)$.

In summary, the above considerations lead to the establishment of a parameter convergence result similar to that found in [5], [8], [9].

Theorem. Let z, z^N be defined as in (6), (7) and let $\bar{q}^N$ be a solution of the problem of minimizing over Q the function J^N given in (8). Then there exists a subsequence $\bar{q}^{N_k}$ converging to q^* in Q and q^* is a solution of the problem of minimizing J over Q subject to (6).

We turn next to a brief summary of some of our numerical findings using the methods outlined above. All of our calculations were carried out using modifications of algorithms and software packages described in [5]. A number of test examples were investigated in which we used an independent numerical method to generate solutions of the system for fixed, known values of the parameters (called "true values" below). These solutions then were used as "data" in the inverse problem as formulated above and estimates of the parameters were sought. In addition to the problems involving multiple time observations at the surface $x = 0$ as described above, we also investigated use of the methods in examples where several discrete spatial observations (at specified x_j in $[0,1]$) were available. Such problems are also important in seismic exploration (i.e., in so-called bore hole problems in which receivers are placed at several locations down a well).

Example 1. We considered (1) with $q_0 = E/\rho$ constant and $s(t,\tilde{q}) = q_3(1 - e^{-5t})e^{q_4 t}$. Observations were given for $t_i = 1/2$, 1 and $x_j = 0$, $1/2$, 1, corresponding to "true" values of $q_0^* = 3$, $q_1^* = -2$, $q_2^* = 1$, $q_3^* = 2$, $q_4^* = -1$. In the iterative scheme to find $\bar{q}^N$, we employed initial guesses of $q_0^0 = 2$, $q_1^0 = -1$, $q_2^0 = 2$, $q_3^0 = 1.5$, $q_4^0 = -.5$. For an approximation index of $N = 8$ (corresponding to $2N+3 = 19$ basis elements for the wave equation written as a vector first order system), we obtained the converged values of $\bar{q}_0^8 = 3.0114$, $\bar{q}_1^8 = -2.0115$, $\bar{q}_2^8 = 1.0035$, $\bar{q}_3^8 = 2.0103$, $\bar{q}_4^8 = -.9934$ with a residual of $J^8(\bar{q}^8) = .224 \times 10^{-8}$.

<u>Example 2.</u> We considered the same problem as in Example 1 (same true values, same initial guesses) except only observations at the surface ($x = 0$) at times $t_i = .25, .5, .75, \ldots, 2.$, were used in the inverse problem. The following results were obtained

N	$\bar{q}_0^N$	$\bar{q}_1^N$	$\bar{q}_2^N$	$\bar{q}_3^N$	$\bar{q}_4^N$
4	2.9381	- 2.0417	.9996	2.0724	- .9775
8	2.9811	- 2.0395	1.0066	2.0494	- .9830

with corresponding residuals of $J^4(\bar{q}^4) = .558 \times 10^{-5}$ and $J^8(\bar{q}^8) = .111 \times 10^{-5}$.

<u>Example 3.</u> Again we considered the problem of Example 1 with the only change being that we took $s(t,\tilde{q}) = q_3(1 - e^{-30t})e^{q_4 t}$. For $N = 8$, we obtained converged values of $\bar{q}_0^8 = 3.0069$, $\bar{q}_1^8 = -1.9452$, $\bar{q}_2^8 = .9906$, $\bar{q}_3^8 = 1.9912$, $\bar{q}_4^8 = -1.0520$, with a residual $J^8(\bar{q}^8) = .4229 \times 10^{-7}$.

<u>Example 4.</u> We present results for a problem with unknown variable elastic modulus E in (1). We take (1) with $\rho \equiv 1$, $s \equiv 0$, and initial data $u(0,x) = e^x$, $u_t(0,x) = -3e^x$. We assume that E is parameterized as $E(x) = 3/2 - 1/\Pi \text{ Arctan } (q_3(x - q_4))$. Data for observations at $x_j = 0, \sqrt{2}, 1$, and $t_i = .16, .33, .5, .66, .83, 1$, were used to obtain the following estimates from initial guesses $q_1^0 = -2$, $q_2^0 = 2$, $q_3^0 = 5$, $q_4^0 = 1$.

N	$\bar{q}_1^N$	$\bar{q}_2^N$	$\bar{q}_3^N$	$\bar{q}_4^N$	$J^N(\bar{q}^N)$
4	- .9909	3.0063	2.9645	.4876	$.86 \times 10^{-4}$
8	- .9999	2.9958	3.0508	.50126	$.28 \times 10^{-5}$
16	- .9999	2.9990	3.0130	.50024	$.19 \times 10^{-6}$
True Values	-1.0	3.0	3.0	.5	

<u>Example 5.</u> Our final example demonstrates that one can obtain good converged values even when the initial guesses are not very close to the true values. We considered (1) with $q_0 = E/\rho$ constant and $s \equiv 0$. Initial data consisted of the functions $u(0,x) = x^3 - 2x^2 + 2x + 1$, $u_t(0,x) = x^4 + x^2 - 2x - 1$. True values were $q_0^* = 1$, $q_1^* = -2$, $q_2^* = 1$, while initial guesses were $q_0^0 = 5$, $q_1^0 = -.1$, $q_2^0 = 5$. The following converged values were obtained using our cubic spline based packages.

N	$\bar{q}_0^N$	$\bar{q}_1^N$	$\bar{q}_2^N$	$J^N(\bar{q}^N)$
6	.9989	- 1.9970	.9989	$.109 \times 10^{-6}$
12	.9995	- 1.9998	.9997	$.603 \times 10^{-7}$

In summary, the cubic spline based parameter estimation techniques proposed in this note appear to have potential for the development of inversion algorithms in seismic signal processing problems. While our discussions here have emphasized the unknown boundary parameters and have been limited mainly to problems with constant elastic parameters, we have already used the ideas in problems with a spatially dependent elastic modulus. Moreover, the methods do not require a specific parameterization of these variable parameters as was done in Example 4 above. In fact, methods which allow one to estimate the parameter function (including its shape) similar to those developed for parabolic systems in [6] are also applicable to these hyperbolic system problems.

ACKNOWLEDGEMENT

The authors would like to express their appreciation to Dr. R. Ewing and Dr. G. Moeckel of Mobil Research and Development Corp. for numerous discussions on problems related to seismic exploration. Special thanks go to George Moeckel for his continued interest, encouragement, and help in our efforts.

REFERENCES

[1] Bamberger, A., G. Chavent , P. Lailly: About the stability of the inverse problem in 1-D wave equations - Application to the interpretation of seismic problems, Appl. Math. Opt. 5 (1979), 1-47.

[2] Banks, H.T.: A survey of some problems and recent results for parameter estimation and optimal control in delay and distributed parameter systems, Proc. Conf. on Volterra and Functional Differential Equations, (VPISU, Blacksburg, June 10-13, 1981), Marcell Dekker, 1982, p. 3-24, (LCDS Tech. Rep. 81-19, July, 1981, Brown Univ.).

[3] Banks, H.T., J.M. Crowley: Parameter estimation for distributed systems arising in elasticity, Proc. Symposium on Engineering Sciences and Mechanics, (National Cheng Kung University, Tainan, Taiwan, Dec. 28-31, 1981), pp. 158-177, LCDS Tech. Rep. 81-24, November, 1981, Brown University.

[4] Banks, H.T., J.M. Crowley: Parameter estimation in Timoshenko beam models, LCDS # 82-14, Brown Univ., June, 1982; J. Astronautical Sci., to appear.

[5] Banks, H.T., J.M. Crowley, K. Kunisch: Cubic spline approximation techniques for parameter estimation in distributed systems, LCDS Tech. Rep. 81-25, Nov., 1981, Brown Univ.; IEEE Trans. Auto. Control, to appear.

[6] Banks, H.T., P.L. Daniel: Estimation of variable coefficients in parabolic distributed systems, LCDS Rep. # 82-22, Sept. 1982, Brown Univ.; IEEE Trans. Auto. Control, submitted.

[7] Banks, H.T., P.L. Daniel, E.S. Armstrong: Parameter estimation for static models of the Maypole Hoop/Column antenna surface, Proc. 1982 IEEE Int'l. Large Scale Systems Symposium, Va. Beach, Va., Oct. 11-13, 1982, pp. 253-255.

[8] Banks, H.T., P. Kareiva: Parameter estimation techniques for transport equations with application to population dispersal and tissue bulk flow models, LCDS # 82-13, Brown University, July, 1982, J. Math. Biology, to appear.

[9] Banks, H.T., K. Kunisch: An approximation theory for nonlinear partial differential equations with applications to identification and control, SIAM J. Control and Optimization, 20 (1982), 815-849.

[10] Dobrin, M.B.: Introduction to Geophysical Prospecting, McGraw-Hill, New York, 1976.

[11] Grant, F.S., G.F. West: Interpretation Theory in Applied Geophysics, McGraw-Hill, New York, 1965.

[12] Kurtz, T.G.: Extensions of Trotter's operator semigroup approximation theorem, J. Functional Anal. 3 (1969), 354-375.

[13] Prenter, P.M.: Splines and Variational Methods, Wiley-Interscience, New York, 1975.

[14] Schultz, M.H.: Spline Analysis, Prentice-Hall, Englewood Cliffs, N.J., 1973.

BOUNDARY CONTROL OF SOME FREE BOUNDARY PROBLEMS

V. Barbu

Seminarul Matematic
Universitatea "Al. I. Cuza"
R-6600 Iaşi, Romania

1. INTRODUCTION

This paper is concerned with first order necessary conditions for certain boundary control problems governed by parabolic variational inequalities of the obstacle type (see problems (P_1) and (P_2)) below.

Throughout in the sequel we shall use the following notations:

1. Let Ω_1 and Ω_2 be two open subsets of R^n with sufficiently smooth boundaries, Γ_1, Γ_2 and such that $\Gamma_1 \cap \Gamma_2 = \emptyset$, $\overline{\Omega}_1 \subset \Omega_2$. Let $\Omega = \Omega_2 \setminus \overline{\Omega}_1$ be the domain of the boundary $\Gamma_1 \cup \Gamma_2$; $[0,T]$ is a real interval, $\Sigma_i = \Gamma_i \times]0,T[$ for $i = 1,2$ and $Q = \Omega \times]0,T[$.
2. a: $H^1(\Omega) \times H^1(\Omega) \to R$ is the bilinear form

$$a(y,z) = \int_\Omega \nabla y(x)\nabla z(x)dx \quad \text{for all} \quad y,z \in H^1(\Omega).$$

By $(.,.)$ we shall denote the scalar product of $L^2(\Omega)$ and the pairing between $H^1(\Omega)$ and $(H^1(\Omega))'$.

3. $H^s(\Omega)$, $W_q^\ell(\Omega)$, $H^{2,1}(Q)$, $W_q^{2,1}(Q)$ and $W^{2-\frac{1}{q},1-\frac{1}{2q}}(\Sigma_1)$ are usual Sobolev spaces on Ω, Q and Σ_1, respectively (see [8]). We shall denote by $||.||_q$ the norm of $X_q = W^{2-\frac{1}{q},1-\frac{1}{2q}}(\Sigma_1)$.

Problem (P_1). Minimize

$$g(y) + \phi(u) + \phi^0(y(T))$$

over all $y \in W_q^{2,1}(Q)$ and $u \in W^{2-\frac{1}{q},1-\frac{1}{2q}}(\Sigma_1)$ subject to

$$(y_t,z-y) + a(y,z-y) \geq (f,z-y) \quad \text{for all } z \in K(t),\ t \in [0,T] \tag{1.1}$$

$$y(x,0) = y_0(x),\ x \in \Omega \tag{1.2}$$

$$u \in U \tag{1.3}$$

where $f \in L^q(Q)$, $q > (n+2)/2$ and

$$U = \{u \in W^{2-\frac{1}{q},1-\frac{1}{2q}}(\Sigma_1);\ u \geq 0 \text{ on } \Sigma_1,\ u(\sigma,0) = y_0(\sigma) \text{ for } \sigma \in \Gamma_1\}. \tag{1.4}$$

$$K(t) = \{y \in H^1(\Omega);\ y \geq 0 \text{ a.e. on } \Omega,\ y = u \text{ on } \Gamma_1,\ y = 0 \text{ on } \Gamma_2\} \tag{1.5}$$

$$y_0 \in W_q^{2-\frac{2}{q}}(\Omega);\ y_0 \geq 0 \text{ a.e. on } \Omega,\ y_0 = 0 \text{ on } \Gamma_2. \tag{1.6}$$

$$g\colon L^2(Q) \to R^+ \text{ is Fréchet differentiable.} \tag{1.7}$$

$$\phi\colon W^{2-\frac{1}{q},1-\frac{1}{2q}}(\Sigma_1) \to \overline{R} =]-\infty,+\infty] \text{ is convex ,} \tag{1.8}$$

lower semicontinuous and $0 \in \text{int}\,\{u;\phi(u) < +\infty\}$.

$$\phi^0\colon L^2(\Omega) \to R \text{ is convex and continuous.} \tag{1.9}$$

<u>Problem</u> (P_2). Minimize

$$g(y) + \psi(u) + \phi^0(y(T))$$

over all $y \in H^{2,1}(Q)$ and $u \in L^2(\Sigma_1)$ subject to

$$(y_t,z-y) + a(y,z-y) + \alpha \int_{\Gamma_1} (y-y^1)(z-y)d\sigma \geq (f,z-y) + \int_{\Gamma_1} u(z-y)d\sigma \tag{1.10}$$

for all $z \in K$, $t \in [0,T]$.

$$y(x,0) = y_0(x) \text{ a.e. } x \in \Omega;\ a \leq u \leq b \text{ a.e. in } \Sigma_1,\ a \geq 0 \tag{1.11}$$

where $f \in L^2(Q)$, $y_0 \in K$ and

$$K = \{y \in H^1(\Omega);\ y \geq 0 \text{ a.e. on } \Omega;\ y = 0 \text{ in } \Gamma_2\} \tag{1.12}$$

$$y^1 \in L^2(\Sigma_1);\ y^1 \geq 0 \text{ a.e. on } \Sigma_1. \tag{1.13}$$

$$\psi\colon L^2(\Omega) \to R \text{ is continuous and convex.} \tag{1.14}$$

As regard functions g and ϕ^0 we shall assume that conditions (1.7) and (1.9) are satisfied.

It is well known (see for instance [5], [7]) that Eqs.(1.1), (1.2) can be formally written as an "obstacle" problem of the form

$$y \geq 0, \quad y_t - \Delta y \geq f \quad \text{in } Q \tag{1.15}$$

$$y_t - \Delta y = f \quad \text{in } \{y > 0\} \tag{1.16}$$

$$y(x,0) = y_0(x), \qquad x \in \Omega. \tag{1.17}$$

$$y = u \quad \text{in } \Sigma_1;\ y = 0 \text{ in } \Sigma_2. \tag{1.18}$$

Similarly, Eqs. (1.10), (1.11) can be written as (1.15) ~ (1.17) and

$$\frac{\partial y}{\partial \nu} + \alpha(y-y^1) = u \quad \text{in } \Sigma_1;\ y = 0 \quad \text{in } \Sigma_2. \tag{1.19}$$

The typical situations are that where

$$g(y) = ||y-y^0||^2_{L^2(Q)} \quad ; \quad y^0 \text{ given in } L^2(Q) \tag{1.20}$$

$$\phi^0(y) = ||y-\xi||^2_{L^2(\Omega)} \quad ; \quad \xi \text{ given in } L^2(\Omega) \tag{1.21}$$

$$\phi(u) = \gamma||u||^2_q \quad \text{if} \quad ||u||_q \leq r; \quad \phi(u) = +\infty \quad \text{if} \quad ||u||_q > r \tag{1.22}$$

$$\psi(u) = \gamma||u||^2_{L^2(\Sigma_1)} \qquad \text{where } \gamma > 0. \tag{1.23}$$

The choices (1.20), (1.21) of the functions g and ϕ^0 correspond to the least square approach of the controllability of the state y of system (1.1), (1.2) respectively, (1.10), (1.11). In systems (1.1), (1.10) the incidence set $E_u = \{(x,t) \in Q;\ y(x,t) = 0\}$ is a free boundary. The control problem of the free boundary E_u can be expressed in few words as follows: given a smooth manifold $\{(x,t;t = \sigma(x)\} = E_0$ find $u \in L^2(\Sigma_1)$ such that $E_u = E_0$.

Several numerical procedures for this problem which in general is improperly posed can be found in literature (see for instance [6]). The least square approach to this problem leads to a control problem of the type (P_1) or (P_2) where (see [9])

$$g(y) = \int_Q \chi_E(x,t)|y(x,t)|^2 dxdt \quad \text{or} \quad g(y) = \int_Q \left|\frac{\varepsilon}{y+\varepsilon} - \chi_E\right|^2 dxdt \tag{1.24}$$

and χ_E is the characteristic function of a given measurable subset E of Q which contains E_o and is "sufficiently close" to E_o.

Control problems of this type arise in heat conduction and diffusion theory (see [5]). For instance the one phase Stefan problem

$$\theta_t - \Delta\theta = 0 \qquad \text{in } \{\ell(x) < t\} \tag{1.26}$$

$$\theta = 0 \qquad \text{in } \{\ell(x) \geq t\} \tag{1.27}$$

$$\nabla_x\theta \cdot \nabla_x\ell = -\rho \qquad \text{in } \{\ell(x) = t\} \tag{1.28}$$

$$\theta = v \ \text{ in } \Sigma_1; \quad \theta = 0 \ \text{ in } \Sigma_2 \tag{1.29}$$

$$\theta(x,0) = 0, \quad x \in \Omega \tag{1.30}$$

can be written in the form (1.1), (1.2) (see [4]) where y is given by the Baiocchi substitution

$$y(x,t) = \int_{\ell(x)}^{t} \theta(x,s)ds \quad \text{for } \ell(x) < t; \ y = 0 \quad \text{for } \ell(x) \geq t \tag{1.31}$$

and $f = -\rho$, $u(x,t) = \int_0^t v(x,s)ds$.

Problem (1.26) ~ (1.30) is the description, typically, of the melting of a body of ice Ω maintained at 0° on the boundary Γ_2 and at v^o on Γ_1. The equation of the interface which separates the solid and liquid regions is $\ell(x) = t$. In terms of v and y defined by Eq. (1.31), problem (1.10), (1.11) describes the situation when the heat flux is concentrated on Γ_1, i.e., instead of (1.29), we have

$$\frac{\partial\theta}{\partial\nu} + \alpha(\theta-\theta_1) = v \ \text{ in } \Sigma_1; \quad \theta = 0 \ \text{ in } \Sigma_2. \tag{1.32}$$

The temperature control in the liquid region or the control of the free boundary $t = \ell(x)$ lead to problems of the form (P_1) and (P_2) with cost functionals of the form (1.20) ~ (1.25). In the sequel we assume familiarity with basic concepts and results of convex analysis.

2. FIRST ORDER NECESSARY CONDITIONS FOR PROBLEM (P_1)

Following the standard terminology, the control $u^* \in X_q = W^{2-\frac{1}{q},1-\frac{1}{2q}}(\Sigma_1)$ which minimizes the functional $g(y_u) + \phi(u) + \phi^o(y_u(T))$ in a neighbourhood $\{u \in U; \|u-u^*\|_q \leq r\}$ of u^* is called local optimal control of problem (P_1). (y_u is the corresponding solution to (1.1), (1.2)). The pair (y^*,u^*) where $y^* = y_{u^*}$ is called local optimal pair of problem (P_1).

Theorem 1. Let (y^*,u^*) be a local optimal pair in problem (P_1). Then there exists a function $p \in L^\infty(0,T;L^2(\Omega)) \cap L^2(0,T;H_0^1(\Omega)) \cap BV([0,T];H^{-s}(\Omega))$; $s > n/2$ such that $p_t + \Delta$ is a bounded Radon measure on Q, $\frac{\partial p}{\partial \nu} \in X_q^*$ and

$$(p_t + \Delta p - \nabla g(y^*))y^* = 0 \quad \text{in } Q \tag{2.1}$$

$$p(T) + \partial\phi^0(y^*(T)) \ni 0 \quad \text{in } \Omega \tag{2.2}$$

$$p(y_t^* - \Delta y^* - f) = 0 \quad \text{a.e. in } Q \tag{2.3}$$

$$(\frac{\partial p}{\partial \nu} - \xi)(u^* - v) \geq 0 \quad \text{for all } v \in U;\ \xi \in \partial\phi(u^*). \tag{2.4}$$

By $\partial\phi$: $X_q \to X_q^*$ (the dual space of X_q) and $\partial\phi^0$: $L^2(\Omega) \to L^2(\Omega)$ we have denoted the subdifferentials of ϕ and ϕ^0, respectively. By $BV([0,T];H^{-s}(\Omega))$ we have denoted the space of functions p: $[0,T] \to H^{-s}(\Omega)$ of bounded variation on $[0,T]$.

In a particular case a similar result has been previously given in [1] and by a different approach in [9].

The main ingredient of the proof is an approximation result for the state system (1.1), (1.2).

For $\varepsilon > 0$ and $u \in U$ consider the boundary value problem

$$\begin{aligned} & y_t - \Delta_y + \beta^\varepsilon(y) = f && \text{in } Q \\ & y(x,0) = y_0(x) && \text{in } \Omega \\ & y = u \ \text{ in } \Sigma_1;\ y = 0 && \text{in } \Sigma_2 \end{aligned} \tag{2.5}$$

where

$$\beta^\varepsilon(r) = \varepsilon^{-1} \int_{\varepsilon^{-1}r}^{\infty} (r - \varepsilon\theta)\rho(\theta)d\theta \quad \text{for } r \in R \tag{2.6}$$

and ρ is a C_0^∞-mollifier on R. The function β^ε is infinitely differentiable, Lipschitzian and monotonically increasing. We shall denote by y_u^ε the solution to (2.5).

Lemma 1. There exists $C > 0$ independent of ε and u such that

$$||y_u^\varepsilon||_{W_q^{2,1}(Q)} \leq C(||u||_q + ||f||_{L^q(Q)}) \quad \text{for all } u \in U. \tag{2.7}$$

Moreover, if $u_\varepsilon \to u$ weakly in X_q then $y^\varepsilon_{u_\varepsilon} \to y_u$ weakly in $W^{2,1}_q(Q)$ where y_u is the unique solution to (1.1), (1.2).

Proof. For each $u \in X_q$ such that $u(\sigma,0) = y_0$ on Γ_1 the boundary value problem

$$\begin{aligned} &\zeta_t - \Delta\zeta = 0 && \text{in } Q \\ &\zeta = u \text{ in } \Sigma_1;\ \zeta = 0 && \text{in } \Sigma_2 \\ &\zeta(x,0) = y_0(x), && x \in \Omega \end{aligned} \tag{2.8}$$

has a unique solution $\zeta_u \in W^{2,1}_q(Q)$ ([8]) satisfying the estimate

$$||\zeta_u||_{W^{2,1}_q(Q)} \le C(||u||_q + 1) \quad \text{for all } u \text{ in } X_q. \tag{2.9}$$

Let $z^\varepsilon_u \in W^{2,1}_q(Q)$ be the solution to

$$\begin{aligned} &(z^\varepsilon_u)_t - \Delta z^\varepsilon_u + \beta^\varepsilon(z^\varepsilon_u + \zeta_u) = f \quad \text{in } Q \\ &z^\varepsilon_u(x,0) = 0 \qquad x \in \Omega \\ &z^\varepsilon_u = 0 \qquad \text{in } \Sigma . \end{aligned} \tag{2.10}$$

If $u \ge 0$ in Σ_1 then $(z^\varepsilon_u + \zeta_u)^- = 0$ in Σ_1 and by a standard device (see for instance [8]) it follows that

$$||\beta^\varepsilon(z^\varepsilon_u + \zeta_u)||_{L^q(Q)} \le C||f||_{L^q(Q)} . \tag{2.11}$$

Hence

$$||z^\varepsilon_u||_{W^{2,1}_q(Q)} \le C_1||f||_{L^q(Q)} . \tag{2.12}$$

Since $y^\varepsilon_u = z^\varepsilon_u + \zeta_u$ is a solution to (2.5) the latter implies (2.7). If $u_\varepsilon \to u$ weakly in X_q then by (2.9), (2.12) and (2.13) we see that on a subsequence again denoted ε,

$$\begin{aligned} &\zeta_{u_\varepsilon} \to \zeta_u \text{ weakly in } W^{2,1}_q(Q) \text{ and} \\ &z^\varepsilon_{u_\varepsilon} \to z_u \text{ strongly in } L^q(Q) \cap C([0,T];L^2(\Omega)). \end{aligned} \tag{2.13}$$

$$\beta^\varepsilon(y^\varepsilon_{u_\varepsilon}) \to \gamma \quad \text{weakly in } L^q(Q) \tag{2.14}$$

where

$$(z_u)_t - \Delta z_u + \gamma = f \quad \text{a.e. in } Q$$

$$z_u(x,0) = 0, \quad z_u = 0 \qquad \text{in } \Sigma$$

and

$$(\gamma, z_u + \zeta_u - r) \geq 0 \quad \text{a.e. in } Q \quad \text{for all } r \geq 0.$$

We may conclude therefore that $y_u = z_u + \zeta_u$ is a solution to (1.1), (1.2) and

$$||y_u||_{W_q^{2,1}(Q)} \leq C(||u||_q + ||f||_{L^q(Q)}) \quad \text{for all } u \text{ in } U. \tag{2.15}$$

The uniqueness of the solution y to (1.1), (1.2) is immediate.

Now let (y^*,u^*) be the local optimal solution to problem (P_1). In other words,

$$g(y^*) + \phi(u^*) + \phi^o(y^*(T)) \leq g(y_u) + \phi(u) + \phi^o(y_u(T)) \tag{2.16}$$

$$\text{for all } u \in U, \ ||u-u^*||_q \leq r.$$

For each $\varepsilon > 0$ consider the approximating control problem

$$(P_1^\varepsilon) \quad \min \{g(y_u^\varepsilon) + \Phi(u) + (2\delta)^{-1}||u-u^*||_q^2 + \phi_\varepsilon^o(y_u(T))\}$$

where $\Phi: X_q \to]-\infty,+\infty]$ is defined by

$$\Phi(u) = \phi(u) \quad \text{if } u \in U; \quad \Phi(u) = +\infty \quad \text{if } u \notin U. \tag{2.17}$$

$$\phi_\varepsilon^o(y) = \inf \{(2\varepsilon)^{-1}||y-z||^2_{L^2(\Omega)} + \phi^o(z); \ z \in L^2(\Omega)\}. \tag{2.18}$$

δ is a positive constant to be fixed later.

Since by Lemma 1 the map $u \to y_u^\varepsilon$ is compact from U to $C([0,T];L^2(\Omega))$ and Φ is weakly lower semicontinuous on X_q (because it is convex and lower semicontinuous) we may infer that problem (P_1) has at least one (global) optimal solution $(y_\varepsilon,u_\varepsilon)$; $y_\varepsilon = y_{u\varepsilon}^\varepsilon$. Hence

$$g(y_\varepsilon) + \Phi(u_\varepsilon) + (2\delta)^{-1}||u^*-u_\varepsilon||_q^2 + \phi_\varepsilon^o(y_\varepsilon(T)) \leq \tag{2.19}$$

$$\leq g(y_{u^*}^\varepsilon) + \Phi(u^*) + \phi^o(y_{u^*}^\varepsilon(T)) \leq C \quad \text{for } \varepsilon > 0.$$

Without loss of generality we may assume that Φ and ϕ^o are positive. Then (2.19) yields

$$||u^*-u_\varepsilon||_q \leq (2\delta C)^{1/2} < r \quad \text{for all} \quad \varepsilon > 0 \tag{2.20}$$

if δ is sufficiently small.

Lemma 2. For $\varepsilon \to 0$,

$$u_\varepsilon \to u^* \quad \text{strongly in } X_q \tag{2.21}$$

$$y_\varepsilon \to y^* \quad \text{weakly in } W_q^{2,1}(Q) \text{ and strongly in } C([0,T];L^2(\Omega)) \tag{2.22}$$

$$\beta^\varepsilon(y_\varepsilon) \to f - y_t^* - \Delta y^* \quad \text{weakly in } L^2(Q). \tag{2.23}$$

Proof. By Lemma 1 and estimate (2.19) we see that on a subsequence we have

$$u_\varepsilon \to u_1 \quad \text{weakly in } X_q$$

$$y_\varepsilon \to y_{u_1} \quad \text{weakly in } W_q^{2,1}(Q) \text{ and strongly in } C([0,T];L^2(\Omega))$$

and

$$g(y_{u_1}) + \Phi(u_1) + (2\delta)^{-1}||u^*-u_1||_q^2 + \phi^0(y_{u_1}(T)) \leq$$

$$\leq g(y^*) + \Phi(u^*) + \phi^0(y^*(T)).$$

Since by (2.20) $||u^*-u_1||_q < r$, the latter implies that $u_1 = u^*$ and (2.21), (2.22) follow. As regards (2.23) it follows by (2.14).

Now let $p_\varepsilon \in H^{2,1}(Q) \cap L^2(0,T;H_0^1(\Omega))$ be the solution to boundary value problem

$$\begin{aligned} &(p_\varepsilon)_t + \Delta p_\varepsilon - (\beta^\varepsilon)'(y_\varepsilon)p_\varepsilon = \nabla g(y_\varepsilon) && \text{in } Q \\ &p_\varepsilon(T) = -\nabla\phi_\varepsilon^0(y_\varepsilon(T)) && \text{in } \Omega \\ &p_\varepsilon = 0 && \text{in } \Sigma. \end{aligned} \tag{2.24}$$

Since u_ε is an optimal control for problem (P_1) and the functions g, ϕ_ε^0 are Fréchet differentiable, a little calculation involving (2.24) leads to

$$\Phi^0(u_\varepsilon,v) + \delta^{-1}F(u_\varepsilon-u^*)(v) \geq \frac{\partial p_\varepsilon}{\partial \nu}(v) \quad \text{for all } v \text{ in } X_q$$

where $\Phi^0\colon X_q \times X_q \to R$ is the directional derivative of Φ and $F\colon X_q \to X_q^*$ is the duality mapping of X_q. The latter yields

$$\frac{\partial p_\varepsilon}{\partial \nu} \in \partial\Phi(u_\varepsilon) + \delta^{-1}F(u_\varepsilon - u^*), \quad \text{a.e. in } \Sigma_1. \tag{2.25}$$

Next by (2.24) we see that

$$||p_\varepsilon(t)||_{L^2(\Omega)} + \int_0^T ||p_\varepsilon(t)||^2_{H_0^1(\Omega)} dt + \int_Q (\beta^\varepsilon)'(y_\varepsilon)|p_\varepsilon| dxdt \leq C. \tag{2.26}$$

Hence $\{(p_\varepsilon)t\}$ is bounded in $L^1(0,T;H^{-s}(\Omega))$ where $s > n/2$ and by the Helly theorem there exists $p \in BV([0,T];H^{-s}(\Omega))$ such that on a subsequence

$$p_\varepsilon(t) \to p(t) \text{ strongly in } H^{-s}(\Omega) \quad \text{for all} \quad t \in [0,T]. \tag{2.27}$$

Next by (2.26) we may assume that

$$p_\varepsilon \to p \quad \text{weakly in } L^2(0,T;H_0^1(\Omega)) \text{ and weak star in } L^\infty(0,T;L^2(\Omega)). \tag{2.28}$$

On the other hand, for every $\eta > 0$, $\exists\, C(\eta) > 0$ such that (see [7], Lemma 5.1, Chap.1)

$$||p_\varepsilon(t) - p(t)||_{L^2(\Omega)} \leq \eta||p_\varepsilon(t) - p(t)||_{H_0^1(\Omega)} + C(\eta)||p_\varepsilon(t) - p(t)||_{H^{-s}(\Omega)}.$$

Hence

$$p_\varepsilon \to p \text{ strongly in } L^2(Q) \tag{2.29}$$

and

$$p_\varepsilon(t) \to p(t) \quad \text{weakly in } L^2(\Omega) \quad \text{for all} \quad t \in [0,T]. \tag{2.30}$$

Finally, by (2.26) we may infer that there exists a bounded measure μ_p on Q such that

$$(\beta^\varepsilon)'(y_\varepsilon)p_\varepsilon \to \mu_p \quad \text{weak star.} \tag{2.31}$$

Now letting ε tend to zero in (2.24) we see that

$$p_t + \Delta p - \mu_p = \nabla g(y^*) \quad \text{in } Q \tag{2.32}$$

$$p(T) + \partial\phi^0(y^*(T)) \ni 0. \tag{2.33}$$

Applying Green's formula in (2.24) it follows that (see [1])

$$\left| \int_{\Sigma_1} \frac{\partial p_\varepsilon}{\partial \nu} \kappa \, d\sigma dt \right| \leq C \| \kappa \|_q \quad \text{for all} \quad \kappa \in X_q$$

(because $W_q^{2,1}(Q) \subset C(\overline{Q})$). Hence $\{\frac{\partial p_\varepsilon}{\partial \nu}\}$ is bounded in X_q^* and letting ε tend to zero in (2.25) we find that

$$\frac{\partial p}{\partial \nu} \in \partial \Phi(u^*) \quad \text{in } \Sigma_1. \tag{2.34}$$

Next by definition of β^ε we have

$$|p_\varepsilon \beta^\varepsilon(y_\varepsilon) - (\beta^\varepsilon)'(y_\varepsilon) p_\varepsilon y_\varepsilon| \leq \varepsilon |p_\varepsilon (\beta^\varepsilon)'(y_\varepsilon) y_\varepsilon| \quad \text{a.e. in } Q.$$

Then arguing as in [2], [3] it follows that

$$p_\varepsilon \beta^\varepsilon(y_\varepsilon) \to 0 \quad \text{a.e. in } Q \tag{2.35}$$

$$(\beta^\varepsilon)'(y_\varepsilon) p_\varepsilon y_\varepsilon \to 0 \quad \text{strongly in } L^1(Q) \tag{2.36}$$

Then by (2.23), (2.29) and (2.35) we see that

$$p(y_t^* - \Delta y^* - f) = 0 \quad \text{a.e. in } Q$$

and since $W_q^{2,1}(Q)$ it compactly imbedded in $C(\overline{Q})$ (because $q > (n+2)/2$) it follows by (2.21) and (2.31) that $\mu_p y^* = 0$ in Q. We have therefore shown that Eqs. (2.1), (2.2), (2.3) hold. As regards (2.4) if follows by (2.34) taking in account the fact that by (2.17) and assumption (1.8),

$$\partial \Phi(u) = \partial \phi(u) + N(u) \quad \forall \quad u \in U$$

where $N(u) \subset X_q^*$ is the cone of normals to U at u. Thus the proof of Theorem 1 is complete.

3. FIRST ORDER NECESSARY CONDITIONS FOR PROBLEM (P_2)

The main result is

Theorem 2. Let (y^*, u^*) be a local optimal pair of problem (P_2). Then there exists $p \in L^2(0,T;H^1(\Omega)) \cap L^\infty(0,T;L^2(\Omega)) \cap BV([0,T];\ H^{-s}(\Omega))$, $s > n/2$ such that $p_t + \Delta p$ is a bounded measure on Q and

$$(p_t + \Delta p)_a = \nabla g(y^*) \quad \text{a.e. in } \{y^* > 0\} \tag{3.1}$$

$$p(T) + \partial\phi^0(y^*(T)) \ni 0 \tag{3.2}$$

$$\frac{\partial p}{\partial \nu} + \alpha p = 0 \quad \text{in } \Sigma_1; \quad p = 0 \quad \text{in } \Sigma_2 \tag{3.3}$$

$$p(y_t^* - \Delta y^* - f) = 0 \quad \text{a.e. in } Q \tag{3.4}$$

$$p \in \partial\psi(u^*) + \eta \qquad \text{a.e. in } \Sigma_1 \tag{3.5}$$

where $\eta = 0$ a.e. in $\{\alpha < u^* < b\}$, $\eta \leq 0$ a.e. in $\{u^* = a\}$ and $\eta \geq 0$ a.e. in $\{u^* = b\}$.

In Eq. (3.1) we have denoted the absolutely continuous part of the measure $p_t + \Delta p$ and it amounts to saying that there exists a bounded measure ν on $\overline{Q}$ such that $\nu_a = 0$ on $\{y^* > 0\}$ and

$$- \int_Q p\kappa_t dxdt - \int_0^T a(p,\kappa)dt - \alpha \int_{\Sigma_1} p\kappa\, dsdt + \nu(\kappa) =$$

$$= - \int_\Omega p(T)\kappa(x,T)dx + \int_Q \nabla g(y^*)\kappa\, dxdt$$

for all $\kappa \in C^1(\overline{Q})$ such that $\kappa(x,0) = 0$ for $x \in \Omega$.

Since the proof is similar to that of Theorem 1 it will be outlined only.

For every $\varepsilon > 0$ the boundary value problem

$$\begin{aligned} & y_t - \Delta y + \beta^\varepsilon(y) = f && \text{in } Q \\ & y(x,0) = y_0(x) && \text{in } \Omega \\ & \frac{\partial y}{\partial \nu} + \alpha(y - y^1) = u \quad \text{in } \Sigma_1; && y = 0 \quad \text{in } \Sigma_2 \end{aligned} \tag{3.6}$$

has a unique solution $y_u^\varepsilon \in L^2(0,T;H^1(\Omega)) \cap C([0,T];L^2(\Omega))$ with $(y_u^\varepsilon)_t \in L^2(0,T;(H^1(\Omega))')$ (this follows by standard existence results for nonlinear parabolic equations [7]). Multiplying Eq. (3.6) where $y = y_u^\varepsilon$ by y_u^ε and $\beta^\varepsilon(y_u^\varepsilon)$ and integrating over Q we get the estimates

$$||y_u^\varepsilon(t)||_{L^2(\Omega)} + \int_0^t ||y_u^\varepsilon(s)||^2_{H^1(\Omega)} ds + ||\beta^\varepsilon(y_u^\varepsilon)||^2_{L^2(Q)} \leq \tag{3.7}$$

$$\leq C(1 + ||u||^2_{L^2(\Sigma_1)}).$$

Hence if $u^{\varepsilon} \to u$ weakly in $L^2(\Sigma_1)$ then $y^{\varepsilon}_{u\varepsilon} \to y_u$ weakly in $L^2(0,T;H^1(\Omega))$, $(y^{\varepsilon}_{u\varepsilon})_t \to (y_u)_t$ weakly in $L^2(0,T;(H^1(\Omega))')$ where y_u is the solution to (1.10), (1.11). Moreover, we have

$$y^{\varepsilon}_{u\varepsilon} \to y_u \quad \text{strongly in } L^2(Q) \text{ and weak star in } L^{\infty}(0,T,L^2(\Omega)).$$

Consider the problem

$$(P^{\varepsilon}_2) \quad \min \{g(y^{\varepsilon}_u) + \Psi(u) + \phi^{o}_{\varepsilon}(y^{\varepsilon}_u(T)) + (2\delta)^{-1}||u-u^*||^2_{L^2(\Sigma_1)}\}$$

where $\delta > 0$ is sufficiently small and

$$\Psi(u) = \psi(u) \quad \text{if } u \in U_o; \quad \Psi(u) = +\infty \quad \text{if } u \,\overline{\in}\, U_o \tag{3.9}$$

$$U_o = \{u \in L^2(\Sigma_1); \quad a \leq u \leq b \quad \text{a.e. in} \quad \Sigma_1\}. \tag{3.10}$$

Let $(y^{\varepsilon},u^{\varepsilon})$ be an optimal pair of the problem (P^{ε}_2) and let $p^{\varepsilon} \in H^{2,1}(Q)$ be the solution to

$$\begin{aligned} &p^{\varepsilon}_t + \Delta p^{\varepsilon} - (\beta^{\varepsilon})'(y^{\varepsilon})p^{\varepsilon} = \nabla g^{\varepsilon}(y^{\varepsilon}) \quad \text{in } Q \\ &\frac{\partial p^{\varepsilon}}{\partial \nu} + \alpha p^{\varepsilon} = 0 \quad \text{in } \Sigma_1; \quad p = 0 \quad \text{in } \Sigma_2 \\ &p^{\varepsilon}(T) + \nabla\phi^{o}(y^{\varepsilon}(T)) = 0. \end{aligned} \tag{3.11}$$

One finds that

$$p^{\varepsilon} \in \partial\Psi(u^{\varepsilon}) + \delta^{-1}(u^{\varepsilon}-u^*) \quad \text{in } \Sigma_1 \tag{3.12}$$

and

$$||p^{\varepsilon}(t)||_{L^2(\Omega)} + ||p^{\varepsilon}||_{L^2(0,T;H^1(\Omega))} + ||(\beta^{\varepsilon})'(y^{\varepsilon})p^{\varepsilon}||_{L^1(Q)} \leq C. \tag{3.13}$$

Arguing as in the proof of Lemma 2 it follows that

$$u^{\varepsilon} \to u^* \quad \text{strongly in } L^2(\Sigma_1) \tag{3.14}$$

and therefore

$$y^{\varepsilon} \to y^* \quad \text{strongly in } C([0,T];L^2(\Omega)) \cap L^2(0,T;H^1(\Omega)) \tag{3.15}$$

$$\beta^{\varepsilon}(y^{\varepsilon}) \to f - y^*_t - \Delta y^* \quad \text{weakly in } L^2(Q).$$

By (3.13) it follows that there exists a function

$p \in BV([0,T];\ H^{-s}(\Omega)\ \cap L^2(0,T;H^1(\Omega)) \cap L^\infty(0,T;L^2(\Omega))$ such that

$$p^\varepsilon \to p \quad \text{weakly in } L^2(0,T;H^1(\Omega)) \text{ and strongly in } L^2(Q) \tag{3.17}$$

$$p^\varepsilon(t) \to p(t) \quad \text{strongly in} \quad H^{-s}(\Omega) \text{ and weakly in } L^2(\Omega) \quad \text{for all} \quad t \in [0,T]. \tag{3.18}$$

As in the proof of Theorem 1, (3.16) and (3,18) yield

$$p(y_t^* - \Delta y^* - f) = 0 \quad \text{in } Q.$$

By (3.17) is follows that $\{p^\varepsilon\}$ is bounded in $L^2(0,T;H^{1/2}(\Gamma))$ and consequently compact in $L^2(0,T;L^2(\Gamma))$. Thus we may pass to limit in (3.11) to get

$$p_t + \Delta p - \nu_p = \nabla g(y^*) \quad \text{a.e. in } Q$$

$$\frac{\partial p}{\partial \nu} + \alpha p = 0 \quad \text{in } \Sigma_1; \quad p = 0 \quad \text{in } \Sigma_2$$

where ν_p is the weak star limit of $(\beta^\varepsilon)'(y^\varepsilon)p^\varepsilon$. By (2.36) we get (3.1) as claimed (see [1], [2]). Finally, letting ε tend to zero in (3.12) one finds (3.5) thereby completing the proof.

Remark. If $n = 1$ then $y^* \in C(\overline{Q})$ and therefore Eq. (3.1) reduces to

$$(p_t + \Delta p - \nabla g(y^*))y^* = 0 \quad \text{in } Q. \tag{3.19}$$

4. FINAL REMARKS

1. Along with the corresponding state systems (1.1), (1.2) and (1.10), (1.11) respectively, the optimality systems (2.1), (2,3) and (3.1), (3.4) can be represented as quasi variational inequalities in Q. For instance by (2.1), (2.2), (2.3) we see that if $f \neq 0$ then the dual extremal arc p in problem (P_1) is the solution to boundary value problem

$$\begin{aligned} &p_t + \Delta p = \nabla g(y^*) && \text{in } E(y^*) \\ &p = 0 && \text{in } \Sigma(y^*) \cup \Sigma \\ &p(T) \in -\partial\phi^0(y^*(T)) && \text{in } E(y^*) \end{aligned} \tag{4.1}$$

where $E(y^*) = \{(x,t) \in Q; y^*(x,t) > 0\}$ and $\Sigma(y^*) = \{(x,t) \in Q; y^*(x,t) = 0\}$. If y^* is known and sufficiently smooth then problem (4.1) can be solved by standard procedure.

Similarly by (3.1) ~ (3.3) we see that the dual extremal arc p of problem (P_2) satisfies

$$(p_t + \Delta p)_a = \nabla g(y^*) \quad \text{in} \quad E(y^*)$$

$$p = 0 \ \text{in} \ E(y^*); \ \frac{\partial p}{\partial \nu} + \alpha p = 0 \ \text{in} \ \Sigma_1, \ p = 0 \ \text{in} \ \Sigma_2 \tag{4.2}$$

$$p(T) + \partial\phi^0(y^*(T)) \ni 0 \quad \text{in} \quad E(y^*).$$

For the numerical calculation of optimal control u* we may therefore use the following algorithm (for problem (P_1)). Starting with u_0 arbitrary we solve inductively the following sequence of variational inequalities; i = 0,1,...

$$\left.\begin{aligned} &((y_i)_t - \Delta y_i - f)y_i = 0, \quad y_i \geq 0 \text{ in } Q \\ &y_i(x,0) = y_0(x), \quad x \in \Omega; \\ &y_i = u_i \quad \text{in } \Sigma_1; \quad y_i = 0 \quad \text{in } \Sigma_2 \end{aligned}\right\} \tag{4.3}$$

$$\left.\begin{aligned} &(p_i)_t + \Delta p_i = \nabla g(y_i) \quad \text{in } \{y_i > 0\} \\ &p_i = 0 \quad \text{in } \{y_i = 0\} \cup \Sigma \\ &p_i(T) = -\partial\phi^0(y_i(T)) \quad \text{in } \{y_i \geq 0\} \end{aligned}\right\} \tag{4.4}$$

$$u_{i+1} = (\partial\Phi)^{-1}\left(\frac{\partial p_i}{\partial \nu}\right) \quad \text{in } \Sigma_1 \tag{4.5}$$

where Φ is defined by (2.17).

2. As noticed in introduction the above results can be applied to obtain necessary conditions of optimality for control problems governed by the Stefan free boundary problem.

For other results in this context we refer the reader to [10] and [11].

3. Our assumptions on g are unnecessary restrictive. Instead of assumptions (1.7) we may suppose that

$$g: L^2(Q) \to R \quad \text{is continuous and convex.} \tag{4.6}$$

Theorems 1 and 2 remain valid in this situation with ∇g replaced by ∂g (the subdifferential of g). The proofs are exactly the same except that in problem (P_1) and (P_2) respectively g is replaced by

$$g_\varepsilon(y) = \inf \left\{\frac{||y-z||^2}{2\varepsilon} L^2(Q) + g(y);\ y \in L^2(Q)\right\}.$$

REFERENCES

[1] Barbu, V.: Necessary conditions for boundary control problems governed by parabolic variational inequalities, An.St.Univ.Al.I.Cuza T.XXVI (1980), 47-66.

[2] Barbu, V.: Necessary conditions for nonconvex distributed control problems governed by elliptic variational inequalities, J.Math.Anal.Appl. Vol. 80,2 (1981), 566-597.

[3] Barbu, V.: Necessary conditions for distributed control problems governed by parabolic variational inequalities, SIAM J.Control Optimiz. 19 (1981), 64-86.

[4] Duvaut, G.: Résolution d'un problème de Stefan, in: New Variational Techniques in Math.Phys. C.I.M.E, Cremonese 1974, 84-102.

[5] Duvaut, G., J.L. Lions: Inequalities in Mechanics and Physics, Springer Verlag 1976, Berlin-Heidelberg-New York.

[6] Jochum, P: The numerical realization of Gauss-Newton's procedure for the inverse Stefan problem, in: Methods and Techniques of Mathematical Physics, E.Brosowski and E.Martensen eds., Verlag Peter D. Lang, Frankfurt am Main 1980.

[7] Lions, J.L.: Quelques Méthodes de Résolution des Problèmes aux Limites Non-lineaires, Dunod Gauthier Villars, Paris 1969.

[8] Ladyzenskaia, O.A., V.A. Solonnikov, N.N. Uraltzeva: Linear and Quasilinear Equations of Parbolic Type (Russian), Moskow 1967.

[9] Moreno, C., Ch. Saguez: Dependence par rapport aux données de la frontière libre associée a certaines inequations variationnelles d'evalution, INRIA Rapport de Recherche 298, May 1978.

[10] Saguez, Ch.: Controle optimal de systemes à frontière libr, Thèse, L'Université de Technologie de Compiègne, 1981.

[11] Zhou Meike, D. Tiba: Optimal control for the Stefan problem (to appear).

FINITE DIMENSIONAL COMPENSATORS FOR NONLINEAR INFINITE DIMENSIONAL SYSTEMS

M.J. Chapman, A.J. Pritchard
Control Theory Centre
University of Warwick
Coventry CV4 7AL, England

In this paper we show how it is possible to develop finite dimensional compensators which stabilize nonlinear infinite dimensional systems. Since our methods rely heavily on linear theory, we will begin by reviewing some of the linear results.

1. INFINITE DIMENSIONAL COMPENSATION

Consider the state space model

$$\dot{x} = Ax + Bu, \quad x(0) = x_0 \tag{1.1}$$

with $x \in X$, a Hilbert space, and $u(.) \in U = L^1_{loc}[0,\infty; \mathbb{R}^m]$. Assume A is the infinitesimal generator of a C_0-semigroup $S(t)$ on X and $B \in L(\mathbb{R}^m, X)$. We interpret (1.1) in the mild form

$$x(t) = S(t)x_0 + \int_0^t S(t-s)Bu(s)ds. \tag{1.2}$$

The output equation is

$$y(t) = Cx(t) \tag{1.3}$$

where $C \in L(X, \mathbb{R}^p)$ and it is required to design a control based on the output of a compensator of the form

$$\dot{w} = \bar{F}w - \bar{G}y, \quad w(0) = w_0 \tag{1.4}$$

$$u = \bar{K}w \tag{1.5}$$

where $\bar{G} \in L(\mathbb{R}^p, W)$, $\bar{K} \in L(W, \mathbb{R}^m)$ and $\bar{F}$ is the generator of a C_0-semigroup on the Hilbert space W. The extended system on $X \oplus W$ is

$$\begin{bmatrix} \dot{x} \\ \dot{w} \end{bmatrix} = \begin{bmatrix} A & B\bar{K} \\ -\bar{G}C & \bar{F} \end{bmatrix} \begin{bmatrix} x \\ w \end{bmatrix}, \quad \begin{bmatrix} x(0) \\ w(0) \end{bmatrix} = \begin{bmatrix} x_0 \\ w_0 \end{bmatrix}. \tag{1.6}$$

In practice the compensator (1.4), (1.5) may be required to satisfy several performance specifications. The most fundamental is that of stability, both to perturbations to the state (internal stability) and to errors in the modelling (structural stability). We concentrate on these specifications in this paper. The standard approach for satisfying the internal stability requirement is to construct a stabilizing state feedback control, if such exists, and to combine it with a state estimator.

Theorem 1.1. Let (A,B) be exponentially stabilizable and (C,A) exponentially detectable both with respect to a decay rate less than $-\alpha$, $\alpha > 0$. So there exists $K \in L(X, \mathbb{R}^m)$, $G \in L(\mathbb{R}^p, X)$ such that

$$||S^K(t)|| \leq Me^{-\alpha t}, \qquad ||S^G(t)|| \leq \hat{M}e^{-\alpha t}, \quad t \geq 0 \tag{1.7}$$

where $S^K(t)$, $S^G(t)$ are the C_0-semigroups generated by $A + BK$ and $A + GC$ respectively. Then the compensator on $X = W$, defined by

$$\dot{w} = (A + BK + GC)w - Gy \tag{1.8}$$

$$u = Kw \tag{1.9}$$

gives rise to an extended system (1.6) having a decay rate less than $-\alpha$.

Of course if X is infinite dimensional the above compensator will also be infinite dimensional and so must be approximated.

2. FINITE DIMENSIONAL APPROXIMATIONS

In this section we examine the possibility of finding a finite dimensional compensator in the form (1.4), (1.5) with $W \cong \mathbb{R}^n$, approximating (1.8), (1.9). We will use the following well known result [5].

Lemma 2.1. Let A be the infinitesimal generator of a C_0-semigroup $S(t)$ on a Banach space X, satisfying

$$||S(t)|| \leq Me^{-\alpha t}, \quad \alpha \in \mathbb{R}.$$

If $B \in L(X)$, then $(A + B)$ generates a C_0-semigroup $S^B(t)$, with

$$||S^B(t)|| \leq Me^{-(\alpha - M||B||)t}$$

Theorem 2.2. Suppose there exists $K \in L(X, \mathbb{R}^m)$, $G \in L(\mathbb{R}^p, X)$ and a finite dimensional subspace V of D(A), such that:

a) $R(G)$ = Range $G \subset V$

b) The semigroup $\overline{S}(t)$, generated by

$$\overline{A} = \begin{bmatrix} A + GC & 0 \\ -\overline{G}C & \overline{A + BK} \end{bmatrix}$$

on $X \oplus V$, satisfies

$$||\overline{S}(t)|| \leq Me^{-\alpha t}$$

where $\overline{A+BK} = P_V(A + BK)|_V$, $\overline{G} = P_V G$ with P_V the orthogonal projection onto V.

c) The operator $\overline{E} \in L(V,X)$ defined by

$$\overline{E} = (I - P_V)(A + BK)|_V$$

satisfies

$$||\overline{E}||_{L(V,X)} < \frac{\alpha}{M} .$$

Then there exists a stabilizing compensator with order equal to dim V.

Proof. Let W be a finite dimensional vector space isomorphic to V and $i \in L(W,X)$ the associated injection, so $R(i) = V$.

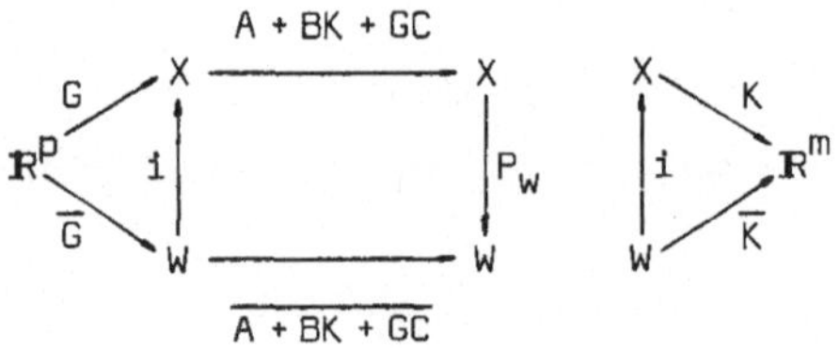

We define operators $\overline{G}$, $\overline{K}$, $\overline{F} = \overline{A + BK + GC}$ so that the above diagrams commute, and $iP_W = P_V$. Now consider the compensator defined by (1.4), (1.5). Setting $e(t) = x(t) - iw(t)$ we find $e(t)$ is the mild solution of

$$\begin{aligned} \dot{e} &= Ax + BKiw - iP_W(A + BK + GC)iw + GCx \\ &= (A + GC)x + BKiw - iP_W(A + BK)iw - GCiw . \end{aligned}$$

Let $\overline{E} \in L(W,x)$ be given by

$$\overline{E} = (I|_X - iP_W)(A + BK)i$$

then

$$\dot{e} = (A + GC)e + \overline{E}w .$$

Also

$$\begin{aligned} \dot{w} &= P_W(A + BK + GC)iw - \overline{G}C(e + iw) \\ &= P_W(A + BK)iw - \overline{G}Ce . \end{aligned}$$

This gives rise to the extended system on $X \oplus W$

$$\begin{bmatrix} \dot{e} \\ \dot{w} \end{bmatrix} = \begin{bmatrix} A + GC & 0 \\ -\overline{G}C & P_W(A+BK)i \end{bmatrix} \begin{bmatrix} e \\ w \end{bmatrix} + \begin{bmatrix} 0 & \overline{E} \\ 0 & 0 \end{bmatrix} \begin{bmatrix} e \\ w \end{bmatrix}$$

and the result follows on application of Lemma 2.1.

<u>Remarks 2.3.</u> (i) If $(A + BK)V \subset V$, then $\overline{A + BK} = (A + BK)|_V$ and $\overline{E} = 0$ so (c) is trivially satisfied. The above theorem then reduces to that given in Schumacher [9]. In this case the Theorem is essentially concerned with taking the controllable part of (1.8), (1.9) with K and G chosen to make it finite dimensional. The result as we state it is a perturbed version and $\overline{E}$ may be thought of as a measure of how close V is to being (A + BK) invariant. This idea is similar in nature to the concept of almost (A,B) invariant subspaces, (see Willems [11]).

(ii) Condition (c) will clearly be satisfied if V is A invariant (i.e. a finite span of (generalized) eigenfunctions of A) and $||(I - P_V)BK|| < \frac{\alpha}{M}$. Moreover since B has a finite dimensional range we may make $||(I - P_V)BK||$ arbitrary small if we can choose V correspondingly large. This will be the case if A has a complete set of eigenfunctions. Unfortunately the constants M, α depend on V. This motivates the problem of finding constants M, α independent of the choice of A invariant subspace V.

Suppose the semigroup $S_e(t)$ generated by

$$A_e = \begin{bmatrix} A + GC & 0 \\ -GC & A + BK \end{bmatrix}$$

on $X \oplus X$, satisfies

$$||S_e(t)|| \le Me^{-\alpha t} . \tag{2.1}$$

<u>Lemma 2.4.</u> Let A + BK generate $S^K(t)$ on X and $P_V(A + BK)|_V$ generate $P_V S^K(t)|_V$, then if (2.1) holds

$$||\overline{S}(t)|| \le Me^{-\alpha t} .$$

<u>Proof.</u> Suppose (A + GC) generates $S^G(t)$ on X, then the conditions of the lemma imply

$$\overline{S}(t) = \begin{bmatrix} S^G(t) & 0 \\ \overline{S}_{21}(t) & P_V S^K(t)|_V \end{bmatrix} \quad \text{on } X \oplus V$$

where $\overline{S}_{21}(t) = P_V S_{21}(t)$, with

$$S_{21}(t)x = - \int_0^t S^K(t-s)GCS^G(s)x \, ds .$$

Moreover

$$S_e(t) = \begin{bmatrix} S^G(t) & 0 \\ S_{21}(t) & S^K(t) \end{bmatrix}$$

and hence

$$\begin{pmatrix} I & 0 \\ 0 & P_V \end{pmatrix} S_e(t) \begin{bmatrix} x \\ v \end{bmatrix} = \bar{S}(t) \begin{bmatrix} x \\ v \end{bmatrix}$$

where $\begin{bmatrix} x \\ v \end{bmatrix}$ is considered as lying in $X \oplus X$ on the left hand side and $X \oplus V$ on the right. So

$$||\bar{S}(t)|| \leq ||S_e(t)|| \leq Me^{-\alpha t} .$$

We also have the following lemma

Lemma 2.5. If $X = V \oplus M$ with $V \subset D(A)$ and $(A+BK)(D(A)\cap M) \subset M$, then $P_V(A+BK)|_V$ generates $P_V S^K(t)|_V$.

Proof. With repect to the decomposition $X = V \oplus M$

$$A + BK = \begin{bmatrix} P_V(A+BK)|_V & 0 \\ P_M(A+BK)|_V & P_M(A+BK)|_M \end{bmatrix} \quad \text{on } V \oplus (D(A) \cap M) .$$

Since $V \subset D(A)$ it follows from the Closed Graph Theorem that

$$\hat{A} = \begin{bmatrix} P_V(A+BK)|_V & 0 \\ 0 & P_M(A+BK)|_M \end{bmatrix} \text{on } V \oplus (D(A) \cap M)$$

is a bounded perturbation of $A + BK$ and hence generates a C_o-semigroup

$$\hat{S}(t) = \begin{bmatrix} \hat{S}_{11}(t) & \hat{S}_{12}(t) \\ \hat{S}_{21}(t) & \hat{S}_{22}(t) \end{bmatrix} \quad \text{on } V \oplus M .$$

Let $\lambda \in \rho(\hat{A})$, the resolvent of $\hat{A}$, then it is easy to see that

$$R(\lambda,\hat{A})V \subset V, \qquad R(\lambda,\hat{A})M \subset M \cap D(A).$$

But

$$R(\lambda,\hat{A})x = \int_0^\infty e^{-\lambda t} \hat{S}(t)x \, dt \qquad \forall \; x \in X, \;\; \lambda \in \rho(\hat{A}) .$$

So

$$\int_0^\infty e^{-\lambda t}\,\hat{S}_{12}(t)m\,dt = \int_0^\infty -e^{-\lambda t}\,\hat{S}_{21}(t)v\,dt = 0 \qquad \forall\ m \in M,\ v \in V,\ \lambda \in \rho(\hat{A}).$$

Hence $\hat{S}_{12}(t) = \hat{S}_{21}(t) = 0$ and this implies $P_V(A + BK)|_V$ generates $\hat{S}_{11}(t)$ and $P_M(A + BK)|_M$ generates $\hat{S}_{22}(t)$. It is now clear that

$$S^K(t) = \begin{pmatrix} \hat{S}_{11}(t) & 0 \\ S^K_{21}(t) & \hat{S}_{22}(t) \end{pmatrix}$$

where

$$S^K_{21}(t)v = \int_0^t \hat{S}_{22}(t-s)P_M(A + BK)|_V\,\hat{S}_{11}(s)v\,ds$$

and the result follows.

Combining these two lemmas with Theorem 2.2 gives

Theorem 2.6. Suppose there exists $K \in \mathcal{L}(X, \mathbb{R}^m)$, $G \in \mathcal{L}(\mathbb{R}^p, X)$ and that X can be decomposed as $X = V \oplus M$ with V a finite dimensional A invariant subspace of $D(A)$ such that

a) $(A + BK)(M \cap D(A)) \subset M$

b) $R(G) \subset V$

c) The semigroup $S_e(t)$ generated by

$$A_e = \begin{bmatrix} A + GC & 0 \\ -GC & A + BK \end{bmatrix}$$

on $X \oplus X$ satisfies

$$||S_e(t)|| \leq Me^{-\alpha t}, \qquad \alpha > 0$$

d) $||(I - P_V)BK|| < \alpha/M$.

Then there exists a stabilizing compensator with order equal to dim V.

The dual theorem to Theorem 2.2 is

Theorem 2.7. Suppose there exists $K \in \mathcal{L}(X, \mathbb{R}^m)$, $G \in \mathcal{L}(\mathbb{R}^p, X)$ and a subspace T of $D(A)$ with finite codimension satisfying

a) $\ker K \supset T$

b) The semigroup $\tilde{S}(t)$ generated by

$$\tilde{A} = \begin{bmatrix} A + BK & -B\overline{K} \\ 0 & \overline{A + GC} \end{bmatrix}$$

on $X \oplus T^{\perp}$ satisfies

$$||\tilde{S}(t)|| \leq Me^{-\alpha t}, \qquad \alpha > 0$$

where $\overline{A + GC} = P_{T^\perp}(A + GC)|_{T^\perp}$, $\overline{K} = K|_{T^\perp}$ with $P_{T^\perp}$ the orthogonal projection onto $T^\perp$.

c) The operator $\tilde{E} \in L(X, T^\perp)$, defined by

$$\tilde{E} = P_{T^\perp}(A + GC)P_T$$

satisfies

$$||\tilde{E}|| < \alpha/M .$$

Then there exists a stabilizing compensator with order equal to the codimension of T.

We omit the proof ot this theorem, but note that the compensator as defined by (1.4),(1.5) is given by

$$\overline{G} = P_{T^\perp}G, \ \overline{K} = K|_{T^\perp}, \quad \overline{F} = P_{T^\perp}(A + BK + GC)|_{T^\perp} .$$

Suppose now that we have chosen $K \in L(X, \mathbb{R}^m)$ so that $(A + BK)$ is exponentially stable with decay rateless than $-\alpha$. Let us try to construct an observer to exponentially estimate Kx of the form

$$\dot{w} = Fw - \overline{G}y + \overline{H}u$$

with $w \in U$ where $U = R(K)$, $\overline{G} \in L(\mathbb{R}^p, U)$, $\overline{H} \in L(\mathbb{R}^m, U)$. Set $e = Kx - w$, then

$$\dot{e} = KAx + KBu - Fw + \overline{G}Cx - \overline{H}u .$$

Let $\overline{H} = KB$ and since $R(\overline{G}) \subset R(K)$, let $\overline{G} = KG$ with $G \in L(\mathbb{R}^p, X)$. Assume

$$K(A + GC) = FK + E \quad \text{on } D(A)$$

with $E = L(X,U)$, then

$$\begin{aligned} \dot{e} &= (FK + E)x - Fw \\ &= Fe + Ex \quad . \end{aligned}$$

Now set $F = K(A + GC)K_R^{-1}$ where $K_R^{-1}: U \to T^\perp$ is the pseudoinverse of K, and so $E = K(A + GC)[I - K_R^{-1}K]$. The compensator with u = w is therefore isomorphic to the one constructed in Theorem 2.7 where T = ker K. (The isomorphism is given by $K|_{T^\perp}: T^\perp \to U$). Of course such a choice for T may not always be possible.

We finish this section by quoting a dual theorem to Theorem 2.6.

Theorem 2.8. Suppose there exists $K \in L(X, \mathbb{R}^m)$, $G \in L(\mathbb{R}^p, X)$ and X can be decomposed as $X = V \oplus M$ with V a finite dimensional subspace of D(A) and $A(M \cap D(A)) \subset M$, such that:

a) $(A + GC)V \subset V$

b) $\ker K \supset M$

c) The semigroup $\tilde{S}_e(t)$ generated by

$$\tilde{A}_e = \begin{bmatrix} A + BK & -BK \\ & \\ 0 & A + GC \end{bmatrix}$$

on $X \oplus X$ satisfies

$$||\tilde{S}_e(t)|| \leq Me^{-\alpha t}, \qquad \alpha > 0$$

d)

$$||P_V GC(I - P_V)|| < \alpha/M.$$

Then there exists a stabilizing compensator with order equal to dim V.

3. STABILIZABILITY VIA SPECTRUM DECOMPOSITION

We make the following definitions

$$\mathbb{C}_\alpha^+ = \{\lambda \in \mathbb{C}: \text{Re } \lambda \geq -\alpha\}, \quad \alpha > 0$$

$$\mathbb{C}_\alpha^- = \{\lambda \in \mathbb{C}: \text{Re } \lambda < -\alpha\}$$

$$\sigma_u(A) = \sigma(A) \cap \mathbb{C}_\alpha^+ ; \qquad \sigma_s(A) = \sigma(A) \cap \mathbb{C}_\alpha^- .$$

Spectrum decomposition assumption 3.1. [5]

If the set $\sigma_u(A)$ is bounded and separable from the set $\sigma_s(A)$ in such a way that a simple rectifiable closed curve can be drawn enclosing an open set containing $\sigma_u(A)$ in its interior and $\sigma_s(A)$ in its exterior, then A is said to satisfy the spectrum decomposition assumption.

If this assumption holds then

$$X = X^u \oplus X^s$$

with $AX^s \subset X^s$, $AX^u \subset X^u$, and

$$\sigma(A|_{X^s}) = \sigma_s(A); \qquad \sigma(A|_{X^u}) = \sigma_u(A) .$$

We will also assume that $A|_{X^s}$ satisfies the following

Spectrum determined growth assumption 3.2.

Suppose $A|_{X^s}$ generates a C_o-semigroup T_t^s such that

$$||T_t^s|| \leq Me^{-\alpha' t} \quad \text{for all } \alpha' < \alpha$$

then we say $A|_{X^s}$ satisfies the spectrum determined growth assumption.

Let P be the projection onto X^u along X^s

$$P = \frac{1}{2\pi i} \int_{\Gamma} (\lambda I - A)^{-1} d\lambda$$

where Γ is a curve of the form given in Assumption 3.1 enclosing $\sigma_u(A)$. Suppose the pair $(A|_{X^u}, PB)$ is exponentially stabilizable and $(C|_{X^u}, A|_{X^u})$ is exponentially detectable both with respect to decay rate, $-\alpha$. Then if we choose $K_o \in L(X^u, \mathbb{R}^m)$, $G_o \in L(\mathbb{R}^p, X^u)$ such that $(A|_{X^u} + PBK_o)$ and $(A|_{X^u} + G_oC|_{X^u})$ are exponentially stable both with decay rate, $-\alpha$, so will $(A + BK)$, $(A + GC)$ be exponentially stable with rate $-\alpha$, where

$$K(x_u \oplus x_s) = K_o x_u, \qquad Gy = G_o y \oplus \{0\}.$$

Now if we assume X^u is finite dimensional and the eigenvectors of A are complete we may take as a possible candidate for V (see Sec. 2) the span of k generalised eigenvectors of A satisfying $V \supset X^u$. It then follows that $R(G) \subset V$. Furthermore, $\ker K \supset X^s$ and $X^s \cap D(A)$ is A invariant, so $X^s \cap D(A)$ is A + BK invariant. The completeness of the eigenvectors of A guarantees that if V is chosen sufficiently large we can ensure

$$||(I - P_V)BK|| < \alpha/M$$

in Theorem 2.6 (d) and hence the existence of a finite dimensional compensator. We remark that these considerations can be generalised to the case of unbounded B and C operators which occur in boundary control and point sensing, [1].

4. A BOUND FOR COMPENSATOR ORDER

An important problem is the calculation of an upper bound for dim V needed to satisfy (d) in Theorem 2.6. In order to do this we need to calculate the constants M, and α. First, however we consider the following problem which is important in its own right:

Suppose we are given a controllable pair (A,B) where A, B are nxn and nxm matrices, then how do we choose a state feedback matrix F such that

(i) A + BF is stable

(ii) A + BF is structurally stable with repect to errors in the matrix A.

Since (A,B) is controllable then (A,B) is stabilizable and we may for instance choose an F via a linear quadratic optimal control, or via some pole placement algorithm. Any such F will guarantee (ii) holds for a sufficiently small neighbourhood of perturbations about A. Our considerations here are to somehow find an F which maximizes this neighbourhood, i.e. we require $\sigma(A' + BF) \subset \mathbb{C}_0^-$ for all A' such that $||A' - A||_{L(\mathbb{R}^n)} \leq d$ and we want to choose F to maximize d. We do not claim to have the answer to this problem but a clue to its resolution may be found by considering the following simple example

$$\dot{x}_1 = x_1 + \epsilon x_2$$

$$\dot{x}_2 = u \quad .$$

For $\epsilon \neq 0$ the above system is controllable and so may be stabilized with arbitrary decay rate, but a perturbation $-\epsilon x_2$ destroys the stability. This indicates that loss of controllability is important and in some sense the distance of the pair (A,B) from the proper algebraic variety of uncontrollable pairs is important. Motivated by these considerations we obtain an estimate for d in the following way:

Let $E(T)$ be the controllability grammian

$$E(T) = \int_0^T e^{-As}BB'e^{-A's}ds$$

and define

$$\overline{E} = \int_0^T (T-s)e^{-As}BB'e^{-A's}ds \quad .$$

If (A,B) is controllable then both E and $\overline{E}$ are strictly positive definite. Now consider the functional

$$V(x) = \langle x, \overline{E}^{-1}x\rangle$$

where the inner product is on $\mathbb{R}^n$. Let $y = \overline{E}^{-1}x$, then

$$\dot{V}(x(t)) = \langle(A\overline{E} + \overline{E}A')y(t),y(t)\rangle + 2\langle BF\overline{E}y(t),y(t)\rangle.$$

But

$$A\overline{E} + \overline{E}A' = -\int_0^T (T-s)\frac{d}{ds}[e^{-As}BB'e^{-A's}]ds$$

$$= BB' - E(T).$$

Hence

$$\dot{V}(x(t)) = -<\mathcal{E}(T)y(t),y(t)> + <B(2F\overline{\mathcal{E}} + B')y(t),y(t)> .$$

We will choose $F = -\sqrt{2}\, B'\mathcal{E}^{-1}$, then

$$\dot{V}(x(t)) = -<\mathcal{E}(T)y(t),y(t)> .$$

Note

$$V(x(t)) = <\overline{\mathcal{E}}y(t),y(t)> \leq T<\mathcal{E}(T)y(t),y(t)> = -T\dot{V}(x(t)) .$$

Hence

$$V(x(t)) \leq e^{-t/T}V(x_0) .$$

Let us assume that

$$\alpha(T)||x||^2 \leq <\mathcal{E}(T)x,x> \leq \beta(T)||x||^2 .$$

Then

$$||x(t)||^2 = <\mathcal{E}\,\overline{\mathcal{E}}^{1/2}y(t),\overline{\mathcal{E}}^{1/2}y(t)> \leq T\beta(T)V(x(t))$$

$$\leq T\beta(T)V(x_0)e^{-t/T} .$$

But

$$||x_0||^2 = <\mathcal{E}\,\overline{\mathcal{E}}^{1/2}y_0,\overline{\mathcal{E}}^{1/2}y_0> \geq (T-\rho)\alpha(\rho)V(x_0)$$

for any $\rho \in [0,T]$, hence

$$||x(t)||^2 \leq \frac{T\beta(T)}{(T-\rho)\alpha(\rho)}\, e^{-t/T}||x_0||^2 .$$

So

$$||e^{(A+BF)t}|| \leq Me^{-\alpha t}, \quad \alpha > 0$$

where we may take

$$M^2 = \frac{T\beta(T)}{(T-\rho)\alpha(\rho)}, \quad \alpha = \frac{1}{2T} .$$

Thus by Lemma 2.1, the system will be stable for any A' with

$$||A' - A|| < \left[\frac{(T-\rho)\alpha(\rho)}{4T^3\beta(T)}\right]^{1/2} .$$

The term on the right hand side of the above inequality must be optimised with respect to both T and ρ.

Let us now consider the role of these results in providing a bound for the compensator order. From Theorem 2.6 we require constants M, α so that

$$||S_e(t)|| \leq Me^{-\alpha t}$$

or equivalently we can work with the semigroup $\tilde{S}_e(t)$ of Theorem 2.8.

Lemma 4.1. Suppose A_{11} generates a C_o-semigroup $S^1(t)$ on X_1 and A_{22} generates a C_o-semigroup $S^2(t)$ on X_2 and $A_{21} \in L(X_1,X_2)$. Then

$$\begin{bmatrix} A_{11} & 0 \\ A_{21} & A_{22} \end{bmatrix}$$

generates a C_o-semigroup S(t) on $X_1 \oplus X_2$. Moreover if

$$||S^i(t)|| \leq M_i \, e^{-\alpha_i t} \quad , \quad i = 1,2$$

and we set

$$||(x_1,x_2)||_{X_1 \oplus X_2} = \max \{||x_1||_{X_1}, ||x_2||_{X_2}\},$$

then

$$||S(t)|| \leq Me^{-\alpha t}$$

where

$$\alpha = \left\{ \begin{array}{ll} \min \{\alpha_1,\alpha_2\} & \text{if } \alpha_1 \neq \alpha_2 \\ \min \{\alpha_1,\alpha_2\} - \epsilon & \text{if } \alpha_1 = \alpha_2 \ \epsilon > 0 \end{array} \right\}$$

$$M = \left\{ \begin{array}{ll} \max \{M_2, \dfrac{M_1M_2||A_{21}||}{|\alpha_1 - \alpha_2|}, M_1\} & \text{if } \alpha_1 \neq \alpha_2 \\ \max \{M_2, \dfrac{M_1M_2||A_{21}||}{\epsilon e}, M_1\} & \text{if } \alpha_1 = \alpha_2 \end{array} \right\} .$$

Proof: The first part is an immediate consequence of Lemma 2.1. Let x_1,x_2 be mild solutions of

$$\dot{x}_1 = A_{11}x_1$$

$$\dot{x}_2 = A_{21}x_1 + A_{22}x_2 .$$

So

$$x_2(t) = S^2(t)x_2(0) + \int_0^t S^2(t-s)A_{21}x_1(s)ds .$$

Then

$$||x_2(t)|| \leq M_2 e^{-\alpha_2 t}||x_2(0)|| + \int_0^t M_1 M_2 ||A_{21}|| e^{-\alpha_2(t-s)} e^{-\alpha_1 s} ds ||x_1(0)||$$

$$\leq \begin{cases} M_2 e^{-\alpha_2 t}||x_2(0)|| + M_1 M_2 ||A_{21}|| \left|\dfrac{e^{-\alpha_1 t} - e^{-\alpha_2 t}}{\alpha_1 - \alpha_2}\right| ||x_1(0)|| & \alpha_1 \neq \alpha_2 \\ M_2 e^{-\alpha_2 t}||x_2(0)|| + M_1 M_2 ||A_{21}|| t\, e^{-\alpha_1 t}||x_1(0)|| & \alpha_1 = \alpha_2 . \end{cases}$$

The above inequality together with

$$||x_1(t)|| \leq M_1 e^{-\alpha_1 t}||x_1(0)||$$

yields the result.

We remark that the choice of $||.||_{X_1 \oplus X_2}$ in the above lemma was illustrative and similar lemmas may be proved using other norms. Now suppose that the conditions of Sect. 3 hold with X^u finite dimensional. Then we may choose K and G of the form

$$K = [K_o, 0] \quad , \qquad G = \begin{bmatrix} G_o \\ 0 \end{bmatrix}$$

with respect to the decomposition $X = X^u \oplus X^s$, where K_o and G_o are matrices. It follows that

$$A + BK = \begin{bmatrix} A^u + PBK_o & 0 \\ (I - P)BK_o & A^s \end{bmatrix}$$

and we may choose

$$K_o = -1/2 \, (PB)^* \overline{E}^{-1}$$

where

$$\overline{E} = \int_0^T (T-s) e^{-A^u s} PB(PB)' e^{-A^{u'} s} ds \quad .$$

Using Lemma 4.1 constants M_1, α_1 may be found such that

$$||S^K(t)|| \leq M_1 e^{-\alpha_1 t}$$

providing we can find a simlar bound for the semigroup $S^s(t)$ generated by A^s. A similar argument then goes through for A + GC and a final application of Lemma 4.1 provides constants M, α bounding the semigroup $S_e(t)$ (or $\tilde{S}_e(t)$). A bound for the compensator order may then be obtained by means of condition d) in Theorem 2.6 or 2.8.

5. NONLINEAR SYSTEMS AND STRUCTURAL STABILITY

We consider the system

$$\dot{x} = Ax + Nx + Bu, \quad x(0) = x_0 \tag{5.1}$$

$$y = Cx \tag{5.2}$$

together with the finite dimensional linear compensator

$$\dot{w} = \overline{F}w - \overline{G}y, \quad w(0) = w_0 \tag{5.3}$$

$$u = \overline{K}w \tag{5.4}$$

where A, B, C, $\overline{F}$, $\overline{G}$, $\overline{K}$ are as before and N is a nonlinear operator with $N(0) = 0$. Proceeding formally we write the extended system on $X \oplus W$ as

$$\begin{bmatrix} \dot{x} \\ \dot{w} \end{bmatrix} = \begin{bmatrix} A & B\overline{K} \\ -\overline{G}C & \overline{F} \end{bmatrix} \begin{bmatrix} x \\ w \end{bmatrix} + \begin{bmatrix} Nx \\ 0 \end{bmatrix}, \quad \begin{bmatrix} x(0) \\ w(0) \end{bmatrix} = \begin{bmatrix} x_0 \\ w_0 \end{bmatrix} \tag{5.5}$$

or

$$\dot{z} = A z + N_e z, \quad z(0) = z_0 \in Z = X \oplus W. \tag{5.6}$$

It is necessary to provide conditions that guarantee there is a unique stable solution of (5.6) and to do this we follow the approach of Ichikawa and Pritchard [4]. We look for a solution in $E = L^r[0,\infty;Z]$ where $r \geq 1$.

First we define an operator

$$(\phi z)(t) = S_e(t)z_0 + \int_0^t S_e(t-s)N_e z(s)ds \tag{5.7}$$

and we will place conditions on the various operators such that ϕ has a fixed point

(A) $\quad ||S_e(.)z_0||_E \leq \overline{\beta}||z_0||_Z, \quad \forall \; z_0 \in Z.$

Condition (A) assumes that the compensator has been designed so that $S_e(t)$ is an exponentially stable semigroup.

Now define the operator valued map $L(.)$ by

$$L(t)\overline{x} = \int_0^t S_e(t-s) \begin{bmatrix} \overline{x}(s) \\ 0 \end{bmatrix} ds$$

and let $\overline{E} = L^s[0,\infty;\overline{X}]$, then our second condition is

(B) $\quad ||L(.)\overline{x}||_E \leq \beta||\overline{x}(.)||_{\overline{E}}, \quad \forall \; \overline{x} \in \overline{E}.$

In general most nonlinearities have a "roughing" action in that if $x \in X$, $Nx \in \overline{X}$ where in general $\overline{X} \supset X$. Condition (B) is then imposed to bring the second term of (5.7) back from $\overline{E}$ to E. The fixed point theorem we intend using is the Banach theorem so we assume that N is locally Lipschitz

(C) $\quad ||Nx - N\hat{x}||_{\overline{E}} \leq k(||x||_{\underline{E}}, ||\hat{x}||_{\underline{E}})||x - \hat{x}||_{\underline{E}}$

for all $x, \hat{x}$ in $B_a^{\underline{E}} = \{x: ||x(.)||_{\underline{E}} \leq a\}$ with $\underline{E} = L^r[0,\infty;X]$.

It then follows that

$$||(\phi z)(.)||_E \leq \overline{\beta}||z_0||_Z + \beta k(||x||_{\underline{E}},0)||x||_{\underline{E}} \quad .$$

So that ϕ will map B_a^E into itself if

(D) $\quad \overline{\beta}||z_0|| + \beta k(||x||_{\underline{E}},0)a \leq a$

for all $x \in B_a^{\underline{E}}$. Moreover ϕ is a contraction if

$$\begin{aligned} ||\phi z - \phi\hat{z}||_E &= ||L(.)Nx - L(.)N\hat{x}||_E \\ &\leq \beta||Nx - N\hat{x}||_{\overline{E}} \\ &\leq \beta k(||x||_{\underline{E}}, ||\hat{x}||_{\underline{E}})||x - \hat{x}||_{\underline{E}} \end{aligned}$$

where

(E) $\quad \beta k(||x||_{\underline{E}}, ||\hat{x}||_{\underline{E}}) \leq K < 1$

for all $x, \hat{x} \in B_a^{\underline{E}}$.

If in (5.1) N is a true nonlinearity, by which we mean that the linear part has been represented in the operator A, then we may assume

$$k(\theta_1,\theta_2) \to 0 \quad \text{as } \theta_1,\theta_2 \to 0 \ .$$

Hence by choosing "a" small enough it is always possible to satisfy condition (E), then for such a choice of K condition (D) will be satisfied if

$$||z_0|| \leq \frac{a(1 - K)}{\overline{\beta}} \quad . \tag{5.8}$$

Clearly this condition must be optimized with respect to a to obtain the largest set of initial states. However since x_0 and hence z_0 is unknown we are not able to check this condition directly. To do this we assume that the linearized system (5.1) is continuously initially observable between X and some output set of functions $\mathcal{Y}$. (This can always be achieved if the map

$$(\mathcal{E}x)(t) = CS(t)x_0$$

is one to one, by letting $\mathcal{Y}$ = Range ($\mathcal{E}$)). Then the extended system with

$$\mathcal{Y}_e(t) = \begin{bmatrix} y(t) \\ w(t) \end{bmatrix} = \begin{bmatrix} C & 0 \\ 0 & I \end{bmatrix} z(t) = C_e z(t)$$

will also be continuously initially observable. This means that the map $\mathcal{E}_e : Z \to \mathcal{Y}_e = \mathcal{Y} \oplus \mathcal{Y}_0$ is invertible (where $\mathcal{Y}_0$ is the set of possible output functions of the observer). Now

$$\mathcal{Y}_e(t) = (\mathcal{E}_e z_0)(t) + C_e \int_0^t S_e(t-s)N_e z(s)ds \ .$$

Hence if we assume $C_e L(.)Nx \in \mathcal{Y}_e$ with

(F) $\quad ||C_e L(.)\overline{x}||_{\mathcal{Y}_e} \leq \gamma ||\overline{x}||_{\overline{E}} \ , \qquad \forall \ \overline{x} \in \overline{E}$

then

$$z_0 = \mathcal{E}_e^{-1}[y_e(.) - C_e L(.)Nx] \ .$$

So

$$||z_0||_Z \leq ||\mathcal{E}_e^{-1}||[||y_e||_{\mathcal{Y}_e} + \gamma k(||x||_{\underline{E}},0)||x||_{\underline{E}}]$$

$$\leq ||\mathcal{E}_e^{-1}||[||y_e||_{\mathcal{Y}_e} + \frac{\gamma k}{\beta} a] \ .$$

Thus if

$$||y_e||_{\mathcal{Y}_e} \leq \frac{a(1 - K[1 + \frac{\gamma\overline{\beta}}{\beta}||\mathcal{E}_e^{-1}||\,])}{\overline{\beta}||\mathcal{E}_e^{-1}||} \qquad (5.9)$$

then (5.8) will certainly hold.

Conditions (A) - (E) guarantee that the finite dimensional compensator stabilizes the nonlinear system (5.1) as well as its linearization. In a way this is a kind of structural stability result, saying that the system is stableto a class of nonlinear perturbations. It is interesting therefore to see what the conditions (A) - (E) imply when in fact N is linear. In this case we take L^∞ spaces for E, $\overline{E}$, and $\underline{E}$, then (A) obviously holds with $\overline{\beta} = M$ and (B) holds with

$$\beta = \frac{M}{w}$$

where

$$||S_e(t)|| \leq Me^{-wt}, \quad w > 0$$

(C) holds with $k(.,.) = ||N||$ and so (D) and (E) require

$$M||z_0|| + \frac{M}{w}||N||a \leq a \tag{5.10}$$

and $\frac{M}{w}||N|| \leq K < 1$.

By choosing a sufficiently large we can always satisfy (5.10) if

$$\frac{M}{w}||N|| \leq K < 1$$

and this is just the condition we obtained in Lemma 2.1.

REFERENCES

[1] Curtain, R.F.: Finite dimensional compensator design for parabolic systems with point and boundary input, IEEE Trans. Automat. Control Vol. AC-26 (1982), 98-104.

[2] Curtain, R.F.: Compensators for infinite dimensional linear systems: a survey, to appear in the special issue of the Journal of the Franklin Institute on Distributed Parameter Systems.

[3] Curtain, R.F., A.J. Pritchard: Infinite dimensional linear systems theory, Lectures Notes in Control and Information Sciences, Vol. 8, Springer Verlag, 1978.

[4] Ichikawa, A., A.J. Pritchard: Existence, uniqueness and stability of nonlinear evolution equations, J. Math. Anal. and Appl., Vol. 68, No. 2, (April 1979), 454-476.

[5] Kato, T.: Perturbation theory of linear operators, Springer Verlag, 1966.

[6] Pritchard, A.J., J. Zabczyk: Stability and stabilizability of infinite dimensional systems, SIAM Review, Vol. 23 (1981), 25-52.

[7] Slemrod, M.: Asymptotic behaviour of C_0-semigroups as determined by the spectrum of the generator, Indiana Univ. Math. J. 25 (1976), 783-791.

[8] Schumacher, J.M.: Compensator synthesis using (C,A,B) pairs, IEEE Trans. Automat. Control, Vol. AC-25 (1980), 1133-1138.

[9] Schumacher, J.M.: A direct approach to compensator design for distributed paramter systems, to appear in SIAM J. Control and Optimization.

[10] Triggiani, R., A.J. Pritchard, Stabilizability in Banach space, Control Theory Centre Report No. 35, University of Warwick., U.K.

[11] Willems, J.C.: Almost A(mod B)-Invariant subspaces, Asterisque 75-76(1980),239-248.

[12] Yosida, K.: Functional analysis, Springer Verlag 1965.

FINITE DIMENSIONAL COMPENSATORS FOR SOME HYPERBOLIC SYSTEMS WITH BOUNDARY CONTROL

R. F. Curtain

Mathematics Institute
Rijksuniversiteit Groningen
Postbus 800
NL-9700 AV Groningen, Netherlands

1. INTRODUCTION

The possiblility of finite dimensional compensators for infinite dimensional systems was first realized by Schumacher in [11],[12], who designed finite dimensional stabilizing schemes via dynamic output feedback for a large class of systems, including parabolic and delay systems. The main restriction was that the control action and the observation be implemented by bounded operators B and C. For parabolic systems this restriction was eliminated by Curtain in [4], who used a different compensator scheme, but still by dynamic output feedback. Hyperbolic systems such as the wave equation cannot be made exponentially stable by finite dimensional state feedback [8], essentially because of the way the eigenvalues cluster along vertical asymptotes. The two schemes [12],[4] work by shifting finitely many eigenvalues to stabilize the system and it is essential that the original system has finitely many eigenvalues to the right of $Re(s) = -a$. This is not satisfied by the wave equation, but it is satisfied by some hyperbolic systems used in modelling, for example [6] and [9]. For such hyperbolic systems the approach of Schumacher is applicable and as noted in [5] the construction of Curtain in [4] can also be applied to stabilize these systems, provided that B and C are bounded. For more background on this, the reader is referred to [5].

In fact point observations for hyperbolic systems usually result is a bounded C operator and so the interest lies with boundary control. We show here how by reformulating the boundary control problem as done by Fattorini in [7], and by augmenting the system, we arrive at a type of system treated in [4] and [11] with a bounded 'B'operator. Thus we obtain a finite dimensional compensator for the augmented system and we show how this can be interpreted as an integral dynamic output feedback compensator for the original system. Finally we illustrate this approach with two examples of hyperbolic systems with boundary control action which have appeared in literature, [6], [9].

2. FINITE DIMENSIONAL COMPENSATORS FOR BOUNDED SYSTEMS

We consider the following linear infinite dimensional system on the Banach space Z

$$\dot{z} = Az + Bu \;; \quad z(0) = z_0 \tag{2.1}$$

$$y = Cz \tag{2.2}$$

where A is the infinitesimal generator of the strongly continuous semigroup T(t) on Z, U and Y are finite dimensional input and output spaces and $B \in L(U,Z)$, $C \in L(Z,Y)$. If $z \in D(A)$ and $u \in C^1(0,t;U)$, then (2.1) has the unique solution

$$z(t) = T(t)z + \int_0^t T(t-s)Bu(s)ds. \tag{2.3}$$

In our applications u is always smooth, so we shall use the differential form without further comment.

The compensator for (2.1), (2.2) has the form

$$\dot{w} = Mw + Ly \tag{2.4}$$

$$u = Qw \tag{2.5}$$

where $w(t) \in W$, finite dimensional, and L, M and Q are suitable matrices. Combining (2.1), (2.2) and (2.4), (2.5) we obtain the following extended system operator A_e on the extended state space $Z \oplus W$

$$A_e = \begin{pmatrix} A & BQ \\ LC & M \end{pmatrix} \quad . \tag{2.6}$$

A_e generates a strongly continuous semigroup $T_e(t)$ on $Z \oplus W$. The problem is to choose W, Q, L and M so that A_e is exponentially stable: i.e. there exist positive constants M and ω such that $||T_e(t)|| \leq Me^{-\omega t}$. This was done in [4] and [12] under certain fundamental assumptions, which we list below.

Assumption 1. A has a discret spectrum and there are finitely many eigenvalues in $Re(s) > -\delta$ for all $\delta > 0$.

This induces a natural state space decomposition [10]. Define

$$\sigma_u(a) = \sigma(A) \cap \{s: Re(s) \geq -\delta\}$$

$$\sigma_s(A) = \sigma(A) \cap \{s: Re(s) < -\delta\}$$

and let Γ be a simple rectifiable closed curve enclosing $\sigma_u(A)$ in its interior and $\sigma_s(A)$ in its exterior. Then define

$$Z^u = PZ, \; Z^s = (I - P)Z; \quad Z = Z^u \oplus Z^s$$

where $P = \frac{1}{2\pi i}\int_{\Gamma}(sI - A)^{-1}ds$ is a bounded projection in Z. Note that $A^u = A/Z^u$ is bounded. Let $A^s = A/Z^s$, then $\sigma(A^s) = \sigma_s(A)$, $\sigma(A^u) = \sigma_u(A)$.

P reduces T(t), by which we mean that P and (I - P) commute with A and T(t) and $T^u(t) = PT(t)$ is the semigroup generated by A^u. $T^u(t) = \exp(A^u t)$. $T^s(t) = (I - P)T(s)$ is the semigroup generated by A^u. $T^u(t) = \exp(A^u t)$. $T^s(t) = (I - P)T(s)$ is the semi-group generated by A^s. With respect to this decomposition we shall write

$$A = \begin{pmatrix} A^u & 0 \\ 0 & A^s \end{pmatrix}, \quad B = \begin{pmatrix} B^u \\ B^s \end{pmatrix}, \quad C = (C^u C^s).$$

<u>Assumption 2.</u> The eigenfunctions of A are complete in Z.

<u>Assumption 3.</u> A^s satisfies the spectrum determined growth assumption:

$$\sup \operatorname{Re} \sigma(A^s) = \lim_{t\to\infty} \frac{\log||T^s(t)||}{t}.$$

<u>Assumption 4.</u> (A^u,B^u,C^u) is minimal for some decomposition $\delta > 0$.

We remark that assumption 4 holds if (A,B) is approximately controllable and initially observable [3].

We now extend the construction in [4] to this more general system operator A. Under assumption 4, we can always find $F_0 \in L(Z^u,U)$, $G_0 \in L(Y,Z^u)$ such $\sigma(A^u+B^uF_0)$ and $\sigma(A^u+G_0C^u)$ are in $Re(s) < -\delta-\epsilon$, $\epsilon > 0$. Then with $F = (F_0 0)$ and $G = \begin{pmatrix} G_0 \\ 0 \end{pmatrix}$, we know that

$$\sigma(A + BF) = \sigma(A^u + B^uF_0) \cup \sigma(A^s)$$

$$\sigma(A + GC) = \sigma(A^u + G_0C^u) \cup \sigma(A^s)$$

and under assumption 3. the semigroups generated by A + BF and A + GC are exponentially stable with decay rate δ. ([12] and appendix). Suppose now that there are n eigenvalues in $Re(s) \geq -\delta$ which we wish to move. Choose initially $W = R^n$ and let R be the isomorphism between Z^n and W, and let P_n be the projection from Z to Z^u. Choosing $Q = FR^{-1}$, $L = -RG$ and $M = R(A + P_nBF + GC)R^{-1}$, produces a well defined extended system operator A_e (2.6). A_e is isomorphic to $\tilde{A}_e$:

$$\tilde{A}_e \Rightarrow \begin{pmatrix} A + BF & -BFR^{-1}T^{-1} \\ TRG(C-C^u) & TR(A+GC)R^{-1}T^{-1} \end{pmatrix} \tag{2.7}$$

where $T: w \to \begin{pmatrix} R\,w \\ w \end{pmatrix}$ maps W onto the space $M = \{\begin{pmatrix} x \\ x\,R \end{pmatrix}, x \in Z^u\}$.

In fact, $A_e = H\tilde{A}_eH^{-1}$, where

$$H = \begin{pmatrix} I & 0 \\ RP_n & -T \end{pmatrix}, \quad H^{-1} = \begin{pmatrix} I & 0 \\ RP_n & -T \end{pmatrix}.$$

Now

$$\tilde{A}_e = A_1 + \begin{pmatrix} 0 & 0 \\ TRG(C-C^u) & 0 \end{pmatrix} \tag{2.8}$$

where

$$A_1 = \begin{pmatrix} A + BF & -BFR^{-1}T^{-1} \\ 0 & TR(A+GC)R^{-1}T^{-1} \end{pmatrix}$$

has discrete spectrum, since $-BFR^{-1}T^{-1}$ is a degenerate perturbation [10].

$$\sigma(A_1) = \sigma(A + BF) \cup \sigma(TR(A + GC)R^{-1}T^{-1})$$

$$= \sigma(A^u + B^uF_o) \cup \sigma(A^u + G_oC^u) \cup \sigma(A^s) \;.$$

Furthermore, A_1 generates a strongly continuous semigroup $T_1(t)$, whose decay rate is δ ([12] and the appendix).

$\begin{pmatrix} 0 & 0 \\ TRG(C-C^u) & 0 \end{pmatrix}$ is a degenerate perturbation of A_1 and so $\sigma(\tilde{A}_e)$ is also discrete [10]. By a standard perturbation result [3], $\tilde{T}_e(t)$ generated by $\tilde{A}_e$ satisfies the estimate

$$\tilde{T}_e(t) \leq M \exp t(-\delta + M||TRG(C-C^u)||) \tag{2.9}$$

where $T_1(t) \leq M \exp(-\delta t)$ is the semigroup generated by A_1. So we see that if $||C - C^u||$ is sufficiently small, $\tilde{T}_e(t)$ and $T_e(t)$ will be exponentially stable. If this is not the case for the initial decomposition, then one can choose a decomposition for a larger δ, which will make $||C - C^u||$ smaller, since C has finite rank and under assumption 2. One has here the option to "freeze" $F = (F_o 0)$. $G = \begin{pmatrix} G_o \\ 0 \end{pmatrix}$ at the original choices; however, wo do not wish to go into practical design considerations here. The conclusion is that for a sufficiently large n one can design a finite dimensional compensator for the original system under assumptions 1 - 4. In practice, one can check the stability of the extended system using the Weinstein-Aronzajn result concerning degenerate perturbations given in [10]. This amounts to finding roots of a polynomial and is fully discussed in [12]. For numerical results on compensators for parabolic systems, see [1]; there the order of the compensator was found to be equal to or one greater than the number of eigenvalues relocated. For details concerning Schumachers' results, see [12] or the survey [5].

3. MATHEMATICAL FORMULATION OF BOUNDARY CONTROL FOR HYPERBOLIC SYSTEMS

It is well known that several boundary control problems can be reformulated by a transformation of a system with inhomogeneous boundary conditions into an equivalent homogeneous system [7].

Consider the following system on a Banach space Z

$$\dot{z} = \mathfrak{a}z \tag{3.1}$$

$$\tau z = Ru \tag{3.2}$$

where $\mathfrak{a}$ is a closed operator on Z and τ is a linear operator with $D(\mathfrak{a}) \subseteq D(\tau)$ and the restriction of τ to $D(\mathfrak{a})$ is continuous with respect to the graph norm of $\mathfrak{a}$. Typically $\mathfrak{a}$ is a partial differential operator acting on its boundary. We suppose that $u(t) \in U$, a finite dimensional input space, and $R \in L(U,R^p)$.

We define the associated operator A on Z by

$$D(A) = \{z \in D(\mathfrak{a})/\tau z = 0\} \text{ and } Az = \mathfrak{a}z \text{ in } D(A) \tag{3.3}$$

and we assume that A is the infinitesimal generator of a strongly continuous semigroup on Z. Our final assumption is that there exists a $B \in L(U,Z)$ so that

(i) $Bu \in D(\mathfrak{a})$ (3.4)

(ii) $\tau(Bu) = Ru$ for all $u \in U$.

Under these assumptions, $N = \mathfrak{a}B \in L(U,Z)$, and the following homogeneous system is well defined

$$\dot{v} = Av - B\dot{u} + Nu \tag{3.5}$$

$$v(0) = v_0$$

and has the unique solution

$$v(t) = T(t)v_0 + \int_0^t T(t-s)Nu(s)ds - \int_0^t T(t-s)B\dot{u}(s)ds \tag{3.6}$$

provided $v_0 \in D(A)$ and $\dot{u}(s)$ is continuously differentiable. It is then easily verified that

$$x(t) = v(t) + Bu(t) \tag{3.7}$$

is a solution of (3.1), (3.2), and conversely, with of course $v_0 = z_0 - Bu(0)$. (We shall choose $u(0) = 0$).

We remark that if $z_0 \notin D(A)$ and $\dot{u}$ is only integrable, we can still identify the mild solution of (3.5) with solution of (3.1), (3.2). In our applications we can always take $z_0 \in D(A)$ and u will be at least continuous.

We now introduce the extended system for $v = \begin{pmatrix} u \\ v \end{pmatrix}$ on the state space $U \oplus Z$. Then

$$\dot{\tilde{v}} = \begin{pmatrix} 0 & 0 \\ N & A \end{pmatrix} \tilde{v} + \begin{pmatrix} I \\ -B \end{pmatrix} \tilde{u} \tag{3.8}$$

where $\tilde{u} = \dot{u}$. Then we have

$$z = (B\ I)\tilde{v} \tag{3.9}$$

and if we have the observation for the original system

$$y = Cz \tag{3.10}$$

we can reformulate this as

$$y = \tilde{C}\tilde{v} \quad \text{with} \quad \tilde{C} = (CB\ C). \tag{3.11}$$

We now define our compensator on W, a finite dimensional space for (3.8), (3.11) to be

$$\dot{w} = Mw + Ly \;; \qquad \tilde{u} = Qw. \tag{3.12}$$

We can apply the construction of §2 (or the Schumacher construction) to (3.8), (3.11), (3.12) provided assumptions 1 - 4 are satisfied. We translate these assumptions in terms of the original system operators.

$\tilde{A} = \begin{pmatrix} 0 & 0 \\ N & A \end{pmatrix}$ has the same eigenvalues as A, plus $\lambda = 0$ with eigenvector $\begin{pmatrix} 1 \\ 0 \end{pmatrix}$.

If the eigenfunctions of A span Z then those of $\tilde{A}$ span $U \oplus Z$. $\tilde{A}$ generates the strongly continuous semigroup

$$\tilde{T}(t) = \begin{pmatrix} I & 0 \\ S(t) & T(t) \end{pmatrix} \tag{3.13}$$

where $S(t) = \int_0^t T(t-s)N ds$ and it is clear that $\tilde{A}$ satisfies assumptions 1 - 3 if and only if A does. One can also show that $(\tilde{A},\tilde{B})$ is approximately controllable iff (A,B) is: $\tilde{B} = \begin{pmatrix} I \\ -B \end{pmatrix}$, and $(\tilde{C},\tilde{A})$ is initially observable iff (C,A) is: $\tilde{C} = (CB\ C)$. Thus assumption 4 holds for $(\tilde{A},\tilde{B},\tilde{C})$ iff it holds for (A,B,C).

It remains to interpret the effect of the compensator (3.12) on the original system, but from (3.9) it follows that if $\tilde{v}$ is exponentially stable, so is z with the same decay rate. So our original system (3.1), (3.2) can be stabilized by the integral control

$$u(t) = \int_0^t Qw(s)ds \tag{3.14}$$

where w(t) is given by (3.12). Thus we coin the phrase "integral dynamic output feedback".

4. EXAMPLES

To illustrate the feasibility of the construction outlined in §3 we consider two examples treated in the literature. The first is a prototype model for large scale flexible space structures considered by Gibson and Navid in [9]. There they assumed a distributed control, although it seems likely that boundary control would be more appropriate.

Example 1. Flexible beam

Consider a simple supported beam of unit length, with first natural frequency π^2 and internal dampling equal to $\sqrt{2}$ % of critical damping, then we have the free system

$$\frac{\partial^2 z}{\partial t^2} + \frac{\partial^4 z}{\partial x^4} - .01 \frac{\partial^3 z}{\partial x^2 \partial t} = 0 \tag{4.1}$$

$$z(0,t) = 0 = z(1,t) = z_{xx}(0,t) = z_{xx}(1,t) \tag{4.2}$$

where $z(x,t)$ is the vertical displacement of the beam at time t and at a distance from one end.

Following [3], example 2.16, p. 25, we define the following operator A on $L_2(0,1)$

$$Ah = \frac{d^4h}{dx^4} \; ; \quad D(A) = \begin{cases} h \in L_2(0,1)\colon h, h_x, h_{xx}, h_{xxx}, h_{xxxx} \in L_2(0,1) \text{ and} \\ h(0) = 0 = h(1) = h_{xx}(0) = h_{xx}(1) \; . \end{cases} \tag{4.3}$$

A is self adjoint and positive on $L_2(0,1)$ and has the square root

$$A^{\sqrt{2}}h = \frac{d^2h}{dx^2} \; ; \quad D(A^{\sqrt{2}}) = \begin{cases} h \in L_2(0,1)\colon h, h_x, h_{xx} \in L_2(0,1) \text{ and} \\ h(0) = 0 = h(1) \; . \end{cases} \tag{4.4}$$

We define $Z = D(A^{\sqrt{2}}) \oplus L_2(0,1)$ with the inner product

$$\langle w, \bar{w} \rangle_Z = \langle A^{\sqrt{2}} w_1, A^{\sqrt{2}} w_1 \rangle + \langle w_2, \bar{w}_2 \rangle \; ; \quad w = \begin{pmatrix} w_1 \\ w_2 \end{pmatrix} \tag{4.5}$$

where $\langle , \rangle$ is the usual inner product on $L_2(0,1)$.

The system operator associated with (4.1), (4.2) is now

$$\mathbf{A} = \begin{pmatrix} 0 & I \\ -A & 2\alpha \frac{d^2}{dx^2} \end{pmatrix} ; \quad D(\mathbf{A}) = D(A) \oplus D(A^{\sqrt{2}}), \tag{4.6}$$

and $2\alpha = .01$.

Since $\langle \mathbf{A}w, w \rangle_Z = \langle \mathbf{A}^* w, w \rangle_Z = -2\alpha \| w_2' \|^2$, we deduce that $\mathbf{A}$ generates a strongly continuous contraction semigroup on Z. (Theorem 2.14, [3]). The eigenvalues of $\mathbf{A}$ are $\lambda_n = n^2\pi^2(-\alpha + i\sqrt{1-\alpha^2})$, $\bar{\lambda}_n$; n = 1,2,... and the eigenvectors are

$$\begin{pmatrix} \sin n\pi x \\ \lambda_n \sin n\pi x \end{pmatrix} , \begin{pmatrix} \sin n\pi x \\ \bar{\lambda}_n \sin n\pi x \end{pmatrix}$$

which are complete in Z. Thus A satisfies assumptions 1 and 2. An analysis of the solution of (4.1), (4.2) shows that assumption 3 is also satisfied. In order to check assumption 4 it proves convenient to use the following orthonormal basis for Z:

$$e_n = \sqrt{2}/(n\pi)^2 \begin{pmatrix} \sin n\pi x \\ 0 \end{pmatrix} , \quad f_n = \sqrt{2} \begin{pmatrix} 0 \\ \sin n\pi x \end{pmatrix} \quad ; \quad n = 1,2,\ldots$$

under which A has the following decomposition

$$A = \text{diag } (\Delta_1, \Delta_2, \Delta_3, \ldots) \tag{4.7}$$

where $\Delta_n = \begin{pmatrix} 0 & -n^2\pi^2 \\ n^2\pi^2 & -2\alpha n^2\pi^2 \end{pmatrix}$.

We now consider the controlled version of (4.1) with boundary conditions

$$z(0,t) = 0 = z(1,t) = z_{xx}(0,t); \quad z_{xx}(1,t) = u(t) \tag{4.8}$$

and we suppose that we can measure the displacement at x_o

$$y(t) = z(x_o,t) \tag{4.9}$$

(4.9) defines a bounded observation operator from Z to Y by

$$C \begin{pmatrix} w_1 \\ w_2 \end{pmatrix} = w_1(x_o) \tag{4.10}$$

whereas the boundary condition corresponds to the description (3.1), (3.2), where

$D(\sigma) = H \oplus D(A^{1/2})$, and

$$H = \begin{cases} h \in L_2(0,1): h_x, h_{xx}, h_{xxx}, h_{xxxx} \in L_2(0,1) \\ \text{and } h(0) = 0 = h(1), h_{xx}(0) = 0 \end{cases}$$

and on their common domain $\sigma = A$.

We now have the boundary operator τ defined by

$$\tau \begin{pmatrix} w_1 \\ w_2 \end{pmatrix} = w_1''(1) \quad \text{and} \quad R = I.$$

To show that $\tau/D(\sigma)$ is continuous with respect the graph norm of σ, we consider first

$$\tau_2: Z \to Z \quad \text{given by} \quad \tau_2 \begin{pmatrix} w_1 \\ w_2 \end{pmatrix} = \begin{pmatrix} w_1'' \\ 0 \end{pmatrix} .$$

Then

$$||\tau_2 \begin{pmatrix} w_1 \\ w_2 \end{pmatrix}||_Z^2 = \langle A^{1/2} w_1'', A^{1/2} w_1'' \rangle = ||w_1^{(4)}||^2 \le (1-\alpha)||\sigma w||_Z^2$$

since

$$||\sigma w||_Z^2 = (1 - 4\alpha^2)||w_2''||^2 + ||w_1^{(4)}||^2 + 2\alpha < w_1^{(4)}, w_2'' >$$

$$\geq (1 - \alpha - 4\alpha^2)||w_2''||^2 + (1 - \alpha)||w_1^{(4)}||^2 .$$

So τ_2 is σ-bounded and the evaluation map at $x = 1$ for $f \in H^2(0,1)$ is continuous and so τ is continuous as required.

The required $b \in Z$ satisfying (3.4)(i) and (ii) is given by

$$b = \begin{pmatrix} 1/6(x^3-x) \\ 0 \end{pmatrix} \tag{4.11}$$

and

$$N = \sigma b = \begin{pmatrix} 0 & I \\ -\frac{d^4}{dx^4} & 2\alpha \frac{d^2}{dx^2} \end{pmatrix} \begin{pmatrix} 1/6(x^3-x) \\ 0 \end{pmatrix} = 0 .$$

Thus

$$z = (B\ I)\tilde{v} \quad , \text{ where} \tag{4.12}$$

$$\dot{\tilde{v}} = \begin{pmatrix} 0 & 0 \\ 0 & A \end{pmatrix} \tilde{v} + \begin{pmatrix} I \\ -b \end{pmatrix} \tilde{u} \ ; \quad \tilde{u} = \dot{u} \tag{4.13}$$

$$y = (Cb\ C)\tilde{v} . \tag{4.14}$$

In order to verify assumption 4, we decompose b and C with respect to the basis $\{e_n, f_n\}$ and obtain

$$b = \sum_{n=1}^{\infty} \alpha_n e_n \tag{4.15}$$

where $\alpha_n = \sqrt{2}(n\pi)^2 \int_0^1 1/6(x^3-x)\sin n\pi x \, ds = \sqrt{2}(-1)^n/(n\pi)$

and

$$Cw = \langle c_0, w\rangle_Z \ ; \quad c_0 = \sum_{n=1}^{\infty} \sqrt{2} \frac{\sin n\pi x_0}{(n\pi)^2} e_n . \tag{4.16}$$

To check the controllability of (A^u, b^u), we form the controllability matrix

$(b^u : A^u b^u : (A^u)^2 b^u : \dots\dots\dots) = C(A^u, b^u)$

$$= \begin{pmatrix} \alpha_1 & 0 & -\mu_1^2\alpha_1 & 2\alpha\mu_1^3\alpha_1 & -(1+2\alpha^2)\mu_1^4\alpha_1 & \cdot \\ 0 & \mu_1\alpha_1 & -2\alpha\mu_1^2\alpha_1 & \mu_1^3\alpha_1(1+2\alpha)^2 & -4\alpha^3\mu_1^3\alpha_1 & \cdot \\ \alpha_2 & 0 & -\mu_2^2\alpha_2 & 2\alpha\mu_2^3\alpha_2 & -(1+2\alpha^2)\mu_2^4\alpha_2 & \cdot \\ 0 & \mu_2\alpha_2 & -2\alpha\mu_2^2\alpha_2 & \mu_2^3\alpha_2(1+2\alpha^2) & -4\alpha^3\mu_2^3\alpha_2 & \cdot \\ \alpha_3 & 0 & -\mu_3^2\alpha_3 & \cdot & \cdot & \cdot \\ 0 & \mu_3\alpha_3 & -2\alpha\mu_3^2\alpha_3 & \cdot & \cdot & \cdot \\ \cdot & \cdot & \cdot & \cdot & \cdot & \cdot \\ \cdot & \cdot & \cdot & \cdot & \cdot & \cdot \end{pmatrix} \tag{4.17}$$

where $\mu_n = n^2\pi^2$.

It is clear that det $C(A^u,b^u)$ can be factored

$$\det C(A^u,b^u) = \alpha_1^2\alpha_2^2 \dots \alpha_p^2\mu_1\mu_2 \dots \mu_p P(\alpha,\mu_1\dots\mu_p)$$

where $P(\alpha,\mu_1\dots\mu_p)$ is a combination of $(\mu_i-\mu_j)$ products and the other terms, clearly nonzero. Thus det $C(A^u,b^u) \neq 0$ iff $\alpha_1,\alpha_2,\dots,\alpha_p$ are nonzero, and from (3.28) we see that they are all nonzero and so (A^u,b^u) is always controllable, or (A,B) is spectrally controllable. Using a dual argument we see that (C,A) is spectrally observable if x_0 is irrational: $\sin n\pi x \neq 0$ for all n. Thus all the assumptions are satisfied and (4.1), (4.8), (4.9) can be stabilized using a finite dimensional compensator of the form (3.12) and the control in (4.8) has the form

$$u(t) = \int_0^t Qw(s)ds \, . \tag{4.18}$$

Similarly one can show that this is also possible with a control via $z_{xx}(0) = u(t)$ for then you obtain $b \in L(U,Z)$ given by

$$b = \begin{pmatrix} -x/6(x-2)(x-1) \\ 0 \end{pmatrix} \quad \text{and } \alpha_n = -\sqrt{2}/(n\pi).$$

The following example is taken from [6], where one was interested in reducing the stress in a locomotive train. This was formulated as an optimization problem and there was no reason to worry about stability. It is, however, an interesting example to analyse.

Example 2. A distributed model for a locomative train

We consider the system

$$\frac{\partial^2 z}{\partial t^2} = a\frac{\partial^2 z}{\partial x^2} + a\frac{\partial z}{\partial x^2 \partial t} \tag{4.19}$$

$$a\left(\frac{\partial z}{\partial x} + \frac{\partial^2 z}{\partial x \partial t}\right)(0,t) = u_o(t) \tag{4.20}$$

$$a\left(\frac{\partial z}{\partial x} + \frac{\partial^2 z}{\partial x \partial t}\right)(1,t) = u_1(t) \; .$$

Following [6], we choose $Z = L_2(0,1) \oplus L_2(0,1)$ as the state space with the inner product

$$<w,\overline{w}>_Z \; = a<w_1,\overline{w}_1> + <w_2,\overline{w}_2> \; ; \; w = \begin{pmatrix} w_1 \\ w_2 \end{pmatrix} \tag{4.21}$$

where $<,>$ is the usual inner product on $L_2(0,1)$.

Then the following operator $\mathbb{A}$ generates a strongly continuous contraction semigroup on Z

$$\mathbb{A} = \begin{pmatrix} 0 & \frac{d}{dx} \\ a\frac{d}{dx} & a\frac{d^2}{dx^3} \end{pmatrix} \tag{4.22}$$

$$D(\mathbb{A}) = \left\{ \begin{array}{l} w \in Z: w_1, w_2, \frac{dw_1}{dx}, \frac{dw_2}{dx}, \frac{d^2 w_2}{dx^2} \in L_2(0,1) \\ \text{and } w_1 + \frac{dw_2}{dx} = 0 \text{ at } x = 0,1. \end{array} \right.$$

has eigenvalues

$$\lambda_n^+ = \frac{-a(n\pi)^2 + \sqrt{a^2(n\pi)^4 - 4a(n\pi)^2}}{2} \; ; \qquad {}^{-}_{n} = \frac{-a(n\pi)^2 - \sqrt{a^2(n\pi)^4 - 4a(n\pi)^2}}{2}$$

with eigenvectors $\begin{pmatrix} 0 \\ 1 \end{pmatrix}$, $v_n^+ = \begin{pmatrix} n\pi \sin n\pi x \\ -\lambda_n^+ \cos n\pi x \end{pmatrix}$, $v_n^- = \begin{pmatrix} n\pi \sin n\pi x \\ -\lambda_n^- \cos n\pi x \end{pmatrix}$.

Thus as in example 1, $\mathbb{A}$ satisfies assumptions 1 - 3 of §2, at least for $\delta > -1$. The eigenvalues have a limit point at -1 and so we must keep $\delta > -1$. It is still possible to improve the stability of the system for $\delta > -1$.

For convenience we use the orthonormal basis $\{e_n, f_n\}$ for Z,

$$e_n = \sqrt{2/a} \begin{pmatrix} \sin n\pi x \\ 0 \end{pmatrix}, \quad f_n = \begin{pmatrix} 0 \\ \sqrt{2} \cos n\pi x \end{pmatrix}, \quad f_o = \begin{pmatrix} 0 \\ 1 \end{pmatrix},$$

under which $\mathbb{A}$ has the following decomposition

$$\mathbb{A} = \text{diag}(0, \Delta_1, \Delta_2, \ldots) \tag{4.23}$$

where

$$\Delta_n = \begin{pmatrix} 0 & \sqrt{a}\, n\pi \\ -\sqrt{a}\, n\pi & -an^2\pi^2 \end{pmatrix} \; .$$

The operator version of the uncontrolled system (4.9), (4.20) on Z is

$$\dot{\tilde{z}} = A\tilde{z} \ ; \qquad \tilde{z} = \begin{pmatrix} z_x \\ z_t \end{pmatrix} \tag{4.24}$$

and the operator $\sigma = A$ on $D(A)$ and

$$D(\sigma) = \{w \in Z: w_1, w_2, \frac{dw_1}{dx}, \frac{dw_2}{dx}, \frac{d^2w_2}{dx^2} \in L_2(0,1)\}.$$

The boundary operator τ is defined by

$$\tau w = \begin{pmatrix} aw_1 + a\frac{dw_2}{dx}(0) \\ aw_1 + a\frac{dw_2}{dx}(1) \end{pmatrix} . \tag{4.25}$$

We must first check that $\tau/D(\sigma)$ is continuous with respect to the graph norm of σ. Now

$$\sigma w = \begin{pmatrix} \frac{dw_2}{dx} \\ a\frac{dw_1}{dx} + a\frac{d^2w_2}{dx^2} \end{pmatrix}$$

and

$$\begin{aligned} ||\sigma w||_Z^2 &= a||w_2'||^2 + a^2||w_1'||^2 + a||w_2''||^2 + 2a^2 <w_1', w_2''> \\ &= a||w_2'||^2 + a^2||w_1'||^2 + a||w_2''||^2 + a^2||w_1'+w_2''||^2 - a^2||w_1'||^2 - a^2||w_2''||^2 \\ &= a^2||w_1' + w_2''||^2 + a^2||w_2''||^2 . \end{aligned}$$

The operator τ_2 on Z defined by

$$\tau_2 w = a\begin{pmatrix} w_1 + w_2' \\ 0 \end{pmatrix} \qquad \text{satisfies}$$

$$||\tau_2 w||_Z^2 = a^3||w_1 + w_2'||^2 \leq \text{const.}||w_1' + w_2''||^2$$

and so τ_2 is σ-bounded.

The evaluation map at the boundary for $f \in H^2(0,1)$ is continuous and so τ is continuous with respect to the graph norm of σ. We now choose $B \in \mathcal{L}(U,Z)$ satisfying (3.4)(i) and (ii):

$$B = 1/a \begin{pmatrix} 1-x & 0 \\ 0 & x^2/2 \end{pmatrix} \tag{4.26}$$

and $N = \sigma B = 1/a \begin{pmatrix} 0 & x \\ -a & a \end{pmatrix}$.

Then the solution of (4.19) is given by

$$\begin{pmatrix} \frac{\partial z}{\partial x} \\ \frac{\partial z}{\partial t} \end{pmatrix} = (B\ I)\tilde{v}$$

where $\tilde{v}$ satisfies the following equation on $U \oplus Z$

$$\dot{\tilde{v}} = \begin{pmatrix} 0 & 0 \\ N & A \end{pmatrix} \tilde{v} + \begin{pmatrix} I \\ -B \end{pmatrix} \tilde{u} \; ; \quad \tilde{u} = \begin{pmatrix} \dot{u}_o \\ \dot{u}_1 \end{pmatrix} . \tag{4.27}$$

Suppose we measure the velocity at x_o, then the observation is

$$y = \frac{\partial z}{\partial t}(x_o,t) = (CB:\ C)\tilde{v} = \tilde{\tilde{C}}\tilde{v} \tag{4.28}$$

and

$$Cz = z_2(x_o), \; z \in Z.$$

This is unfortunately unbounded in our framework, so we shall replace it by an average round x_o.

$$\underline{C}z = \begin{pmatrix} 0 \\ \frac{1}{2\epsilon} \int_{x_o-\epsilon}^{x_o+\epsilon} z_2(x)dx \end{pmatrix} . \tag{4.29}$$

We need to check if $(A^u, B^u, \underline{C}^u)$ is minimal. Write

$$b_o(x) = \begin{pmatrix} 1-x \\ 0 \end{pmatrix} = \sum_{n=1}^{\infty} \alpha_n e_n \cong (o \;\; \alpha_1 \;\; o \;\; \alpha_2 \;\; \ldots\ldots\ldots)^T ; \; \alpha_n = \sqrt{2/a} \;.\; 1/n\pi$$

and check det $C(A^u, b_o^u)$ as in example 1.

$$C(A^u, b_o^u) = \begin{pmatrix} 0 & 0 & 0 & 0 & . \\ \alpha_1 & 0 & -a\alpha_1\pi^2 & a^2\alpha_1\pi^4 & . \\ 0 & -\alpha_1\sqrt{a}\pi & a\sqrt{a}\alpha_1\pi^3 & a\sqrt{a}\pi^3\alpha_1 - a^2\sqrt{a}\pi^5\alpha_1 & . \\ \alpha_2 & 0 & -4\alpha_2 a\pi^2 & 8a\alpha_2\pi^3 & . \\ 0 & -\alpha_2 2\pi\sqrt{a} & 4\alpha_2\sqrt{a}\pi^3 & 8a^3\pi^3\alpha_2 - 16\alpha_2 a\sqrt{a}\pi^4 & . \\ \alpha_3 & . & . & . & . \\ 0 & . & . & . & . \\ . & . & . & . & . \\ . & . & . & . & . \\ . & . & . & . & . \end{pmatrix}$$

and we see that this is not controllable due to the row of zeroes, but that the rest has full rank since $\alpha_i \neq 0$. For

$$b_1(x) = \begin{pmatrix} 0 \\ x^2/2 \end{pmatrix} = \sum_{n=0}^{\infty} \beta_n f_n \cong (\beta_o \;\; o \;\; \beta_1 \;\; o \;\; {}_2 \;\ldots)^T ; \; \beta_o = 1/3, \; \beta_n = \frac{(-1)^n\sqrt{2}}{(n\pi)^2} ,$$

we get a similar configuration, but no row of zeroes, and $C(A^u, b^u)$ will have full rank since $\beta_i \neq 0$ for all i.

Now $\underline{C}z = \langle c_o, z \rangle$ yields

$$c_0 = \sum_{n=0}^{\infty} \gamma_n f_n \;;\quad \gamma_0 = 1,\quad \gamma_n = \frac{\sqrt{2}}{n\pi\epsilon} \cos n\pi x_0 \sin n\pi\epsilon,$$

so we see that $(\underline{C}^u, A^u)$ will be observable.

This is an interesting example because it does not satisfy assumption 1. for all δ, due to the limit point at -1. The construction of §2 can still be applied to improve the stability for $\delta > -1$. We do not now have any prior guarantee that this will work in this case, however, because we cannot increase the dimension of W, the compensator order without limit. What is also interesting is the fact that you cannot stabilize by control at x = 0 alone, but you can by boundary control at x = 1 alone.

5. CONCLUSIONS

We have presented a compensator scheme for a class of hyperbolic systems with boundary control by transforming the system into one with bounded control action. At this stage one is free to use either the Schumacher design [12] or the one outlined in §2. Experience with parabolic systems in [1] showed that the design presented here is much simpler computationally and the results with the two methods seemed equally good. In theory the stability of the constructed extended system can be checked using the Weinstein-Aronzajn result as was successfully done for parabolic systems in [1] and [12]. However, this may prove difficult numerically for the hyperbolic case and simulation of the extended system might be a better approach.

REFERENCES

[1] Bontsema, J.: Finite dimensional compensators for parabolic systems, MSc. thesis, Rijksuniversiteit Groningen 1982.

[2] Curtain, R.F.: Finite dimensional compensators for parabolic systems with point and boundary input, IEEE Trans.Automatic Control Vol. AC-26 (1982), 99-104.

[3] Curtain, R.F., A.J. Pritchard: Infinite dimensional linear systems theory, Lecture Notes in Control and Information Sciences, vol.8, Springer Verlag, 1978.

[4] Curtain, R.F.: Finite dimensional compensators for parabolic distributed systems with unbounded control and observation, Report TW 234, Rijksuniversiteit Groningen, 1982.

[5] Curtain, R.F.: Compensators for infinite dimensional linear systems: a survey, (to appear in the special issue of the Journal of the Franklin Institute on Distributed Parameter Systems.)

[6] Davis, J.H., B.M. Barry: A distributed model for stress control in multiple locomotive trains, Appl.Math.Optim. 3 (1977), 163-190.

[7] Fattorini, H.O.: Boundary control systems, SIAM J. Control 6 (1968), 349-385.

[8] Gibson, J.S.: A note on stabilization of infinite dimensional linear oscillators by compact linear feedback, SIAM J. Control and Optim. 18 (1980), 311-316.

[9] Gibson, J.S., M. Navid: Optimal Control of Flexible Structures, Proc. 20th IEEE Conference on Decision and Control, 1981, 700-701.

[10] Kato,T: Perturbation theory for linear operators, Springer Verlag, 1966.

[11] Schumacher, J.M.: A direct approach to compensator disign for distributed parameter systems, to appear in SIAM J. Control and Optimization.

[12] Schumacher, J.M.: Dynamic feedback in finite and infinite dimensional linear systems, Mathematical Centre Tracts, No. 143, Mathematish Centrum, Amsterdam 1981.

APPENDIX

For completeness we state a result by Schumacher in [12] which we have appealed to in §2.

Lemma.

Suppose that A_{11} and A_{22} are generators of strongly continuous semigroups on Banach spaces X_1 and X_2, respectively, with growth constants ω_1 and ω_2. Suppose also that $A_{21} \in L(X_1, X_2)$, then the operator on $X_1 \oplus X_2$ defined by

$$A = \begin{pmatrix} A_{11} & 0 \\ A_{21} & A_{22} \end{pmatrix}$$

generates a semigroup whose growth constant equals max (ω_1, ω_2). Analagously for

$$\tilde{A} = \begin{pmatrix} A_{11} & A_{12} \\ 0 & A_{22} \end{pmatrix} \quad \text{, where } A_{12} \in L(X_2, X_1).$$

DIRECT SOLUTION OF THE BELLMANN EQUATION FOR A STOCHASTIC CONTROL PROBLEM

G. Da Prato

Scuola Normale Superiore
Piazza Dei Cavaliere 7
I-56100 Pisa, Italy

1. INTRODUCTION

We are here concerned with the backward Cauchy problem:

$$\left.\begin{aligned} &\psi_t(t,x) - 1/2|\psi_x(t,x)|^2 + (Ax,\psi_x(t,x)) \\ &+ 1/2(\psi_{xx}(t,x)Bx,Bx) + V(t,x) = 0 \\ &\psi(T,x) = \phi_o(x) \end{aligned}\right\} \tag{1}$$

where ψ and V map $[0,T] \times H$ into $\mathbb{R}$ ($T > 0$, H Hilbert space) and A,B are linear operators (generally unbounded). Moreover ψ_t and ψ_x represent the derivatives of ψ respectively in t and in x. Eq. (1) is the Bellman equation of the following control problem:

$$\text{Minimize} \quad E\left\{\int_0^T (V(t,\xi) + 1/2|u(t)|^2)dt + \phi_o(\xi(T))\right\} \tag{2}$$

over all $u \in M^2_W(0,T;H)$ subject to the state equation:

$$d\xi = (A\xi + u)dt + B\xi dW_t, \qquad \xi(0) = \xi_o. \tag{3}$$

Here W is a Brownian motion (which we take unidimensional for simplicity) in a probability space (Ω,ε,P) and $M^2_W(0,T;H)$ represents the set of all H-valued adapted processes X such that

$$E\int_0^T |X(s)|^2 ds < +\infty$$

(here E means the expectation).

Under suitable assumptions (roughly speaking $V(t,.)$ and ϕ_o are assumed to be convex, of class C^2 and with a polynomial growth) we are able to prove existence and uniqueness of a regular solution to Eq. (1).

Moreover, by a verification theorem, we can show that ψ is the value function of problem (2) - (3) and that the optimal control u^* is related to the optimal state ξ^* by the synthesis formula:

$$u^*(t) = -\psi_x(t,\xi^*(t)). \tag{4}$$

If in addition V and ϕ_o are quadratic,

$$\left.\begin{aligned} V(t,x) &= 1/2(M(t)x,x) \\ \phi_o(x) &= 1/2(P_o x,x) \end{aligned}\right\} \tag{5}$$

then we can find a solution of Eq. (1) of the form

$$\psi(t,x) = 1/2(P(t)x,x)$$

where P(t) is the solution of Riccati equation

$$\left.\begin{aligned} & P'(t) - P^2(t) + A^*P(t) + P(t)A - B^*P(t)B + M(t) = 0 \\ & P(T) = P_o \ . \end{aligned}\right\} \tag{6}$$

In section 2 we recall some results for the state equation (3) (see [6], [7]) and in section 3 we study Eq. (1) using similar methods as in [2] and [3]. Finally in Section 4 we give synthesis results.

2. STATE EQUATION

We assume here

i) A is self-adjoint negative
ii) B is a linear bounded operator of $D((-A)^{1/2})$ into H. (7)
iii) There exist $\eta \in [0,1[$ and $\lambda \in \mathbb{R}$ such that
$$\eta(Ax,x) + 1/2|Bx|^2 \leq \lambda |x|^2 \quad \forall\, x \in D(A).$$

We remark that $D((-A)^{1/2})$ coincides with the interpolation space $D_A(1/2,2)$ (see [12]). Under hypotheses (7) we are able to prove that the integral equation

$$\xi(s) = e^{(s-t)A}\xi(t) + \int_t^s e^{(s-\sigma)A}u(\sigma)d\sigma + \int_t^s e^{(s-\sigma)A}B\xi(\sigma)dW\sigma, \quad s \geq t \tag{8}$$

has a unique solution $\xi \in M_W^2(t,T;D((-A)^{1/2}))$ for any $u \in M_W^2(0,T;H)$ and $\xi(t) \in H$ (see [6] Proposition 4.5).

To get strong solution of Eq. (3) let us recall the definition of extrapolation spaces ([8]).

We set $K = (H \times H)/G_A$ where G_A is the graph of A and we define the embedding J of H into K by

$$J: H \to K, \; x \to (0,x)^{\sim}$$

where $(0,x)^{\sim}$ is the coset of $(0,x)$. Moreover we get an "extension" of A in K by setting

$$\left.\begin{aligned} D(\tilde{A}) &= J(H) \\ \tilde{A}(0,x)^{\sim} &= -(x,0)^{\sim} \end{aligned}\right\}$$

Remark that if $x \in D(A)$ we have in fact, $\tilde{A}(0,x)^{\sim} = (0,Ax)^{\sim}$. Now we consider the interpolation space $D_{\tilde{A}}(\vartheta,2)$ which we denote by $D_A(\vartheta-1,2)$ ($D_A(\vartheta-1,2)$ is the extrapolation space). If $x \in D((-A)^{1/2})$ we have $\tilde{A}J(x) \in D_A(-1/2,2)$. In what follows we shall write $\tilde{A} = A$.

<u>Proposition 1.</u> Assume that hypotheses (7) hold and moreover that $u \in M^2_W(t,T;H)$ and $x \in H$. Then there exists a unique $\xi \in M^2_W(t,T;D((-A)^{1/2})$ such that

$$\xi(s) = x + \int_t^s (A\xi(\sigma) + u(\sigma))d\sigma + \int_t^s B\xi(\sigma)dW_\sigma \tag{9}$$

(the function in the first integral is in the space $D_A(-1/2,2)$). Moreover if $u = 0$ we have

$$E|\xi(s)|^2 \leq e^{\lambda(\sigma-t)}|x|^2 . \tag{10}$$

Finally if $u = 0$ and if for $m \geq 2$ (m integer) there exists $\mu_m \in \mathbb{R}$ such that

$$\begin{aligned} &(Ax,x) + 1/2|Bx|^2 + (m-1)|(Bx,x)|^2|x|^{-2} \leq \\ &\leq \mu_m|x|^2 \quad \forall \; x \in D(A), \; x \neq 0 \end{aligned} \tag{11}$$

then

$$E|\xi(s)|^{2m} \leq \exp(\mu_m(s-t))|x|^{2m} . \tag{12}$$

<u>Proof.</u> The first assertion follows from Theorem 4.4 in [7]; the others ones are consequences of the Itô formula.

In the sequel if $u = 0$ we set

$$\xi(s) = G(s,t)x \tag{13}$$

$G(s,t)$, $s \geq t$, is a linear random operator and by (10) we have

$$E|G(s,t)|^2 \leq e^{\lambda(t-s)} \tag{14}$$

moreover if (11) holds then

$$E|G(s,t)|^{2m} \leq \exp(\mu_m(t-s)) \quad \text{for } m \in \mathbb{N} . \tag{15}$$

Example 1. Let $H = L^2(0,1)$, $Au = u_{xx}$ with domain $D(A) = H^2(0,1) \cap H_0^1(0,1)$. Moreover, let $Bu = bu_x$ with $b \in \mathbb{R}$ and domain $D(B) = H_0^1(0,1)$. Then it can be seen that if $|b| < \sqrt{2}$ the hypotheses (7) as well as the hypotheses (11) are verified.

Remark 1. We have assumed that A is self-adjoint only for sake of simplicity (for a more general case see [7]).

2. SOLUTION OF PROBLEM (1)

First of all we remark that by setting $\phi(t,x) = \psi(T-t,x)$ and $g(t,x) = V(T-t,x)$ problem (1) becomes

$$\left.\begin{aligned} &\phi_t(t,x) + 1/2|\phi_x(t,x)|^2 - (Ax,\phi_x(t,x)) + \\ &- 1/2(\phi_{xx}(t,x)Bx,Bx) = g(t,x) \\ &\phi(0,x) = \phi_0(x) . \end{aligned}\right\} \tag{16}$$

We need now some notation. By $C^k(H)$, $k = 0,1...,$ we mean the set of all mappings ϕ: $H \to \mathbb{R}$ continuous and bounded on each ball of H with their derivatives of order less than k.

For any $\phi \in C^k(H)$ and $n \in \mathbb{N}$ we set

$$|\phi|_{k,n} = \sup_{x\in H} |\phi^k(x)|/(1 + |x|^{2n})$$

$$||\phi||_{k,n} = \sup_{\substack{x,y\in H \\ x\neq y}} |\phi^k(x) - \phi^k(y)|/(|x-y| (1 + \sup |x|^{2n},|y|^{2n}))$$

moreover we denote by X and Z the spaces

$$X = \{\phi \in C(H);\ |\phi|_{0,n_0} < +\infty\}$$

$$Z = \{\phi \in C^2(H);\ |\phi|_{0,n_0}, |\phi|_{1,n_1}, |\phi|_{2,n_2}, ||\phi||_{2,n_3} < +\infty\}$$

where $n_0 \geq n_1 \geq n_2 \geq n_3 \geq 0$ are intergers to be fixed later and X and Z are endowed with the norms

$$|\phi|_X = |\phi|_{0,n_0}$$

$$|\phi|_Z = |\phi|_{0,n_0} + |\phi|_{1,n_1} + |\phi|_{2,n_2} + ||\phi||_{2,n_3} .$$

We denote by $B([0,T];Z)$ the set of all mappings ϕ: $[0,T] \times H \to \mathbb{R}$ such that

i) $\phi(t,.) \in Z \ \forall\, t \in [0,T]$ and $\sup_{t\in[0,T]} |\phi(t,.)|_Z < +\infty$

ii) ϕ and ϕ_x are continuous in $[0,T] \times H$, moreover $(\phi_{xx}.\xi,\eta)$ is continuous in $[0,T] \times H$ for any $\xi,\eta \in H$.

We use the space $B([0,T];Z)$ (instead $C([0,T];Z)$) so that also functions as $\phi(t,x) = (e^{tA}x,x)$ can be considered. Finally we denote by K the set of all convex functions $H \to \mathbb{R}$.

To solve problem (16) we use the same procedure as in [2] - [3]. First we consider the linear problem:

$$\left.\begin{aligned} &\phi_t(t,x) - (Ax,\phi_x(t,x)) - 1/2(\phi_{xx}(t,x)Bx,Bx) = 0 \\ &\phi(0,x) = \phi_0(x) \quad . \end{aligned}\right\} \tag{17}$$

Recalling (12) and proceeding as for the proof of Proposition 4 in [3] we get the result:

Proposition 2. Assume that hypotheses (7) and (11) (with $m = n_0$) hold. Then for any $\phi_0 \in Z$ there exists a unique solution ϕ of problem (17). Moreover we have

$$\phi(t,x) = E\phi_0(H(t)x) \tag{18}$$

where

$$H(t) = G(T,T-t) \tag{19}$$

and G is defined by (13).

Remark 2. In the linear quadratic case Eq. (17) reduces to the Riccati equation

$$\left.\begin{aligned} &Q' = A^*Q + QA - B^*QB = 0 \\ &Q(0) = P_0 \end{aligned}\right\} \tag{20}$$

where $\phi_0(t,x) = 1/2(Q(t)x,x)$ and $\phi_0(x) = 1/2(P_0x,x)$. By Proposition 2 we have

$$Q(t) = E(H^*(t)P_0Q(t)) . \tag{21}$$

To get a solution of (16) we consider an approximating problem (see [3]).

$$\left.\begin{aligned} &\phi^\alpha_t + \frac{\phi^\alpha - \phi^\alpha_\alpha}{\alpha} - (Ax,\phi^\alpha_x) - 1/2(\phi^\alpha_{xx}Bx,Bx) = g \\ &\phi^\alpha(0,x) = \phi_o(x) \end{aligned}\right\} \tag{22}$$

where

$$\phi^\alpha_\alpha(t,x) = \inf\{\phi^\alpha(t,y) + \frac{1}{2\alpha}|x-y|^2;\ y \in H\}$$

and ϕ_o, $g(t,.)$ are convex functions as well as $\phi^\alpha(t,.)$. Recalling Proposition 2 we can write Eq. (22) in the following weak form:

$$\begin{aligned} \phi^\alpha(t,x) = {} & e^{-t/\alpha}E\phi_o(H(t)x) + \\ & + E\int_0^t e^{-(t-s)/\alpha}\left(\frac{\phi^\alpha_\alpha}{\alpha} + g\right)(s,H(t-s)x)ds\ . \end{aligned} \tag{23}$$

Moreover by a proceeding similar to the one used in the proof of the Theorem 1 in [3] we get:

<u>Theorem 1.</u> Assume that

i) $n_o \geq 2n_1(1+n_2)$
ii) hypotheses (7) and (11) (with $m = n_o$) hold
iii) $\phi_o \in Z \cap K$, $g \in B([0,T];Z)$ and $g(t,.) \in K \quad \forall\, t \in [0,T]$.

Then the following conclusions hold:

a) Eq. (23) has a unique solution $\phi^\alpha \in B([0,T];Z)$ which is also a classical solution to problem (22). Moreover $|\phi^\alpha(t,.)|_Z$ is bounded uniformly with respect to t and α.
b) There exists ϕ such that $\phi^\alpha \to \phi$ in $C([0,T];C^1(H))$ and moreover $\sup_{t\in[0,T]} |\phi(t,.)|_Z$ is bounded.
c) For all $x \in D(A)$ and t a.e. in $[0,T]$, ϕ is a solution of (17).
d) ϕ is unique.

If ϕ_o and g are quadratic we can study directly the following Riccati equation

$$\left.\begin{aligned} &Q'(t) = A^*Q(t) + Q(t)A - Q^2(t) - B^*Q(t)B + N(t) \\ &Q(0) = P_o \end{aligned}\right\} \tag{24}$$

where $N(t) = M(T-t)$.

By Proposition 2 we can write Eq. (24) in the following weak form:

$$Q(t)x = H^*(t)P_oH(t)x + \int_0^t H^*(t-s)(N(s) + Q^2(s))H(t-s)dsx \qquad \forall\, x \in H. \tag{25}$$

Proceedings as in [5] we get the result (we denote by $\Sigma^+(H)$ the set of hermitian positive operator in H and by $C_s([0,T];\Sigma^+(H))$ the set of all the mappings $[0,T] \to \Sigma^+(H)$ strongly continuous).

Theorem 2. Assume the hypotheses (7) hold and moreover that $P_0 \in \Sigma^+(H)$ and $N \in C_s([0,T];\Sigma^+(H))$. Then Eq. (25) has a unique solution $Q \in C_s([0,T];\Sigma^+(H))$. Moreover $Q(.)x$ is continuously differentiable for any $x \in H$ and for each $t > 0$, $x \in D(A)$ we have $Q(t)Ax \in D(A^*)$ and

$$\left.\begin{aligned} &Q'(t)x = A^*Q(t)x + Q(t)Ax - Q^2(t)x - B^*Q(t)Bx + N(t)x \\ &Q(0) = P_0 x. \end{aligned}\right\} \qquad (26)$$

4. DYNAMIC PROGRAMMING

Theorem 3. Assume that the hypotheses of Theorem 1 hold and let be the solution to problem (1). Then for each $t \in [0,T]$ and $x_0 \in H$ we have

$$\begin{aligned} \psi(t,x_0) = \inf \{E(\int_t^T (V(s,\xi(s)) + \sqrt{2}|u(s)|^2)ds + \\ + \phi_0(\xi(T))); u \in M_W^2(t,T;H), d\xi(s) = (A\xi + u)ds + B\xi dW_s\} \end{aligned} \qquad (27)$$

Moreover the solution ξ^* to the problem

$$\left.\begin{aligned} &d\xi = (A\xi - \psi_x(t,\xi))dt + B\xi dW_t \\ &\xi(0) = x_0 \end{aligned}\right\} \qquad (28)$$

is an optimal trajectory to problem (2) - (3) corresponding to the optimal control u^* given by

$$u^*(t) = - \psi_x(t,\xi^*(t)) \quad \text{a.e.} \quad t \in [0,T]. \qquad (29)$$

Proof. The proof is similar to that in [2] and [3].

Remark 4. In the linear quadratic case the synthesis formula is given by

$$u^*(t) = - P(t)\xi^*(t) \qquad (30)$$

where P is the solution to (6) and ξ^* is the solution to the differential stochastic equation

$$\left.\begin{aligned} &d\xi = (A - P(t))\xi dt + B\xi dW_t \\ &\xi(0) = x. \end{aligned}\right\} \qquad (31)$$

REFERENCES

[1] Curtain, R.F., A.J. Pritchard: Infinite Dimensional Linear Systems Theory, Springer-Verlag 1978.

[2] Barbu, V., G. Da Prato: A direct method for studying the dynamic programming equation for controlled diffusion processes in Hilbert spaces, Numer.Funct.Anal. and Optimiz. 4 (1) (1981), 23-43.

[3] Barbu, V., G. Da Prato: Solution of Bellman equation associated with an infinite dimensional stochastic control problem and synthesis of optimal control, SIAM J. Control and Optimization (to appear).

[4] Barbu, V., G. Da Prato: Hamilton-Jacobi equations in Hilbert spaces, Pitman (to appear).

[5] Da Prato, G.: Quelques résultats d'existence unicité et régularité pour un problème de la théorie du contrôle, J.Maths pures et appl. 52 (1973), 353-375.

[6] Da Prato, G.: Regularity results of a convolution stochastic integral and applications to parabolic stochastic equations in Hilbert spaces, Conferenze Seminario matematico Università di Bari (1982).

[7] Da Prato, G.: Some results on Linear Stochastic Evolution Equations in Hilbert spaces by the semi-groups method, Stochastic Analysis and Applications (to appear).

[8] Da Prato, G., P. Grisvard: Maximal regularity for evolution equations by interpolation and extrapolation, submitted.

[9] Fleming, W.H., R.W. Rishel: Deterministic and stochastic Optimal Control, Springer-Verlag 1975.

[10] Ichikawa, A.: Linear stochastic evolution equations in Hilbert spaces, J. Diff. Equat. 28 (1978), 266-283.

[11] Ichikawa, A.: Dynamic Programming Approach to Stochastic Evolution Equations, SIAM Journal in Control and Optimization 17 (1979), 152-174.

[12] Lions, J.L., J. Peetre: Sur une classe d'espaces d'interpolation, Publ.I.H.E.S. 19 (1964), 5-68.

[13] Metivier, M., T. Pellaumail: Stochastic Integral, Academic Press 1977.

[14] Pardoux, E: Equations aux dérivées partielles stochastiques non linéaires monotones, Thèse, Université Parix XI (1975).

[15] Yosida, K.: Functional Analysis and Semigroups, Springer-Verlag 1965.

DEGENERATE DIFFERENTIAL EQUATIONS AND APPLICATIONS

A. Favini

Istituto Matematico "S. Pincherle"
Piazza di Porta San Donato 5
I-40127 Bologna, Italy

1. INTRODUCTION

Singular systems of differential equations, both in the finite dimensional and infinite dimensional case, are an area of current research by several poeple. We only want to quote the very recent books by S.L. Campbell [1], where applications are also considered as, for instance, singular perturbations, cheap control problems and descriptor systems.

We shall be concerned with two types of problems. The first part of the paper is devoted to obtain existence and uniqueness results for the operational equation

$$BA_1u + A_0u = h, \qquad (1.1)$$

where A_0, A_1 are closed linear operators from F into E, E and F being complex Banach spaces, B is a linear closed operator in E, and h is a given element of E. (Of course, u is the sought solution.)

The "singularity" in (1.1) depends on the fact that A_1 may fail to be invertible.

In [3] Ju.A. Dubinskii proved some results for (1.1) under the basic assumption that the operator B commutes with A_0 and A_1. We treat (1.1) without this assumption; this permits to treat suitable partial differential equations with coefficients depending on time.

The approach we follow is the operational method by G. Da Prato and P. Grisvard [2] and most of these results has been proved in [4].

The second part of the paper shows that the preceding techniques apply to certain two-point problems arising in singular optimal control.

2. THE OPERATIONAL EQUATION

We list the following hypotheses:

(H1) $\sigma(-B)$, the spectrum of B, is contained in $S_{a,\phi} = \{z: |\arg z| < \phi, |z| \geq a\}$, $\phi < \pi$, $a > 0$, $\sigma(P)$, the set of all complex numbers z such that $zA_1 + A_0$

is not invertible, lies outside the sector $S_\phi = \{z: |\arg z| \leq \phi, |z| > 0\}$, $\Gamma_\phi = \partial S_\phi$, $\Gamma_{a,\phi} = \partial S_{a,\phi}$. We use $P(z)$ for $zA_1 + A_0$.

(H2) For each $z \notin S_{a,\phi}$, we have $||(B-z)^{-1}; L(E)|| \leq C(1 + |z|)^{-1}$, where $||x;G||$ denotes the norm of x in the Banach space G; here L(E) denotes the space of all linear bounded operators from E into itself and we shalluse L(E,F) for the corresponding F-valued operators.

(H3) For each $z \notin S_\phi$, $||P(z)^{-1}; L(E;F)|| \leq C(1 + |z|)^h$, $||A_0P(z)^{-1}; L(E)|| \leq C(1 + |z|)^m$, where h, m are integers ≥ -1.

(H4) For each z in $\rho(-B) \cap \rho(P)$, the resolvent sets of B and P(z), respectively, we have

$$||(B-z)^{-1}[B; A_0P(z)^{-1}]x; D(B^k)|| \leq C(1 + |z|)^\alpha ||x;E||, \quad x \in D(B),$$

where k is a non-negative integer, $\alpha \in \mathbb{R}$, $C \in \mathbb{R}^+$, and the bracket $[B_1;B_2]$ denotes the commutator of B_1 and B_2.

(H5) For each z in $\rho(-B) \cap \rho(P)$,

$$||[B; A_0P(z)^{-1}](B-z)^{-1}; L(E;D(B^k))|| \leq C(1 + |z|)^\beta,$$

where k is a non-negative integer, $\beta \in \mathbb{R}$, $C \in \mathbb{R}^+$.

Such hypotheses will be necessary in order to prove our existence and uniqueness results. We begin with the ones relative to existence of a solution.

Theroem 2.1. Suppose that (H1) - (H2), (H3), (H5) hold. If the constant k in (H5) satisfies $k > \max(h,m,\beta)$ and the C there is sufficiently small, then (1.1) has at least a solution for any $h \in D(B^k)$.

Such a solution u is given by

$$u = Sf = (2\pi i)^{-1} \int_{\Gamma_{a,\phi}} z^{-k}P(z)^{-1}(B-z)^{-1}B^k f \, dz,$$

where f is a suitable element of $D(B^k)$. In fact, it is a simple matter to recognize that Sf satisfies $BA_1Sf + A_0Sf = f - VB^kf$, with

$$V = (2\pi i)^{-1} \int_{\Gamma_{a,\phi}} z^{-(k+1)}[B; R_T(z)](B-z)^{-1}dz,$$

where $T = A_1A_0^{-1}$ and $R_T(z) = (zT + 1)^{-1}$.

In virtue of the assumptions $[B; R_T(z)]\,(B-z)^{-1}$ is an operator from E into $D(B^k)$ and hence, we deduce that $1 - VB^k$ is an isomorphism from $D(B^k)$ onto itself i.e., a solution of (1.1) is given by $S(1 - VB^k)^{-1}h$.

A particularly interesting case is provided by m = 0 in (H3). Then it is possible to prove the following result:

Theorem 2.2. Assume (H1), (H2) and the second inequality in (H3) hold with m = 0, $\phi = \pi/2$ (and thus it is easy to recognize that we can assume $\phi > \pi/2$). If

$$||[B; R_T(z)]f;E|| \leq C||f;E||, \quad ||[B;[B; R_T(z)]]f;E|| \leq C||f;E||$$

on $\rho(-B) \cap \rho(P)$, and the constant C in (H2) may be taken sufficiently small, then (1.1) has a solution for any $h \in D(B)$.

In order to weaken the smoothness assumptions on h, we also use the real interpolation spaces $(E_0,E_1)_{\theta,p}$, [8]. To this regard, we prove the following result, extending the corresponding "regular" one in [2].

Theorem 2.3. Assume (H1), (H2), (H3) and (H5) hold with k = 0. If $1 < p < \infty$, $0 < \theta < 1$,

$$||[B; R_T(z)] (B-z)^{-1}; L(V;W)|| \leq C(1 + |z|)^{\gamma}, \tag{2.1}$$

for any $z \in \rho(-B) \cap \rho(P)$, where $V = (E, D(B))_{\theta,p}$, $W = (E; D(B^{k+1}))_{\omega,p}$, $\omega = (\theta+k)(k+1)^{-1}$, $k > \max(h,m-1,\beta,\gamma)$, and the constant C in (2.1) is sufficiently small small, then (1.1) has at least one solution for each $h \in W$. Further, such a solution u satisfies A_0u, $BA_1u \in V$.

Remark 2.1. Since $(B-z)^{-1}$ maps continuously V into itself by interpolation, (2.1) signifies, "grosso modo", that $[B; R_T(z)]x \in D(B^k)$ for each $x \in V$ and $||B^k[B; R_T(z)]x;V|| \leq C(1 + |z|)^{\gamma+1}||x,V||$. This is achieved, for example, if $||B^k[B, R_T(z)]x,E|| \leq C_1(z)||x;E||$, $||B^{k+1}[B; R_T(z)]x,E|| \leq C_2(z)||x,D(B)||$, with suitable $C_i(z)$, in view of the interpolation property for $(E_0,E_1)_{\theta,p}$.

It is to be pointed out that (H5) and (2.1) guarantee the convergence of certain integrals related to the solutions, in view of the fact that convergence of the integral of the norm implies convergence of the integral. Sometimes, we may deal with such integrals directly. Then, for example, we have

Theorem 2.4. Assume (H1), (H2) and

$$||R_T(z); L(E)|| \leq C(1 + |z|)^k$$

for $\text{Re } z \geq -a_0 |\text{Im } z|$, $a_0 > 0$, k a non-negative integer.

Let $\Gamma = \Gamma_{a,\phi}$ for suitable a,ϕ and define

$$S_1 = (2\pi i)^{-1} \int_\Gamma z^{-(k+1)}[B; R_T(z)] (B-z)^{-1} dz.$$

Suppose B^kS_1, $B^{k+1}S_1 \in L(E)$ and that $[B; R_T(z)]$ defines an element of L(E), with $||[B; R_T(z)]; L(E)|| \leq C'(1 + |z|)^k$ on Γ. If the constant C in (H2) may be taken sufficiently small, then the conclusions in Theorem 2.3. hold.

Analogously, by using the same technique, we can prove a result which extends Theorem 2.2.:

Theorem 2.5. Under the assumptions in Theorem 2.2., (1.1) has one solution u for all $h \in (E;D(B))_{\theta,p} = V$, such that $A_o u$, $BA_1 u \in V$.

We now turn to the uniqueness problem.

Theorem 2.6. Assume (H1), (H2), (H4) and the second inequality in (H3). If we suppose k in (H4) to satisfy $k > \max(m,\alpha)$ and the constant C there to be sufficiently small, then (1.1) has at most one solution.

The proof of Theorem 2.6 makes use of the integral

$$\int_{\Gamma_{a,\phi}} z^{-k}(B-z)^{-1}R_T(z)dz$$

and for this we need assumption (H4).

Theorem 2.7. Assume (H1), (H2) and the second inequality in (H3). Moreover, suppose that $[B;\, A_o P(z)^{-1}]$ has a bounded extension from E into $D(B^k)$ for all $z \in \rho(-B) \cap \rho(P)$ such that $||[B;\, R_T(z)];\, L(E;D(B^k))|| \le C'(1 + |z|)^{\alpha-1}$. If the constant C in (H2) may be taken sufficiently small (a non restrictive assumption in the applications to partial differential equations) and $k > \max(m,\alpha)$, then (1.1) has a unique solution for all $h \in D(B^k)$.

We only have to point out that under these assumptions $(B-z)^{-1}$ maps $D(B^k)$ into itself with a bound for its norm given by $C(1 + |z|)^{-1}$ and we can suppose $h \le m$ in (H3). In fact, if $h > m$, from $||A_o P(z)^{-1};\, L(E)|| \le C(1 + |z|)^m$ we deduce $||P(z)^{-1};\, L(E;F)|| \le C(1 + |z|)^m$ and thus we substitute h by m. Hence our preceding theorems apply.

At this point some remarks on the system

$$A_1 Bu + A_o u = h, \tag{2.2}$$

are in order. Here, A_o, A_1, B satisfy assumptions corresponding to the ones in the introduction; that is, A_i $(i = 0,1)$ is a linear closed operator from F into E, B is a closed operator from F into itself and A_o is supposed to have a bounded inverse.

If $S = A_o^{-1}A_1$ is a densely defined closed operator, then, instead of (2.2), we can consider the equation

$$SBu + u = A_o^{-1}h = f. \tag{2.3}$$

Now, (2.3) may be handled in the preceding way if we assume that $[B;S]$ has a bounded extension, for then it is allowed to put (2.3) under the form

$BSu - [B;S]u + u = f.$

Existence and uniqueness assumptions shall concern the operators S and $1 - [B;S]$. This approach can be taken mainly when A_0, A_1 are bounded operators from F into E. But in most applications, the spaces E and F coincide and then we can consider the system $BA_1u + (A_0 - [B;A_1])u = h$, directly, and all our results apply if we substitute $A_0 - [B;A_1]$ for A_0. Observe that (2.2) is in general more difficult to treat than (1.1); this is to be expected also because $A_0^{-1}A_1$ needs not to be closed! On the other hand, we sometimes must make use of equations as (2.2). For example, it is an easy matter to prove the following uniqueness result:

Theorem 2.8. Assume that the equation $A_1Bu + A_0u = f$ has a solution for any f in a dense subspace of F. Then (1.1) has at most one solution.

Example 2.1. Let X be a complex Banach space, and $E = L^p(0,T;X)$, $p \in (1,+\infty)$, $D(B) = \{u \in W^{1,p}(0,T;X):u(0) = 0\} = W_0^{1,p}(0,T,X)$, $(Bu)(t) = u'(t) = du(t)/dt$, with the usual notations for Sobolev spaces.

For each $0 \leq t \leq T$, let $A_i(t)$, $i = 0,1$, be a linear closed operator from a complex Banach space Y into X, with domain $D(A_i(t))$, $D(A_0(t)) \subseteq D(A_1(t))$ everywhere dense in Y, and $A_0(t)^{-1} \in L(X;Y)$, $0 \leq t \leq T$.

If $D(A_i) = \{u \in F = L^p(0,T;Y):u(t) \in D(A_i(t))$ a.e., $A_i(.)u(.) \in E\}$, $(A_iu)(t) = A_i(t)u(t)$, $i = 0,1$, then (1.1) is equivalent to the following problem. Given $h \in L^p(0,T,X)$, find an element $u \in L^p(0,T,Y)$ such that $A_i(.)u(.) \in E$, $i = 0,1$, $d(A_1(.)u(.))/dt \in E$ and

$$\begin{aligned} &d(A_1(t)u(t))/dt + A_0(t)u(t) = h(t), \quad 0 < t < T, \\ &\lim_{t \downarrow 0} A_1(t)u(t) = 0. \end{aligned} \tag{2.4}$$

Such a function u is said to be a strict solution of (2.4).

Conditions on $A_i(t)$ may be given ensuring existence and uniqueness for the strict solution of (2.4). In particular, by use of Hardy inequality [8, p. 262], we see that if $-A_0(t)$ is the infinitesimal generator of a bounded analytic semigroup of linear operators, sufficiently smooth in t, $A_1(t) = t$, then (2.4) has a unique strict solution both for all $h \in W_0^{1,p}(0,T;X)$ and for $h \in W^{\theta,p}(0,T;X)$, (see [2] for notations).

We also can treat concrete differential problems, where $A_i(t)$, $i = 0,1$, is defined by a suitable partial differential operator, with smooth coefficients. For example, $A_1(t)$ may be the multiplication operator by $m(t,x) \geq 0$, $0 \leq t \leq T$, $x \in \Omega$, Ω a bounded domain in $\mathbf{R}^n$.

Example 2.2. Consider the problem

$$d(ty(t))/dt = -tx(t) + y(t) + f(t),\ y'(t) = -x(t) + g(t),\ 0 < t < T,\ y(0) = 0.$$

We have $E = F = L^2(0,T;\mathbb{C}^2)$, $A_0(t)(x,y) = (tx-y,x)$, $A_1(t)(x,y) = (ty,y)$.

It is a simple matter to see that (H2), (H3) hold with $h = m = 1$, but an estimate as in (H4), (H5) is not possible, because this would involve higher smoothness to the x in (H4), (H5). In fact, there is existence only if $f(t) = tg(t)$, and uniqueness fails because both $x(t) = g(t)$, $y(t) = 0$ and $x(t) = g(t) - 1$, $y(t) = t$, solve the problem.

Remark 2.2. A variant of (1.1) is given by the problem

$$\tilde{B}A_1u + A_0u = h, \quad CA_1u = v_0, \tag{2.5}$$

where A_0, A_1, B satisfy the same general hypotheses as in the introduction, while C acts as a bounded operator form E_1, a Banach space continuously imbedded in E, into a Banach space G.

Problem (2.5) is reduced to a problem of the type (1.1) if one supposes that the operator B defined by

$$D(B) = \{u \in D(\tilde{B}) \cap E_1 : Cu = 0\},\ Bu = \tilde{B}u,$$

satsisfies (H1), (H2).

Then, if $v_0 = Cy_0$, $y_0 \in D(\tilde{B}) \cap E_1$, (2.5) may be put under the form $B(A_1u - y_0) + \tilde{B}y_0 + A_0u = h$, or else, if $A_0u = y$, $A_1A_0^{-1} = T$, $B(Ty - y_0) + y = h - \tilde{B}y_0$.

At last, we have the system

$$Bv + y = h - \tilde{B}y_0, \quad Ty - v = y_0.$$

It follows that this is equivalent to the equation in the one unknown v:

$$(TB + 1)v = Th - T\tilde{B}y_0 - y_0, \tag{2.6}$$

an equation of type (2.2), which can be handled chiefly in two ways, according what we saw previously; the first way corresponds to assuming that $[B,T]$ has a suitable bounded extension, and then one considers the equation $BTv + (1 - [B;T])v = \ldots$, or, if we suppose that $Th - T\tilde{B}y_0 - y_0 \in D(B)$, we are allowed to apply B to both the members in (2.6) and obtain the equivalent equation $(BT + 1)w = B(Th - T\tilde{B}y_0 - y_0)$, which is of type (1.1).

Note that solvability of (2.5) implies a relation between y_0, that is v_0, and h. In concrete applications, the operator C may be the trace operator $y \to y(0)$ and thus equation (2.4), with initial condition $v_0 \neq 0$, shall have a solution, in general, only if $h(t)$ and v_0 satisfy certain compatibility assumptions. This, on the other hand, was to be expected in view of simple counter-examples.

3. APPLICATIONS TO A CONTROL PROBLEM

We want to introduce the following control problem, following [6, p. 111].

Let us denote by U a real Hilbert space and $C(t) \in L(U;V)$ for any $0 \leq t \leq T$, continuous in t in the norm of $L(U;V)$, where V is another Hilbert space. Let f be given in $L^2(0,T;V)$. We denote by $x(t,v)$ a strong solution of the problem

$$\begin{aligned} &d(A_1(t)x(t,v))/dt + A_0(t)x(t,v) = f + C(t)v, \\ &\lim_{t\downarrow 0} A_1(t)x(t,v) = 0, \end{aligned} \tag{3.1}$$

where $A_i(t)$, $i = 0,1$, is supposed to be a linear closed operator from W into V, W a Hilbert space. The observation is given by $z(t,v) = H(t)x(t,v)$, where $H(t)$ is a strongly continuous $L(W;Z)$-valued map on $0 \leq t \leq T$, Z Hilbert space. If $N(t)$ is a bounded, strongly continuous in t, operator in U such that $(N(t)u,u)_U \geq 0$, the cost function is given by

$$J(v) = \int_0^T ||H(t)x(t,v);Z||^2 dt + \int_0^T (N(t)v(t),v(t))_U dt.$$

We seek $\inf_{v\in U_{ad}} J(v)$, where U_{ad} is a suitable subspace of $L^2(0,T;U)$.

Of course, we must suppose that (3.1) is well posed for each v in U_{ad} and we have seen that this request puts some restrictions on the regularity of v. In this paper we content ourselves with considering as U_{ad} the space $W_0^{k,p}(0,T;X)$ and assuming that A_0, A_1 satisfy those properties that permit us to apply the preceding results.

Note that the same $J(v)$ may be singular, in view of the hypotheses on $N(t)$.

Of course, following H.O. Fattorini, we shall use the complexified version of (3.1).

It is then possible to approach our problem by means of the methods in [6], which leads to a two-point problem. In fact, we can prove the following result (see also [6, p. 114]):

<u>Theorem 3.1.</u> The optimal control-state functions u, x for the control problem are characterized by

$$\begin{aligned} &d(A_1(t)x(t))/dt + A_0(t)x(t) = f(t) + C(t)u(t), \\ &- A_1(t)^*p'(t) + A_0(t)^*p(t) = H(t)^*H(t)x(t), \qquad 0 < t < T \\ &0 = C(t)^*p(t) + N(t)u(t) = 0, \end{aligned} \tag{3.2...}$$

$$\lim_{t\downarrow 0} A_1(t)x(t) = 0 = \lim_{t\uparrow T} p(t). \qquad (..3.2)$$

In order to treat Problem (3.2), we remark that $u \to -du/dt$, with $u(T) = 0$, is the adjoint of B in Example 2.1 and satisfies the same properties as B; thus it is possible to treat (3.2) as an operational problem of the type

$$BA_1x + A_0x = f + Cu, \qquad (3.3)$$

$$A_3B_1p + A_2p = Kx, \qquad (3.4)$$

$$Qp + Nu = 0. \qquad (3.5)$$

We may treat this problem in some different ways, after (3.4) is reduced to the form considered in (1.1). Of course, if B_1 and A_3 commute, there is no trouble.

It is tempting to consider the problem under the form $\mathcal{B}\mathcal{A}_1Y = -\mathcal{A}_0Y + F$, where $\mathcal{B}(x,p,u) = (Bx, B_1p, u)$, $\mathcal{A}_1(x,p,u) = (A_1x, A_3p, 0)$, $\mathcal{A}_0(x,p,u) = (A_0x - Cu, -Kx + (A_2 - [B_1;A_3])p, Qp + Nu)$, $F = (f,0,0)$, which is of the type investigated in Sec. 2.

But this approach reserves some troubles. First of all, it is in general not trivial to give conditions entailing invertibility of $\mathcal{A}_0$, estimates for $(z\mathcal{A}_1 + \mathcal{A}_0)^{-1}$, and secondly, there are problems relative to the domains and stability of operators.

A different approach for (3.2) has been given in [5] for the regular case, where $A_1 = A_3 = I$, $h = -1$, $m = 0$ and N has a bounded inverse. Then (3.3) - (3.5) is reduced to a system with two equations and two unknowns and the problem is solved in the interpolation spaces, via a perturbation technique. The same method can be applied in these more general cases when $m = 0$; in fact the result we need is Theorem 2.3.

If N is invertible or not, we have two orders of problems, which appear at a first look of (3.3) - (3.5).

First, if we are able to solve (3.3) in the variable x, considering u as a parameter, nobody ensures us that Kx has a regularity implying that a solution of (3.4) exists. Second, we must, after having found this p, seek a solution u of (3.5) and make sure that this u has the smoothness sufficient to solve (3.3): see the results of Sec. 2!

In general, we could proceed according this plan. Let the system be described by

$$x = \Phi(Cu + f), \; p = \Psi(Kx), \; 0 = Qp + Nu, \qquad (3.6)$$

where Φ, Ψ are suitable linear continuous operators acting in certain Banach spaces, $\Phi: A^0 \to A^1$, $\Psi: B^0 \to B^1$. Then (3.6) has a solution if $Q\Psi(K\Phi(Cu + f)) + Qu = 0$ has a solution u for which $Cu + f \in A^0$.

The most interesting case is given by N a bounded non-negative operator in the Hilbert space U. Then we have the direct sum decomposition $U = N(Q) \oplus R(Q)^a$, ([7], p. 352), and thus the preceding equation must be projected on these subspaces.

We refer to a forthcoming paper for a detailed solution of the problem.

REFERENCES

[1] Campbell, S.L.: Singular systems of differential equations I, II, ed. Pitman, 1980/82.

[2] Da Prato, G., P. Grisvard: Sommes d'opérateurs linéaires et équations différentielles opérationnelles, J. Math. Pures Appl. 54 (1975), 305-387.

[3] Dubinskii, Ju.A.: On some differential-operator equations of arbitrary order, Math. Sbornik 90 (132); english transl.: Math. USSR Sbornik 19 (1973), 1-21.

[4] Favini, A.: Degenerate and singular evolution equations in Banach space, forthcoming.

[5] Favini,A., A. Venni: On a two-point problem for a system of abstract differential equations, Numer. funct. anal. & optim. 2(4), (1980), 301-322.

[6] Lions, J.L.: Optimal control of systems governed by partial differential equations, ed. Springer, 1971.

[7] Taylor, A.E., D.C. Lay: Introduction to functional analysis, ed. Wiley, 1980.

[8] Triebel, H.: Interpolation theory, function spaces, differential operators, ed. North Holland, 1978.

THE NUMERICAL SOLUTION OF DIFFERENTIAL EQUATIONS ARISING IN CONTROL THEORY FOR LUMPED AND DISTRIBUTED PARAMETER SYSTEMS

L. Graney

BP Research Centre
Chertsey Road, Sunbury on Thames
Middlesex TW16 7LN, England

1. INTRODUCTION

Process simulators can be classified as steady-state or dynamic simulators, with the latter category being further sub-divided into lumped parameter systems and distributed parameter systems. The types of mathematical equations representing these systems are algebraic equations for steady-state processes, ordinary differential equations for lumped parameter systems and partial differential equations for distributed parameter systems. A detailed description of a method is presented that automatically solves each of these different types of simulators, discussing in particular the work of Graney and Richardson [8]. The method, known as the Numerical Method of Lines, was developed for parabolic partial differential equations but can be applied to elliptic and hyperbolic partial differential equations, to ordinary differential equations and to algebraic equations. An advantage of this method is the simultaneous solution for each component of the process rather than the sequential approach needed by other methods. Various aspects and shortcomings of the method are discussed in detail and an example of how the method can be applied to a chemical reactor control problem is presented together with results of test problems.

The method can be sub-divided into a number of stages and the sections of this paper will be consistent with these stages. The first step in the method is to formulate the problem as a system of ordinary differential equations and to solve these equations numerically: this is covered in Section 2. The solution of the differential equations involves the solution of an associated algebraic equation and this aspect is discussed in Section 3. Section 4 deals with the application of the method to various problems and Section 5 describes some of the packages available and some of the developments that are needed to improve the method.

2. FORMULATION AS A SYSTEM OF O.D.ES.

2.1. Form of partial differential equations

The method of lines was developed to solve parabolic p.d.es. and it will be described in that context; its application to other classes of problems will be discussed separately. The general formulation of a system of parabolic equations can be written as

$$c(x,t)\,\frac{\partial \underset{\sim}{u}}{\partial t} = \underset{\sim}{f}\left(x,t,\underset{\sim}{u},\frac{\partial \underset{\sim}{u}}{\partial x},\frac{\partial^2 \underset{\sim}{u}}{\partial x^2}\right) \tag{1}$$

which is an initial value problem in time, t, and a boundary value problem in space, x.

A specific form of Equation 1 for a single equation is

$$c(x,t)\,\frac{\partial u}{\partial t} = \frac{\partial}{\partial x}\left\{g(x,t,u)\frac{\partial u}{\partial x}\right\} + h(x,t,u)\,\frac{\partial u}{\partial x} + p(x,t,u)\ . \tag{2}$$

The method will be described in terms of equations of this form, with generalisations to Equation 1 being straightforward as other classes of equation also covered by Equation 2, viz:

if $g(x,t,u) = 0$	the equation is hyperbolic
if $c(x,t) = 0$	the equation is a steady-state equation, i.e. elliptic
if $g(x,t,u) = (h(x,t,u) = 0$	the equation reduces to an o.d.e. corresponding to a lumped parameter system

2.2. Reduction to a system of o.d.es.

Let the region of definition of the equation be $[a,b] \times [0,\infty)$ with the solution known at $t = 0$. Dirichlet or Neumann type boundary conditions will be specified along $x = a$ and $x = b$. For hyperbolic problems a boundary condition needs to be specified along either $x = a$ or $x = b$.

The region is sub-divided, in the x direction, into m sub-intervals, not necessarily of the same length, such that $a = x_o < x_1 < \ldots < x_m = b$. Let $x_i - x_{i-1} = \delta x_i$. The solution u(t) is defined along the lines $x = x_i$, $i = 0,1,\ldots,m$ to be $(u_o(t),u_1(t),\ldots,u_m(t)) = \underset{\sim}{u}(t)$ (see Figure 1).

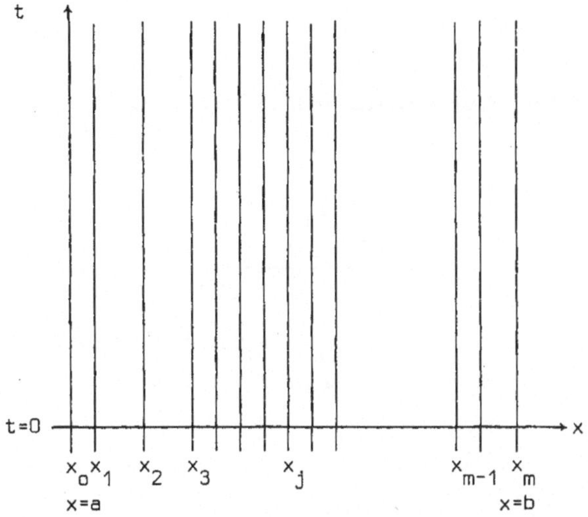

Figure 1 - Sub-Division of Region with Integration with respect to t

The method is based on a variables separable technique and the derivatives in the two directions are approximated differently. The spatial derivatives are considered first. There are a number of ways of approximating these derivatives, the principal ones being by finite differences, by finite elements and by collocation. Finite differences is the most widely used technique as some of the advantages of finite elements are not as important in one space dimension. Collocation methods are being used more widely now, for example PDECOL [12], and higher order methods are being developed.

To simplify the explanation we will consider a uniform mesh in space, i.e. $\delta x_1 = \delta x_2 = \ldots = \delta x_m = \delta x$. The choice of δx is very important as, in the basic form of the method, there is no automatic spatial mesh adjustment based on error estimates of the solution. This point is discussed in more detail later.

The most popular finite differences used in the method of lines are second order centred differences, i.e.

$$\left(\frac{\partial u}{\partial x}\right)_j = \frac{u_{j+1} - u_{j-1}}{2\delta x} + O((\delta x)^2) \tag{3}$$

$$\left(\frac{\partial^2 u}{\partial x^2}\right)_j = \frac{u_{j+1} - 2u_j + u_{j-1}}{(\delta x)^2} + O((\delta x)^2) . \tag{4}$$

For particularly difficult problems we have found it beneficial to impose some damping on the solution of the form

$$(u)_j = \frac{u_{j+1} + u_{j-1}}{2} + O((\delta x)^2) . \tag{5}$$

An example of the usefulness of this approximation will be illustrated in Section 4.

Substitution of Equations 3, 4 and 5 into Equation 1 yields the following ordinary differential equation along the line $x = x_j$:

$$c(x_j,t)\frac{du_j}{dt} = f(x_j,t,1/2(u_{j+1} + u_{j-1}),\frac{u_{j+1} - u_{j-1}}{2\delta x},\frac{u_{j+1} - 2u_j + u_{j-1}}{(\delta x)^2}) . \tag{6}$$

This equation is applied along $x = x_1,x_2,\ldots,x_{m-1}$ with boundary conditions applied along $x = x_0$ and $x = x_m$. Thus the problem can now be formulated as

$$\frac{d\underset{\sim}{u}}{dt} = \underset{\sim}{f}(t,\underset{\sim}{u}) \tag{7}$$

subject to:

$\underset{\sim}{u}(x,0)$ known.

As an illustration the model problem

$$\frac{\partial u}{\partial t} = \frac{\partial^2 u}{\partial x^2}$$

subject to zero Dirichlet boundary conditions and u known at t = 0, can be approximated by

$$\frac{du_1}{dt} = \frac{1}{(\delta x)^2} (- 2u_1 + u_2)$$

$$\vdots$$

$$\frac{du_j}{dt} = \frac{1}{(\delta x)^2} (u_{j-1} - 2u_j + u_{j+1}) \quad j = 2,3,\ldots,m-2$$

$$\vdots$$

$$\frac{du_{m-1}}{dt} = \frac{1}{(\delta x)^2} (u_{m-2} - 2u_{m-1})$$

that is,

$$\frac{du}{dt} = A\underset{\sim}{u} \tag{8}$$

where

$$A = \frac{1}{(\delta x)^2}\begin{bmatrix} -2 & 1 & & & 0 \\ 1 & -2 & 1 & & \\ & \ddots & \ddots & \ddots & \\ & & & & 1 \\ 0 & & & 1 & -2 \end{bmatrix}$$

with $\underset{\sim}{u}(x,0)$ known.

Experience with these approximations reveals an accuracy limitation for difficult problems (see Section 4) and fourth order centred differences, listed below, are preferred in these cases:

$$\left(\frac{\partial u}{\partial x}\right)_j = \frac{1}{12\delta x}\left(u_{j-2} + 8(u_{j+1} - u_{j-1}) - u_{j+2}\right) + O((\delta x)^4) \tag{9}$$

$$\left(\frac{\partial^2 u}{\partial x^2}\right)_j = \frac{1}{12(\delta x)^2}\left(-u_{j-2} + 16(u_{j-1} + u_{j+1}) - u_{j+2} - 30u_j\right) + O((\delta x)^4). \tag{10}$$

As before a damping effect may be appropriate

$$(u)_j = 1/6(4(u_{j-1} + u_{j+1}) - u_{j-2} - u_{j+2}) + O((\delta x)^4)\ . \tag{11}$$

These approximations are applied along $j = 2,3,\ldots,m-2$ with equivalent, but modified, approximations along the penultimate lines $j = 1$ and $j = m-1$, and boundary conditions are applied on $j = 0$ and $j = m$. The problem is again specified in the form of Equation 7.

Clearly these approximations are more accurate than the second order ones but the compensating disadvantage is that the larger bandwidth means more computation is required per time step. Fourth order approximations with smaller bandwidths are available and should be seriously considered, see Chawla [6] for example.

2.3. Solution of resulting o.d.es

Equations 7 are numerically integrated in time with the error in integration controlled by varying the step-size, δt in time and the order of the method used. The difficulty with the o.d.es. generated by the method of lines is that they are generally "stiff". A stiff system of equations is one for which the solution comprises slowly varying terms, together with some rapidly decaying components. For Equations 8, for example, a measure of the stiffness of the equation is

$$S(A) = \frac{\max|\text{eigenvalue of } A|}{\min|\text{eigenvalue of } A|}\ .$$

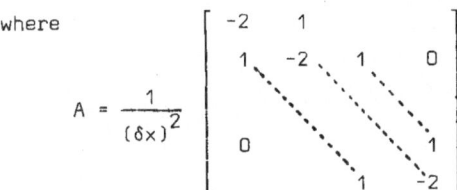

For the model problem using second order approximations, the matrix A has the eigenvalues

$$\lambda_j = -\frac{4}{(\delta x)^2}\sin^2\{\frac{j\pi}{2m}\} \quad j = 1,2,\ldots,m-1 .$$

Thus

$$S(A) = \frac{\sin^2\{\frac{(m-1)\pi}{2m}\}}{\sin^2\{\frac{\pi}{2m}\}} \simeq \frac{4m^2}{\pi^2} .$$

If the spatial direction is divided into 10 sub-intervals $S(A) \simeq 40$ so the equations are only mildly stiff, whereas if 100 sub-intervals are used the stiffness is increased significantly with $S(A) \simeq 4000$.

It is this problem of stiffness that initially denied the method of lines its present position as the most commonly used method for solving parabolic p.d.es. The method has been known for many years but it was not until "stiffly stable" methods had been developed by Gear [7], circa 1970, that the method became a feasible means of solving the problem. Stiffly stable methods are implicit linear multistep methods, using backward differences, developed such that they possess good stability properties for stiff problems. The methods use approximations of the form:

$$\frac{d\underset{\sim}{u}}{dt}(t_{n+1}) \simeq \frac{\underset{\sim}{u}(t_{n+1}) - \sum_{i=1}^{k} \alpha_i \underset{\sim}{u}_{n+1-i}}{\beta_o \delta t} \tag{12}$$

where $\{\underset{\sim}{u}_{n+1-i}\}_{i=1}^{k}$ are known solution values at previous time steps, and the constants k, β_o and $\{\alpha_i\}_{i=1}^{k}$ determine the method being used, see Gear [7].

Substitution of Equation 12 into the differential Equation 7 yields

$$\underset{\sim}{u}(t_{n+1}) = \sum_{i=1}^{k} \alpha_i \underset{\sim}{u}_{n+1-i} + \beta_o \delta t\, \underset{\sim}{f}(t_{n+1}, \underset{\sim}{u}_{n+1}) . \tag{13}$$

The solution at the next point in time, t_{n+1}, is obtained by solving the algebraic Equation 13. The equation is clearly implicit in the solution $\underset{\sim}{u}_{n+1}$ and furthermore is non-linear in $\underset{\sim}{u}_{n+1}$ if the original p.d.e. was non-linear. The solution of this equation is dealt with in Section 3.

Gear's formulation of stiffly stable methods is a modification of Equation 12 in that scaled estimates of the first (k-1) derivatives of the solution at $t = t_n$ are stored instead of the solution values at previous points. The advantage of this approach is that the expression for the estimate of the error in the solution is simplified. Also, should this estimate be unacceptably large, the procedure for reducing the step-size is also simpler.

Stiffly stable methods of orders up to 10 and more are available but in practice methods of no more than four are usually used. There is an interesting comparison between the methods used to approximate the derivatives in the two directions. The approximations in time are of variable order with a variable step-size: the approximations in space are of fixed order, being generally of lower order than that of the approximations in time, and the step-size is of fixed length.

An effect of Gear-type methods is their damping effect on oscillations present in the solution which is consistent with the nature of solutions of parabolic problems. The methods are less successful for problems with oscillatory solutions and usually have difficulty in maintaining accuracy in the results. This point should be noted in the context of hyperbolic problems, where the method can be used but with caution.

An alternative formulation of the problem can be achieved by reversing the roles of the independent variables.

The region of definition is now sub-divided in the time direction with the solution defined as functions of x, $\{u_i(x)\}$ at constant time intervals.

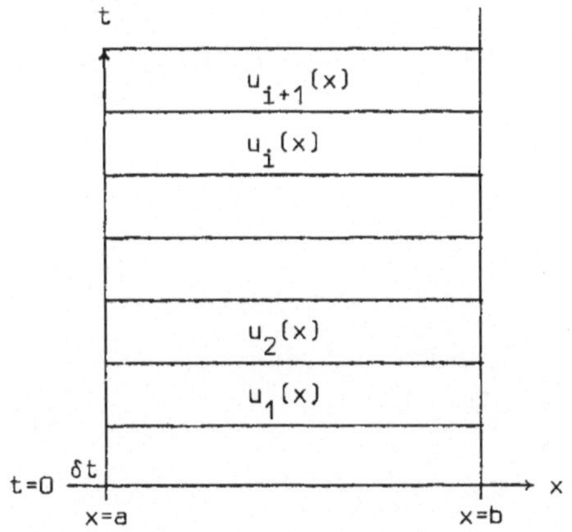

Figure 2 - Sub-Division of Region with Integration with respect to x

The derivative with respect to time is approximated in terms of the $\{u_i(x)\}$ so that the equation to be solved along each of the lines t = $i\delta t$ (assuming a constant time interval) is of the form

$$\frac{d^2u_i}{dx^2} = f(x,\tilde{u}) \quad i = 1,2,...$$

subject to boundary conditions at x = a and at x = b.

This sequence of boundary value problems can be solved by shooting techniques, for example, in which the step-lengths in the space direction are selected automatically by the error control mechanism of the integration procedure. Clearly, in this formulation there is no error control mechanism for the step-length in the time direction.

3. THE ALGEBRAIC PROBLEM

The problem is now formulated algebraically in the form of Equation 14:

$$\underset{\sim}{u}(t_{n+1}) = \sum_{i=1}^{k} \alpha_i \underset{\sim}{u}_{n+1} + \beta_o \delta t \, \underset{\sim}{f}(t_{n+1}, \underset{\sim}{u}_{n+1}) \ . \tag{14}$$

The solution at the next point in time, $t = t_{n+1}$, is obtained by solving this non-linear equation where $\{\underset{\sim}{u}_{n+1-i}\}_{i=1}^{k}$ are known values of the solution at previous points in time. Obviously a different integration technique needs to be used initially to calculate the values $\underset{\sim}{u}_1$ to $\underset{\sim}{u}_{k-1}$.

Denoting the unknown $\underset{\sim}{u}(t_{n+1})$ by $\underset{\sim}{u}$, we can rewrite Equation 14 as

$$\underset{\sim}{f}(\underset{\sim}{u}) \equiv \underset{\sim}{u} - \underset{\sim}{c} - \beta_o \, \delta t \, \underset{\sim}{f}(t_{n+1}, \underset{\sim}{u}) = \underset{\sim}{0} \ . \tag{15}$$

Using Newton's method to solve this yields

$$J_r(\underset{\sim}{u}^{(r+1)} - \underset{\sim}{u}^{(r)}) = -\underset{\sim}{f}((\underset{\sim}{u}^{(r)})$$

$$= - \underset{\sim}{u}^{(r)} + \underset{\sim}{c} + \beta_o \delta t \, \underset{\sim}{f}(t_{n+1}, \underset{\sim}{u}^{(r)}) \tag{16}$$

where the Jacobian J_r is defined by

$$(J_r)_{ij} \equiv \frac{\partial \phi_i}{\partial u_j}(\underset{\sim}{u}^{(r)}) = \frac{\partial u_i}{\partial u_j} - \beta_o \delta t \frac{\partial f_i}{\partial u_j}(\underset{\sim}{u}^{(r)}) \ . \tag{17}$$

The partial derivatives $\frac{\partial f_i}{\partial u_j}$ can be approximated either (a) analytically or (b) by finite differences.

(a) If second order approximations are used for the spatial derivatives the functions f_i, obtained from Equation 7, are dependent upon t, u_{i-1}, u_i and u_{i+1} only, and the associated differential equations are

$$\frac{du_i}{dt} = f_i(t, \underset{\sim}{u}) = f(t, 1/2(u_{i-1} + u_{i+1}), \frac{u_{i+1} - u_{i-1}}{2\delta x}, \frac{u_{i+1} - 2u_i + u_{i-1}}{(\delta x)^2}) \tag{18}$$

$$= f(t, u_i^*, u_x^*, u_{xx}^*) \ . \tag{19}$$

Thus analytic approximations to $\frac{\partial f_i}{\partial u_j}$ can be obtained by using Equations 18 and 19.

For example

$$\frac{\partial f_i}{\partial u_{i-1}} = \frac{\partial f}{\partial u_i^*}\frac{\partial u_i^*}{\partial u_{i-1}} + \frac{\partial f}{\partial u_x^*}\frac{\partial u^*_x}{\partial u_{i-1}} + \frac{\partial f}{\partial u_{xx}^*}\frac{\partial u^*_{xx}}{\partial u_{i-1}}$$

$$= \sqrt{2}\,\frac{\partial f}{\partial u_i^*} - \frac{1}{2\delta x}\frac{\partial f}{\partial u_x^*} + \frac{1}{(\delta x)^2}\frac{\partial f}{\partial u_{xx}^*}$$

i.e. $$\frac{\partial f_i}{\partial u_{i-1}} = \sqrt{2}\,\frac{\partial f}{\partial u} - \frac{1}{2\delta x}\frac{\partial f}{\partial u_x} + \frac{1}{(\delta x)^2}\frac{\partial f}{\partial u_{xx}} . \tag{20}$$

There are similar expressions for $\frac{\partial f_i}{\partial u_i}$ and $\frac{\partial f_i}{\partial u_{i+1}}$. Obviously $\frac{\partial f_i}{\partial u_j} = 0$ if $j \neq i-1,i,i+1$.

(b) Finite difference approximations to $\frac{\partial f_i}{\partial u_j}$ would be of the form

$$\frac{\partial f_i}{\partial u_j}(\underset{\sim}{u}^{(r)}) \simeq \frac{f_i(\underset{\sim}{u}^{(r)} + \varepsilon \underset{\sim}{e}_j) - f_i(\underset{\sim}{u}^{(r)})}{\varepsilon}$$

where $\underset{\sim}{e}_j$ is the jth column of the unit matrix and ε is a constant.

The value of ε should be chosen to be small, so that the approximation is as accurate as possible, but not so small that accuracy is lost due to rounding error cancellation effects.

Another method of solving Equation 15 is to use one of the quasi-Newton class of methods in which an approximation to the Jacobian, or its inverse, is updated at each iteration (see Murray [14]).

To compute the new iterates in the above schemes the linear algebraic Equation 16 has to be solved. The Jacobian is (almost) tridiagonal when second order spatial approximations are used and (almost) pentadiagonal when using fourth order approximations. Advantage must be taken of the sparsity, especially for problems defined over two and three dimensions. The best methods for such problems are Successive Over-Relaxation, Pre-conditioned Conjugate Gradients and Multi-grid methods.

4. EXAMPLES

4.1. Mixed problem

A chemical engineering mixed control problem is illustrated in the following example of Heydweiller et al. [9] which consists of a gas phase tubular reactor, a gas absorber and a completely mixed tank.

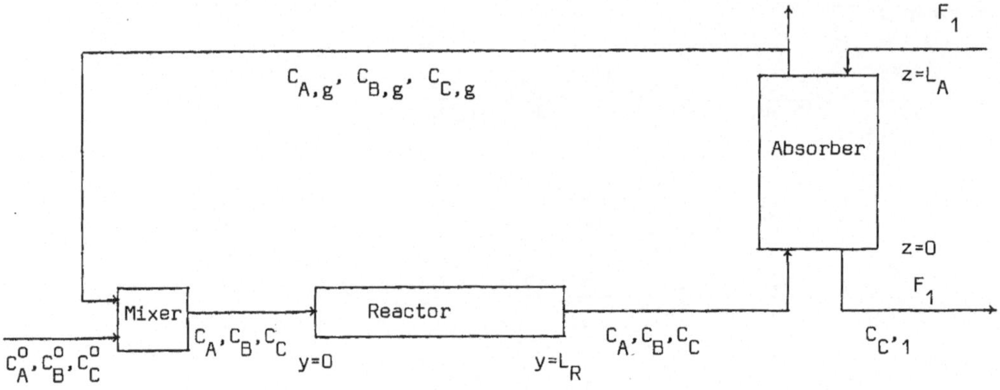

It is assumed that the reactants and products are present in low concentrations, that the gas phase has a constant density and that the whole process is isothermal. A one-dimensional dispersion model of the flow in the reactor leads to the following equations:

$$\frac{\partial C_A}{\partial t} = D\frac{\partial^2 C_A}{\partial y^2} - v\frac{\partial C_A}{\partial y} - kC_A C_B$$

$$\frac{\partial C_B}{\partial t} = D\frac{\partial^2 C_B}{\partial y^2} - v\frac{\partial C_B}{\partial y} - kC_A C_B$$

$$\frac{\partial C_C}{\partial t} = D\frac{\partial^2 C_C}{\partial y^2} - v\frac{\partial C_C}{\partial y} + 2kC_A C_B$$

where C_A, C_B and C_C are the concentrations of the reactants and products, and D, c and k are associated constants.

This model is in terms of parabolic partial differential equations. The equations for the gas absorber, on the other hand, are hyperbolic. The model assumes plug flow in both gas and liquid phases and assumes that the product C is the only component which is significantly absorbed into the liquid phase:

$$\frac{\partial C_{A,g}}{\partial t} = - v_g \frac{\partial C_{A,g}}{\partial z}$$

$$\frac{\partial C_{B,g}}{\partial t} = - v_g \frac{\partial C_{B,g}}{\partial z}$$

$$\frac{\partial C_{C,g}}{\partial t} = - v_g \frac{\partial C_{C,g}}{\partial z} - \frac{k_g a}{\varepsilon_g}(C_{C,g} - C^*_{C,g})$$

$$\frac{\partial C_{C,1}}{\partial t} = v_1 \frac{\partial C_{C,1}}{\partial z} + \frac{k_g a}{\varepsilon_1} (C_{C,g} - C^*_{C,g}) .$$

The third component in the system is the perfectly mixed tank. Dynamic mass balances lead to the following ordinary differential equations:

$$\frac{dC_A}{dt} = \frac{F_f}{V_m} (C_A^o + rC_{A,g}) - \frac{F}{V_m} C_A$$

$$\frac{dC_B}{dt} = \frac{F_f}{V_m} (C_B^o + rC_{B,g}) - \frac{F}{V_m} C_B$$

$$\frac{dC_C}{dt} = \frac{F_f}{V_m} (C_C^o + rC_{C,g}) - \frac{F}{V_m} C_C .$$

With appropriate initial and boundary values these three sets of equations can be solved simultaneously, in a uniform manner, by the method of lines. As mentioned previously the approximations described above are not ideal for hyperbolics and Heydweiller et al. developed approximations that achieve a balance between accuracy and stability. Details are included in their paper.

4.2. Model test problem

A number of features of the method can be illustrated by means of the test problem

$$\frac{\partial u}{\partial t} = \frac{\partial^2 u}{\partial x^2} \qquad 0 \le x \le 4, \quad t > 0 \tag{21}$$

subject to:

$$u(0,t) = 1$$

$$\frac{\partial u}{\partial x}(4,t) = 0$$

$$u(x,0) = 0 .$$

The solution is

$$u(x,t) = 1 - \frac{4}{\pi} \sum_{k=1}^{\infty} \frac{1}{(2k-1)} \exp(-a_k^2 t) \sin(a_k x)$$

where

$$a_k = \frac{(2k-1)\pi}{8} .$$

The problem has been solved, using second order spatial approximations, with 20 and 40 equi-spaced points in the spatial direction.

The error tolerance per step when integrating temporally was taken to be 0.5×10^{-5}. The results are shown in Table 1.

	$\delta x = 0.2$		$\delta x = 0.1$	
t	$\|\|Error\|\|_\infty$	δt	$\|\|Error\|\|_\infty$	δt
0.5	4.37E-4	0.0165	1.09E-3	0.0285
1.0	1.45E-4	0.0292	5.77E-4	0.0509
1.5	1.54E-4	0.0385	6.18E-4	0.0907
3.0	7.93E-5	0.0903	3.13E-4	0.1628
4.0	4.22E-5	0.1163	1.66E-4	0.2775
4.5	2.98E-5	0.1163	1.13E-4	0.2775
5.0	1.20E-5	0.1578	7.44E-5	0.2775
10.0	1.88E-5	0.3883	7.46E-5	0.5420
15.0	2.08E-5	0.4648	7.24E-5	1.1381
20.0	1.50E-5	0.5319	4.58E-5	1.1381
25.0	9.31E-6	0.6234	2.58E-5	1.5195
30.0	5.31E-6	0.7501	9.67E-6	1.5195

Table 1 - Comparison of Errors in Solutions of Model Problem using Different Mesh Lengths

For small values of t ($t \leq 4$) it can be seen that the errors obtained by halving δx are reduced by a factor of 4, approximately. This is consistent with using second order approximations for the spatial derivatives and the errors of these approximations dominating the errors of the higher order time integrator. As the steady-state solution is approached the step-size in time increases whilst the errors reduce and, in fact, become comparable for $\delta x = 0.1$ and $\delta x = 0.2$. Clearly an automatic adaptive mechanism for determining δx, whilst maintaining accuracy, would be beneficial.

The problem was also solved using fourth order spatial approximations and comparisons with second order results are shown in Table 2. δx was taken to be 0.2 and the error tolerance per step was 0.5×10^{-4}.

	Second order method		Fourth order method	
t	$\|\text{Error}\|_\infty$	δt	$\|\text{Error}\|_\infty$	δt
0.1	.532E-2	.00405	.105E-2	.00408
0.2	.271E-2	.00910	.269E-3	.00865
0.3	.178E-2	.01382	.107E-3	.01320
0.4	.134E-2	.01382	.487E-4	.01989
0.5	.108E-2	.02065	.256E-4	.01989
0.6	.897E-3	.02065	.146E-4	.03033
0.7	.772E-3	.03149	.999E-5	.03033
0.8	.677E-3	.03149	.681E-5	.03033
0.9	.597E-3	.03149	.474E-5	.04549
1.0	.526E-3	.04720	.373E-5	.04549
1.5	.257E-3	.07132	.212E-5	.06874
2.0	.113E-3	.07132	.142E-5	.06874
2.5	.985E-4	.10855	.150E-5	.10466
3.0	.943E-4	.10855	.137E-5	.10466

Table 2 - Comparison of Errors using Second and Fourth Order Approximations

As expected the fourth order results are more accurate, with the permitted time-steps δt being of comparable size.

4.3. Burgers' Equation

$$\frac{\partial u}{\partial t} = \nu \frac{\partial^2 u}{\partial x^2} - u \frac{\partial u}{\partial x} \ . \tag{22}$$

This is a well-known diffusion problem, i.e. it is possible in nature. However, if $\nu \to 0$ the equation becomes convective and hyperbolic. A number of workers have had difficulties with this problem for small values of ν (see Ames [1] and Sincovec and Madsen [18]). No difficulty was experienced with the above approach until ν was less than 0.005. For example, consider Equation 22 subjected to the following initial and boundary conditions

$$u(0,t) = u(1,t) = 0$$

$$u(x,0) = \sin \pi x$$

with ν taken to be 0.001. The solution is of the form shown in Figure 3.

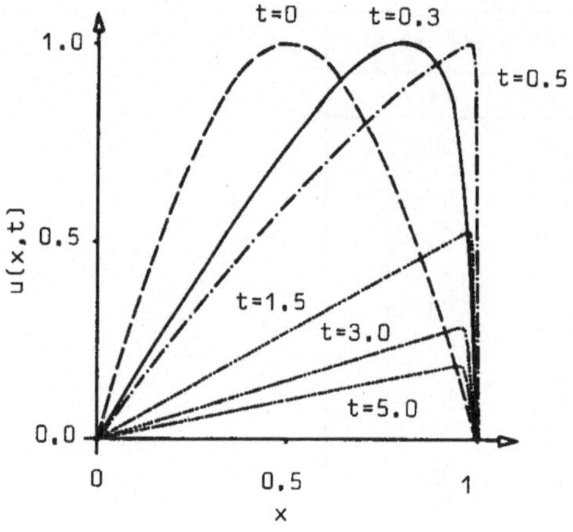

Figure 3 - Solution of Problem 4.3

The initial sine curve is rapidly distorted with the peak moving to the right of the interval. From t = 0.5 onwards the solution decreases in value and the peak moves slightly away from the right hand end of the interval, gradually becoming less sharp. The numerical results obtained using second order spatial differences with 100 mesh intervals are shown in Figures 4 and 5.

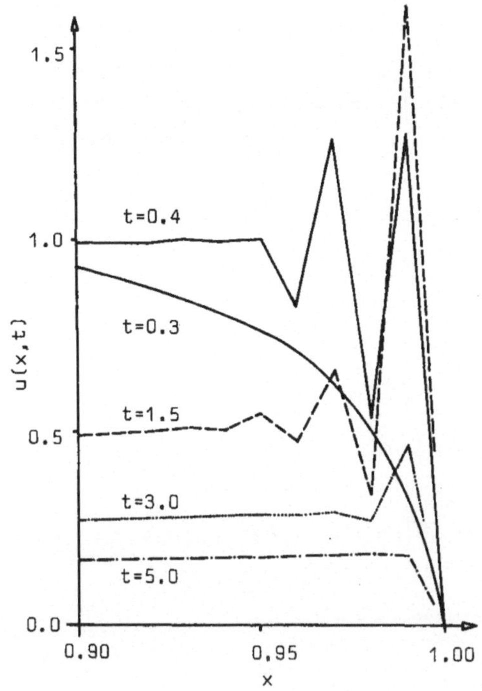

Max(u_{99}) = 4.36 at t = 0.6

CPU time: 4 min 18 s

Figure 4 - Numerical Solution of Problem 4.3 (without damping)

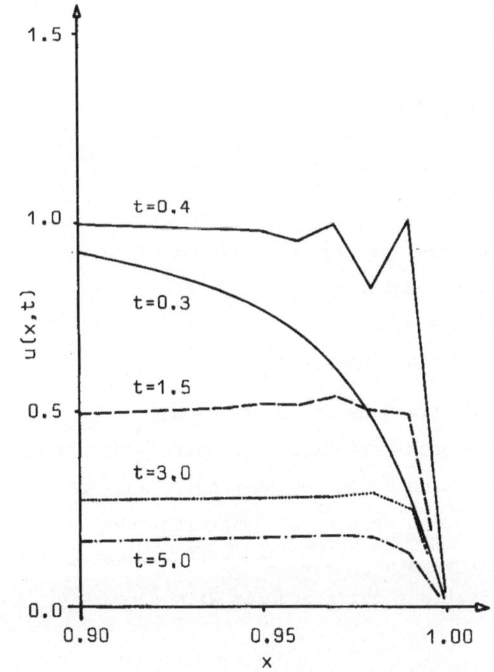

Max(u_{99}) = 1.22 at t = 0.5

CPU time: 6 min 8 s

Figure 5 - Numerical Solution of Problem 4.3 (with damping)

Figure 4 shows the results without the damping effect of Equation 5 whereas the results of Figure 5 have the influence of the damping term. For the undamped case a maximum value of 4.36 was reached at t = 0.6 whilst for the damped case the maximum value was 1.22 at t = 0.5. In both cases the instabilities decreased with time with the improvement being more rapid for the damped solution. Comparable results for the fourth order method are shown in Figure 6. The oscillations apparent in the second order methods were reduced considerably.

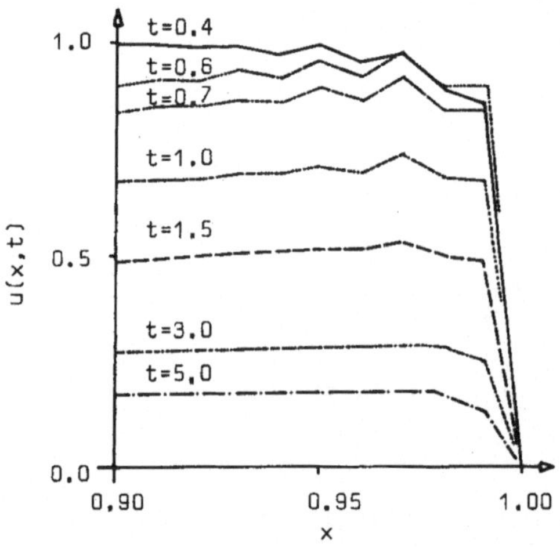

Figure 6 - Fourth Order Solution to Problem 4.3

5. COMMENTS AND CONCLUSIONS

5.1. The numerical method of lines is a flexible, general purpose algorithm well-suited to a wide variety of control problems, including problems of mixed type. The removal of the necessity to perform a stability analysis is particularly beneficial as is the automatic error estimation in the temporal integration.

5.2. The size of the spatial mesh used is important as there is no means of estimating and controlling the errors due to approximations in the spatial direction. It would be useful if the size of the mesh could be chosen automatically such that the estimation of the errors in the space direction were consistent with the specified error tolerance in time. The ability to generate a non-uniform mesh is required as, clearly, a uniform mesh has limitations for a problem with a rapidly varying solution.

5.3. Another development could be towards an adaptive mechanism for varying the spatial mesh with time. For example, as a steady-state situtation is approached it may be possible to increase the mesh whilst still maintaining the required error control. Another example where a time-varying mesh would be expedient is that of a wavefront in the solution moving through the region.

5.4. The choice of approximation in the spatial direction is important. Although second order approximations have the advantage of simplicity of use, fourth order approximations can solve problems more accurately, assuming the same mesh size is used. This is particularly necessary for unstable problems and for problems whose solution involves a wavefront.

5.5. There is a wide literature on the Method of Lines and on the integration of stiff o.d.es. It would not be appropriate to survey these topics in detail here but some of the packages available for use are listed.

Some comprehensive packages can solve a large variety of problems in one or two space dimensions including problems involving p.d.es of different types (i.e. elliptic, parabolic and hyperbolic) and also mixed o.d.e./p.d.e. problems. Two packages that allow the use of high order finite difference approximations are DSS/2 developed by Schiesser [16] and FORSIM VI developed at Chalk River Nuclear Laboratories by Carver et al. [5]. Other packages include PDEL [4] which uses finite difference space discretisations, and POST [17] and DISPL [3] which use Galerkin methods with B-Spline basis functions.

Less comprehensive routines for parabolic problems are also available including:

(i) PDEONE [18] one-dimensional, finite difference approach;
(ii) PDETWO [13] as PDEONE but for two-dimensional problems;
(iii) PDECOL [12] uses collocation in the space direction;
(iv) PDEF1 [2] one-dimensional, Galerkin approach;
(v) NAG Library [15] one-dimensional, finite difference technique;
(vi) TWODEPEP [10] two-dimensional, finite element method.

A useful survey of various packages and routines has been carried out by Machura and Sweet [11].

ACKNOWLEDGEMENT

The author wishes to thank BP International Limited for the time and facilities afforded him in the preparation and presentation of this paper.

REFERENCES

[1] Ames, W.F.: Numerical methods for partial differential equations, Nelson, London, 1969.

[2] Bakker, M.: Report NW 52/77, Mathematisch Centrum, Amsterdam, 1977.

[3] Bennett and Vichnevetsky (eds): "Numerical methods for differential equations and simulation", North-Holland,1978.

[4] Cardenas, A.F., W.J. Karplus: Comm ACM 13 (1970), 184-191.

[5] Carver, M.B., D.G. Stewart, J.M. Blair, W.N. Selander: FORSIM VI Manual, AECL-5821, Chalk River Nuclear Laboratories, Ontario, 1978.

[6] Chawla, M.M.: J. IMA 21 (1976), 83-93.

[7] Gear, C.W.: "Numerical initial value problems in ordinary differential equations", Prentice Hall, New Jersey, 1971.

[8] Graney, L., A.A. Richardson: J. Computat. and Appl. Maths. 7 (1981), 229-236.

[9] Heydweiller, J.C., R.F. Sincovec, L.T. Fan: Computers and Chem. Eng. 1 (1977), 125-131.

[10] International Mathematical and Statistical Library: 7500 Bellaire Boulevard, Houston, Texas.

[11] Machura, M., R.A. Sweet: ACM Trans. on Mathematical Software 6 (1980), 461-488.

[12] Madsen, N.K., R. Sincovec: ACM Trans. on Mathematical Software 5 (1979), 326-351.

[13] Melgaard, D., R. Sincovec, ACM Trans. on Mathematical Software 7 (1981), 106-135.

[14] Murray. W.(ed.): "Numerical methods for unconstrained optimisation", Academic Press, London, 1972.

[15] Numerical Algorithms Group Library: NAG Central Office, 256 Banbury Road, Oxford.

[16] Schiesser, W.E.: DSS/2 Manuals, Lehigh University, Bethlehem, Pennsylvania, 1976.

[17] Schryer, N.: Comp. Sci. Tech. Report 53, Bell Laboratories, Murray Hill, New Jersey, 1977.

[18] Sincovec, R.F., N.K. Madsen: ACM Trans. on Mathematical Software 1 (1975), 232-260.

ON TIME-OPTIMAL BOUNDARY CONTROL OF VIBRATING BEAMS

W. Krabs

Fachbereich Mathematik
Technische Hochschule Darmstadt
Schlossgartenstrasse 7
D-6100 Darmstadt, West Germany

SUMMARY

For the vibration of a homogeneous beam it is shown, for a certain class of initial states at the time $t = 0$, that null-controllability at $t = T$ is possible for every time $T > 0$, if control is applied to one of the boundary conditions on the right-hand side which is of sufficiently high order.

The controls v are assumed to be in $H^1(0,T)$ and to satisfy $v(0) = v(T) = 0$. The last condition guarantees that the beam stays in rest for all $t \geq T$, if v is continued by putting $v(t) = 0$.

It is further shown that restricted null-controllability at $t = T$ is possible by controls v with $||v'||_{L^2(0,T)}$ being bounded by some given constant $M > 0$, if T is sufficiently large. This guarantees the existence of time-minimal controls v_M for which it can be shown that $||v'_M||_{L^2(0,T(M))} = M$ where $T(M)$ is the minimum time such that restricted null-controllability at $t = T$ is possible.

1. THE MODEL

We consider the vibration of a homogeneous beam of length 1 whose deviation $y = y(x,t)$ from the position of rest as a function of the space variable $x \in [0,1]$ and the time $t \in [0,\infty)$ is governed by the differential equation

$$y_{tt} + y_{xxxx} = 0 \quad \text{in } (0,1) \times (0,T) \tag{1.1}$$

for all $T > 0$.

The motion of the beam is assumed to be controlled through one boundary condition on the right-hand side where boundary conditions of the following kind are admitted

$$\frac{\partial^{j_1}}{\partial x^{j_1}} y(0,t) = \frac{\partial^{j_2}}{\partial x^{j_2}} y(0,t) = 0, \tag{1.2...}$$

$$\frac{\partial^{j_3}}{\partial x^{j_3}} y(1,t) = kv(t), \quad \frac{\partial^{j_4}}{\partial x^{j_4}} y(1,t) = (1-k)v(t), \; t \in [0,T], \tag{...1.2}$$

for all $T > 0$ with $k = 0$ or 1 and v being a control function from $[0,\infty)$ into $\mathbb{R}$ which for the beginning is assumed to belong to $H^2[0,T]$ for all $T > 0$.

If the quadruple (j_1,j_2,j_3,j_4) is chosen to be one of the following 6 possibilities

j_1	0	0	0	0	0	0
j_2	1	1	1	1	2	2
j_3	0	0	1	2	0	1
j_4	1	2	3	3	2	3

(1.3)

(which will be assumed in the sequel), then it can be shown that the operator

$$Lz(x) = z^{(4)}(x), \; x \in (0,1),$$

is symmetric and positive definite on

$$D(L) = \{z \in C^{(4)}[0,1] \,|\, z^{(j_1)}(0) = z^{(j_2)}(0) = z^{(j_3)}(1) = z^{(j_4)}(1) = 0\}$$

and therefore has a complete orthonormal sequence $(e_j)_{j\in\mathbb{N}}$ of eigenfunctions in $D(L)$ and corresponding sequence $(\lambda_j)_{j\in\mathbb{N}}$ of real positive eigenvalues with $\lambda_j \to \infty$ as $j \to \infty$ (see, for instance, [1]). Furthermore all the eigenvalues are simple and of the form

$$\lambda_j = [(j-\sigma)\pi + \varepsilon_j]^4, \; j \in \mathbb{N}, \tag{1.4}$$

with $\sigma = -\frac{1}{2}, -\frac{1}{4}, \frac{1}{4}, \frac{1}{2}, 0, \frac{1}{2}$ for the corresponding choice in (1.3), $|\varepsilon_j| < \frac{\pi}{4}$ for all $j \in \mathbb{N}$, and $\varepsilon_j \to 0$ as $j \to \infty$.

Next we assume initial conditions of the form

$$y(x,0) = y_0(x), \; y_t(x,0) = y_1(x) \quad \text{for almost all } x \in (0,1) \tag{1.5}$$

where

$$y_0 \in H(L) = \{z \in L_2(0,1) \,|\, \sum_{j=1}^{\infty} \lambda_j \langle z,z_j\rangle^2 < \infty\}, \tag{1.6}$$

$y_1 \in L^2(0,1)$, and $\langle .,. \rangle$ denotes the scalar product in $L^2(0,1)$.

Let $T > 0$ be given arbitrarily. Then for every choice of $v \in H^2(0,T)$, $y_0 \in H(L)$, and $y_1 \in L^2(0,1)$ there is exactly one generalized solution $y = y(x,t)$ of (1.1), (1.2), (1.5) in the following sense: From the properties of L it follows that there is exactly one solution $r \in C^4[0,1]$ of the boundary value problem

$$\begin{aligned} &Lr(x) = r^{(4)}(x) = 0. \quad x \in (0,1), \\ &r^{(j_1)}(0) = r^{(j_2)}(0) = 0, \\ &r^{(j_3)}(1) = k, r^{(j_4)}(1) = 1 - k. \end{aligned} \tag{1.7}$$

Now we consider the initial boundary value problem

$$\left.\begin{aligned} &y^*_{tt}(x,t) + y^*_{xxxx}(x,t) = - r(x)v''(t) \quad \text{for } x \in (0,1),\ t \in (0,T), \\ &\frac{\partial^{j_1}}{\partial x^{j_1}} y^*(0,t) = \frac{\partial^{j_2}}{\partial x^{j_2}} y^*(0,t) = 0, \\ & \hspace{18em} t \in [0,T], \\ &\frac{\partial^{j_3}}{\partial x^{j_3}} y^*(1,t) = \frac{\partial^{j_4}}{\partial x^{j_4}} y^*(1,t) = 0, \\ &y^*(x,0) = y_0(x) - r(x)v(0),\ y^*_t(x,0) = y_1(x) - r(x)v'(0) \\ &\qquad \text{for almost all } x \in (0,1). \end{aligned}\right\} \tag{1.8}$$

By Theorem 1.1 in Chapter IV of [6] there is exactly one generalized solution y^* of (1.8) satisfying

$$y^* \in C([0,T],H(L)),\ \frac{dy^*}{dt} \in C([0,T],L^2(0,1)). \tag{1.9}$$

If we put

$$y(x,t) = y^*(x,t) + r(x)v(t), \tag{1.10}$$

then y satisfies (1.1) (in the sense of distributions), (1.2), (1.5) and

$$y \in C([0,T],H(L)),\ \frac{dy}{dt} \in C([0,T],L^2(0,1)), \tag{1.11}$$

if $r \in H(L)$. Under this assumption the unique solution of (1.1), (1.2), (1.5) which satisfies (1.11) is given by (1.10). This is an immediate consequence of the fact that there is exactly one solution y^* of (1.8) which satisfies (1.9).

The assumption $r \in H(L)$ is not always fulfilled. For instance, if $j_1 = j_3 = 0$, $j_2 = j_4 = 2$, and $k = 1$, then $r(x) = x$, $x \in [0,1]$, which is not in $H(L)$.
However, for $j_1 = j_3 = 0$, $j_2 = j_4 = 2$, and $k = 0$ one obtains

$$r(x) = \frac{1}{6}\, x(x^2-1) \text{ which is in } H(L).$$

In general it can be shown that (see [2])

$$\langle r,e_j\rangle = \begin{cases} O(j^{-(j_3+1)}) & \text{for } k = 1, \\ O(j^{-(j_4+1)}) & \text{for } k = 0. \end{cases}$$

This implies

$$r \in H(L), \text{ if } \begin{cases} j_3 \geq 2 & \text{for } k = 1, \\ j_4 \geq 2 & \text{for } k = 0. \end{cases}$$

The unique generalized solution y^* of (1.8) which satisfies (1.9) can be explicitly represented in the form

$$\begin{aligned} y^*(x,t) = & \sum_{j=1}^{\infty} [a_j(v)\cos\sqrt{\lambda_j}t + b_j\sin\sqrt{\lambda_j}t]e_j(x) \\ & - \sum_{j=1}^{\infty} \frac{h_j^k}{\sqrt{\lambda_j}} \int_0^t v''(s)\sin\sqrt{\lambda_j}(t-s)ds\ e_j(x) \end{aligned} \tag{1.12}$$

where

$$\begin{aligned} a_j(v) &= \int_0^1 y_0(x)e_j(x)dx - h_j^k v(0), \\ b_j(v) &= \frac{1}{\sqrt{\lambda_j}} \int_0^1 y_1(x)e_j(x)dx - \frac{h_j^k}{\sqrt{\lambda_j}} v'(0), \\ h_j^k &= \langle r,e_j\rangle \end{aligned} \tag{1.13}$$

with $r \in C^4[0,1]$ being the solution of (1.7) for $k = 1$ or $k = 0$. If $r \in H(L)$, we can allow for control functions which are less smooth. In order to see that we apply integration by parts to (1.12) and obtain

$$y^*(x,t) = \hat{y}(x,t) - \sum_{j=1}^{\infty} h_j^k \int_0^t v'(x)\cos\sqrt{\lambda_j}(t-s)ds\ e_j(x)$$

where

$$\hat{y}(x,t) = \sum_{j=1}^{\infty} [a_j(v)\cos\sqrt{\lambda_j}t + b_j(0)\sin\sqrt{\lambda_j}t]e_j(x) \tag{1.14}$$

is the generalized solution of

$$\hat{y}_{tt} + \hat{y}_{xxxx} = 0 \text{ in } (0,1) \times (0,T),$$

the boundary conditions of (1.8), and

$$\hat{y}(x,0) = y_0(x) - r(x)v(0),\ \hat{y}_t(x,0) = y_1(x) \quad \text{for almost all } x \in (0,1)$$

which satisfies

$$\hat{y} \in C([0,T],H(L)), \frac{d\hat{y}}{dt} \in C([0,T],L^2(0,1)).$$

If we define

$$y(x,t) = \hat{y}(x,t) - \sum_{k=1}^{\infty} h_j^k \int_0^t v'(s)\cos\sqrt{\lambda_j}(t-s)ds\, e_j(x) + r(x)v(t), \tag{1.15}$$

then y satisfies (1.1) (in the sense of distributions), (1.2) and (1.5). Furthermore

$$y_t(x,t) = \hat{y}_t(x,t) + \sum_{k=1}^{\infty} h_j^k \sqrt{\lambda_j} \int_0^t v'(s)\sin\sqrt{\lambda_j}(t-s)ds\, e_j(x) \tag{1.16}$$

and (1.11) can be verified.

Result: If $r \in H(L)$ and $v \in H^1(0,T)$, then the unique generalized solution of (1.1), (1.2), (1.5) which satisfies (1.11) is given by (1.15).

The uniqueness follows from the fact that (1.8) has only the trivial solution as generalized solution which satisfies (1.9), if $y_0 = y_1 = 0$ a.e. and $v \equiv 0$.

2. THE PROBLEM OF NULL-CONTROLLABILITY

As in Section 1 we assume that $y_0 \in H(L)$, $y_1 \in L^2(0,1)$ in (1.5) and $r \in H(L)$ for the unique solution $r \in C^4[0,1]$ of (1.7). Let $T > 0$ be given. Then the problem of null-controllability consists of asking for the existence of some $v \in H^1(0,T)$ with $v(0) = v(T) = 0$ such that for the corresponding generalized solution y of (1.1), (1.2), (1.5) which satisfies (1.11) it follows that

$$y(\cdot,T) = y_t(\cdot,T) = 0 \quad \text{a.e. on } (0,1). \tag{2.1}$$

If this is possible, then, by defining $v(t) = 0$ for all $t > T$, it is guaranteed that the corresponding generalized solution of (1.1), (1.2), (1.5) with (1.11) satisfies (2.1) for all $t > T$ instead of T, i.e., the beam stays in rest for all $t \geq T$ and the extended control function v is in $H^2(0,\tilde{T})$ for all $\tilde{T} > 0$. If in addition to the above requirements on v it is assumed that

$$||v'||_{L^2(0,T)} = \left(\int_0^T v'(t)^2 dt\right)^{1/2} \leq M \tag{2.2}$$

for some constant $M > 0$, then we speak of the problem of restricted null-controllability. Let this problem be solvable for some $T > 0$. Then the infimum $T(M)$ for all times $T > 0$ for which restricted null-controllability is possible is well defined and it can be shown that there is a time-minimal control function $v_M \in H^1(0,T(M))$ with $v_M(0) = v_M(T) = 0$ such that, for $T = T(M)$, the corresponding generalized solution y of (1.1), (1.2), (1.5) with (1.11) satisfies (2.1). Moreover,

T(M) is positive unless $y_0 = y_1 = 0$ a.e. on (0,1) (see Section 3).

All these problems have been considered in [2], [4], and [5] where, in the problem of null-controllability, it was assumed that $v \in H^2(0,T)$ and $v(0) = v(T) = v'(0) = v'(T) = 0$ without requiring $r \in H(L)$. The main tool for their solution was the theory of trigonometric moment problems which can also be applied here but in a somewhat simpler fashion. In order to do this we make use of the explicit representation (1.15) of the generalized solution y of (1.1),(1.2), (1.5) with (1.11). This in conjunction with (1.14) and (1.16) shows that, for $v \in H^1(0,T)$ with $v(0) = v(T) = 0$, (2.1) is equivalent to

$$h_j^k \int_0^T v'(t)\cos\sqrt{\lambda_j}(T-t)dt = a_j(0)\cos\sqrt{\lambda_j}T + b_j(0)\sin\sqrt{\lambda_j}T,$$

$$h_j^k \int_0^T v'(t)\sin\sqrt{\lambda_j}(T-t)dt = a_j(0)\sin\sqrt{\lambda_j}T - b_j(0)\cos\sqrt{\lambda_j}T \qquad (2.3)$$

for $j \in \mathbb{N}$ and $k = 1$ or 0 with $a_j(0)$, $b_j(0)$, h_j^k being given by (1.13). By [2], Satz 7.10, all h_j^k, $j \in \mathbb{N}$, are nonzero so that (2.3) can be equivalently put into the form.

$$\int_0^T v'(t)\cos\sqrt{\lambda_j}t\,dt = c_j^1 = a_j(0)/h_j^k,$$
$$\int_0^T v'(t)\sin\sqrt{\lambda_j}t\,dt = c_j^2 = -\,b_j(0)/h_j^k \qquad (2.4)$$

for $j \in \mathbb{N}$ and $k = 1$ or 0.

If we take into account that for every $v \in H^1(0,T)$ with $v(0) = 0$ the statement $v(T) = 0$ is equivalent to

$$\int_0^T v'(t)dt = 0 \qquad (2.5)$$

then we can formulate the following result: If, for some $T > 0$, the function $u \in L^2(0,T)$ is a solution of the trigonometric moment problem

$$\int_0^T u(t)dt = 0,$$

$$\int_0^T u(t)\cos\sqrt{\lambda_j}t\,dt = c_j^1, \qquad (2.6)$$

$$\int_0^T u(t)\sin\sqrt{\lambda_j}t\,dt = c_j^2$$

for $j \in \mathbb{N}$ with c_j^1 and c_j^2 being given by (2.4), then the function

$$v(t) = \int_0^t u(s)ds \tag{2.7}$$

is in $H^1(0,T)$, satisfies $v(0) = v(T) = 0$ and solves the problem of null-controllability, i.e., the corresponding generalized solution y of (1.1), (1.2), (1.5) with (1.11) satisfies (2.1). Conversely, if some $v \in H^1(0,T)$ with $v(0) = v(T) = 0$ solves the problem of null-controllability for some $T > 0$, then $u = v'$ is in $L^2(0,T)$ and solves the trigonometric moment problem (2.6). Similarly, the problem of restricted null-controllability turns out to be equivalent to finding some $u \in L^2(0,T)$ which solves (2.6) and satisfies

$$||u||_{L^2(0,T)} \leq M \tag{2.8}$$

for some $M > 0$. The infimum $T(M)$ of all times $T > 0$ for which restricted null-controllability is possible can therefore be defined as

$$T(M) = \inf \{T > 0 \,|\, \exists u \in L^2(0,T) \text{ which satisfies (2.6) and (2.8)}\}. \tag{2.9}$$

3. SOLUTION VIA MOMENT THEORY

We make the same assumptions concerning y_o, y_1 and r as in Section 1. In addition we require that

$$\sum_{j=1}^{\infty} (c_j^1)^2 + (c_j^2)^2 < \infty \tag{3.1}$$

where c_j^1 and c_j^2 for $j \in \mathbb{N}$ are defined by (2.4). This is a further restriction on y_o and y_1, if the choice of (j_1,j_2,j_3,j_4) in (1.3) and $k = 1$ or 0 has been made. For instance, if $j_1 = j_3 = 0$, $j_2 = j_4 = 2$ and $k = 0$ (where $r(x) = \frac{1}{6} x(x^2-1)$ and hence $h_j^o = 2(-1)^j(j\pi)^{-3}$ for $j \in \mathbb{N}$), the condition (3.1) is satisfies for $y_o \in H^3(0,1)$, $y_o(0) = y_o(1) = y_o''(0) = y_o''(1) = 0$, and $y_1 \in H^1(0,1)$, $y_1(0) = y_1(1) = 0$.

In order to apply the results of [4] and [5] we replace (2.6) by the following complex system:

$$\begin{aligned} &\int_0^T u(t)dt = c_1 = 0, \\ &\int_0^T u(t)e^{2i\omega_j t}dt = c_{2j} = c_j^1 + ic_j^2, \\ &\int_0^T u(t)e^{-2i\omega_j t}dt = c_{2j+1} = c_j^1 - ic_j^2 \end{aligned} \tag{3.2}$$

for $j \in \mathbb{N}$, $i = \sqrt{-1}$, $\omega_j = \frac{1}{2}\sqrt{\lambda_j}$.

From Theorem 4.3 in [4] and the considerations following it we first obtain

Theorem 3.1. Let $T > 0$ be such that

$$\omega_j - \omega_{j-1} \geq \frac{\pi+\epsilon}{T} \quad \text{for all } j \in \mathbb{N} \tag{3.3}$$

for some $\epsilon > 0$ and $\omega_0 = 0$. Then there exists exactly one least norm solution $u \in L^2(0,T)$ of (3.2) which is real and satisfies

$$||u||_{L^2(0,T)} \leq \left(\frac{2A(\epsilon)}{T}\right)^{1/2} \left(\sum_{j=1}^{\infty} (c_j^1)^2 + (c_j^2)^2\right)^{1/2} \tag{3.4}$$

where

$$A(\epsilon) = \frac{\pi(\pi+\epsilon)^2}{2\epsilon(2\pi+\epsilon)} . \tag{3.5}$$

As an immediate consequence we have the

Corollary: There exists some $d > 0$ such that for all $T > \frac{\pi}{d}$ null-controllability is possible.

Proof: From the form (1.4) of the eigenvalues λ_j of L it follows that

$$\omega_j - \omega_{j-1} \geq d \quad \text{for all } j \in \mathbb{N}$$

for some $d > 0$ (where again $\omega_0 = 0$). For instance, for $j_1 = j_3 = 0$, $j_2 = j_4 = 2$ we can choose $d = \pi^2/2$. Therefore (3.3) is satisfied, if we put $\epsilon = dT - \pi$ (which is positive by assumption). Hence by Theorem 3.1 there exists a real solution $u \in L^2(0,T)$ which also solves (2.6) such that v, defined by (2.7), is in $H^1(0,T)$ satisfies $v(0) = v(T) = 0$ and solves the problem of null-controllability.

Another easy consequence of Theorem 3.1 is the

Theorem 3.2. Let (3.3) be satisfied for some $T > 0$, some $\epsilon > 0$, and $\omega_0 = 0$. Then, for every $M > 0$, there is some $T^* > 0$ and some real solution $u = u^* \in L^2(0,T^*)$ of (3.2) for $T = T^*$ which satisfies

$$||u^*||_{L^2(0,T^*)} \leq M. \tag{3.6}$$

Proof: If (3.3) is satisfied for some $T > 0$ and some $\epsilon > 0$, then it is also satisfied for every $T^* \geq T$ and the same ϵ. Therefore, for every $T^* \geq T$, there exists, by Theorem 3.1, a unique least norm solution $u = u^* \in L^2(0,T^*)$ of (3.2) for $T = T^*$ which is real and satisfies (3.4) for $T = T^*$. This implies

$$\lim_{T^* \to \infty} ||u^*||_{L^2(0,T^*)} = 0$$

and hence (3.6) for $T^* \geq T$ sufficiently large. As a consequence we have the following

Corollary: For every $M > 0$ restricted null-controllability is possible.

Proof: From the proof of the Corollary of Theorem 3.1 we infer the existence of some $d > 0$ such that (3.3) is satisfied for all $T > \frac{\pi}{d}$ and $\epsilon = dT - \pi$. Hence the assertion follows from Theorem 3.2 in conjunction with the final arguments in the proof of its Corollary.

If we define, for a given $M > 0$, the infimal time $T(M)$ by (2.9), then by Theorem 4.3 in [3] it follows that there exists some $u = u_M \in L^2(0,T(M))$ with $||u_M||_{L^2(0,T(M))} \leq M$ which solves (2.6) for $T = T(M)$. Moreover, $T(M)$ is positive unless all $c_j^1 = c_j^2 = 0$, $j \in \mathbb{N}$. In turn this implies the existence of a time minimal control function $v_M \in H^1(0,T(M))$ with $v_M(0) = v_M(T) = 0$ (given by $v_M(t) = \int_0^t u_M(s)ds$, $t \in [0,T(M)]$) which solves the problem of restricted null-controllability for $T = T(M)$. Moreover, $T(M)$ is positive unless $y_o = y_1 = 0$ a.e. (which is equivalent with $c_j^1 = c_j^2 = 0$ for all $j \in \mathbb{N}$). From results in [5] it can furthermore be derived that

$$||v_M'||_{L^2(0,T(M))} = M \quad \text{unless } y_o = y_1 = 0 \quad \text{a.e.,} \tag{3.7}$$

which is intuitively conceivable. The proof, however, is not simple and makes use of the fact that, for every choice of $(c_j)_{j\in\mathbb{N}} \in l_2$ and every $T > 0$, there exists a solution $u \in L^2(0,T)$ of (3.2). This can be roughly seen as follows: Let for any $x > 0$,

$\Lambda(x)$ be the number of $\omega_j'x$ with $2\omega_j < x$.

Then from (1.4) and the corresponding properties of the sequence $(\epsilon_j)_{j\in\mathbb{N}}$ we conclude, for every $x > 0$, that

$$\Lambda(x) \leq [\tfrac{1}{\pi}\sqrt{x} + \tfrac{7}{4}]$$

where $[\alpha]$ denotes the largest $k \in \mathbb{N}$ with $k \leq \alpha$. For a given $x > 0$ put

$$j = [\tfrac{1}{\pi}\sqrt{x} + \sigma - \tfrac{1}{4}].$$

Then

$$((j-\sigma) + \tfrac{\pi}{4})^2 \leq x \Rightarrow 2\omega_j = \sqrt{\lambda_j} < x,$$

hence

$$[\tfrac{1}{\pi}\sqrt{x} + \sigma + \tfrac{3}{4}] \leq \Lambda(x) \text{ for sufficiently large } x. \qquad \text{O}$$

As a result we obtain for $x > 0$ sufficiently large and $y > 0$

$$0 \leq \frac{\Lambda(x+y)-\Lambda(x)}{y} \leq \frac{\sqrt{x+y}-\sqrt{x}}{\pi y} + \frac{2+\sigma}{y} = \frac{1}{\pi\sqrt{x+y}+\sqrt{x})} + \frac{2+\sigma}{y}$$

which implies

$$\limsup_{y\to\infty} \limsup_{x\to\infty} \frac{\Lambda(x+y)-\Lambda(x)}{y} = 0.$$

Therefore, by [7], the sequence $(z_j)_{j\in\mathbb{N}\cup\{0\}}$ of functions $z_1(t) = 1$, $z_{2k}(t) = e^{2i\omega_k t}$, $z_{2k+1}(t) = e^{-2i\omega_k t}$, $t \in \mathbb{R}$, $k \geq 1$, is incomplete and, by a result in [8] minimal on every interval $[0,T]$ with $T > 0$, i.e., there exists a sequence $(x^j)_{j\in\mathbb{N}}$ in $L^2(0,T)$, with

$$\int_0^T x^j(t) z_k(t)dt = \begin{cases} 1 \text{ for } j = k, \\ 0 \text{ for } j = k. \end{cases}$$

Let $T > 0$ and $\epsilon > 0$ be given arbitrarily. Then from (1.4) it follows that there exists some $j(T,\epsilon) \in \mathbb{N}$ such that

$$\omega_j - \omega_{j-1} \geq \frac{\pi+\epsilon}{T} \text{ for all } j \geq j(T,\epsilon).$$

Similar to Theorem 3.1 one can show (see [4]) the existence of a solution $\hat{u} \in L^2(0,T)$ of (3.2) for all $j \geq j(T,\epsilon)$. By

$$u = \hat{u} + \sum_{j=1}^{j(T,\epsilon)-1} (c_j - \langle\hat{u},z_j\rangle)x^j$$

we then obtain a solution of (3.2) for all $j \in \mathbb{N}$. This is in general a complex valued function. For the special sequence $(c_j)_{j\in\mathbb{N}}$ in (3.2) we can assume it to be real because the real part of any solution of (3.2) is also a solution in this case.

REFERENCES

[1] Coddington, E.A., N. Levinson: Theory of Ordinary Differential Equations, McGraw-Hill: New York - Toronto - London 1955.

[2] Eichenauer, W.: Über trigonometrische Momentenprobleme und deren Anwendung auf gewisse Schwingungskontrollprobleme, Dissertation, Darmstadt 1982.

[3] Hajek, O., W. Krabs: On a General Method for Solving Time-Optimal Linear Control Problems, Preprint No. 579 des Fachbereichs Mathematik der TH Darmstadt, Januar 1981.

[4] Krabs, W.: On Boundary Controllability of One-Dimensional Vibrating Systems, Math.Meth. in the Appl. Sc. 1 (1979), 322-345.

[5] Krabs, W.: Optimal Control of Processes Governed by Partial Differential Equations, Part II: Vibrations, Zeitschrift für Operations Research 26 (1982), 63-86.

[6] Lions, J.L.: Optimal Control of Systems Governed by Partial Differential Equations, Springer-Verlag: Berlin - Heidelberg - New York 1971.

[7] Redheffer, R.M.: Remarks on Incompleteness of $\{e^{i\lambda_n x}\}$, Non-Averaging Sets and Entire Functions, Proc.Amer.Math.Soc. 2 (1951), 365-369.

[8] Schwartz, L.: Etude de Sommes d'Exponentielles, Herman: Paris 1959.

AN L_2 THEORY FOR THE QUADRATIC OPTIMAL COST PROBLEM OF HYPERBOLIC EQUATIONS WITH CONTROL IN THE DIRICHLET B.C. *)

I. Lasiecka and R. Triggiani

Mathematics Department
University of Florida
Gainesville, Fl. 32611, USA

0. INTRODUCTION

We take the opportunity of this Workshop to announce and proof - in a general outline - some very recent results of work still in progress [6], which give a fully L_2-theory of the quadratic optimal control problem on $[0,T]$ for boundary input (linear) hyperbolic equations of order two. Here, we shall confine ourselves to the most challenging case, which occurs when the $L_2(0,T;\ L_2(\Gamma))$ - boundary control acts through the Dirichlet B.C., as then the regularity of the corresponding solutions is the lowest, as compared to the Neumann or elastic B.C. cases. A more complete exposition will be given in the full paper [6] , which will also include the Neumann or elastic B.C. cases (where the theory is "richer"), and other related topics. The crux of the case that we study is that we penalize both the Dirichlet boundary control and the corresponding solution in the L_2-norm; i.e., in $L_2(0,T;\ L_2(\Gamma))$ and $L_2(0,T;\ L_2(\Omega))$, respectively. This is the distinguishing feature which differentiates the present results from those already existing in the literature, e.g., [1],[7],[10]. In fact, the basic difficulty encountered in the study of our problem is, of course, a question of regularity of the solutions, In face of this, one may either take smoother boundary controls (e.g. $u \in H_0^2([0,T] \times \Gamma)$ as in [7 , p. 325], or $u \in L_2(0,T;\ H^{1/2}(\Gamma))$ as in [1], [10]), or else take $L_2(0,T;\ L_2(\Gamma))$ - Dirichlet controls but penalize the corresponding solutions in a space larger and less smoother than $L_2(0,T;\ L_2(\Omega))$, typically involving $H^{-s}(\Omega)$, for some $s > 0$. It was for this reason that the regularity question was studied per se in our paper [3]: here we managed to prove that the following desired implication holds true: the map from the Dirichlet control into the solution is continuous from $L_2(0,T;\ L_2(\Gamma)) \to L_2(0,T;\ L_2(\Omega))$.

This result made it possible to study the problem in this paper. We first establish existence and uniqueness of the optimal control, and then derive a Riccati Differential equation for the feedback synthesis (pointwise in time a.e.) of

---

*) Presented at the Workshop by the first named author.

the optimal control. (Riccati's synthesis is not investigated in [7] in the hyperbolic boundary case, only in the distributed case, see p. 348). Our approach here is "explicit" in the sense that an operator is first defined in terms of the given dynamics, and only subsequently proved to be a solution of the Riccati Differential equation.

1. PROBLEM FORMULATION AND STATEMENT OF MAIN RESULTS

Let Ω be an open bounded domain in R^n with boundary Γ. Let $A(\xi,\partial)$ be a uniformly strongly elliptic operator of order two in Ω with smooth real coefficients. We consider the mixed hyperbolic problem

$$\frac{\partial^2 y}{\partial t^2}(t,\xi) = -A(\xi,\partial)y(t,\xi) \qquad \text{in } (0,T] \times \Omega \equiv Q$$

$$y(0,\xi) = y_0(\xi); \quad \frac{\partial y}{\partial t}(0,\xi) = y_1(\xi) \qquad \xi \in \Omega \tag{1.1}$$

$$y(t,\sigma) = u(t,\sigma) \qquad \text{in } (0,T] \times \Gamma \equiv \Sigma$$

where the control function $u(t,\sigma)$ acting in the Dirichlet B.C. is assumed to belong to $L_2(0,T; L_2(\Gamma))$. By Fubini Theorem, $L_2(0,T; L_2(\Gamma)) \equiv L_2(\Sigma)$ and $L_2(0,T; L_2(\Omega)) \equiv L_2(Q)$. We assume throughout that the homogeneous problem (i.e. $u \equiv 0$) is uniformly well posed in $L_2(\Omega)$; equivalently [2], that the operator $-A$ consisting of $-A(x,\partial)$ plus homogeneous Dirichlet B.C. is the generator of a strongly continuous (s.c). cosine operator $C(t)$ on $L_2(\Omega)$, $t \in R$. We next associate with (1.1) a quadratic cost functional. (The norms are all L_2-norms over the indicated domains):

$$J(u,y) \equiv \int_0^T |y(t)|_\Omega^2 + |u(t)|_\Gamma^2 dt = |y|_Q^2 + |u|_\Sigma^2$$

on $[0,T]$. The optimal control problem is now: Minimize $J(u,y(u))$ over all $u \in L_2(\Sigma)$ where $y(u)$ is the solution to (1.1) corresponding to u. (C.P.)

In order to establish existence and uniqueness of the above optimal control problem, we need regularity results for (1.1). □

Theorem 1.1. [3]. Let Ω either have C^1-boundary Γ or else be a parallelepiped. Then, for $y_0 \in L_2(\Omega)$ and $y_1 \in H^{-1}(\Omega)$, the map $u \to y(u)$ is a continuous operator $L_2(\Sigma) \to L_2(\Omega)$. □

Remark 1.1. After publishing [3], the authors were able to markedly improve the regularity result of Theorem 1.1. and prove that, in fact, the map $u \to y(u)$ is continuous as an operator $L_2(\Sigma) \to C([0,T]; L_2(\Omega))$ [5]. This sharper regularity result of the mixed problem (1.1) is not strictly needed for the optimal control problem of the present paper, although use of it would permit some simplifications in a few points of our treatement. In any case, we shall need here crucially some properties (see Lemma 2.4 below) that are also used in an essential way in the regularity proof of [5]. □

In light of Theorem 1.1, the functional $J(u,y(u))$ is continuous on $L_2(\Sigma)$; since it is strictly convex, it is weakly lower semicontinuous and by standard arguments in optimization theory, the control problem (C.P.) admits a unique solution, which we shall denote by u^o. The corresponding optimal solution is then y^o. Thus

Theorem 1.2. Under the conclusion of Theorem 1.1, the contol problem (C.P.) admits a unique solution $u^o \in L_2(\Sigma)$ and $y^o \in L_2(Q)$, actually $y^o \in C([0,T]; L_2(\Omega))$ by Remark 1.1. □

In order to formulate our main results, some preliminary background material is needed. It is well known that the operator $\mathcal{A}$ given by

$$\mathcal{A} = \begin{vmatrix} 0 & I \\ -A & 0 \end{vmatrix}, \quad \mathcal{D}(\mathcal{A}) = \mathcal{D}(-A) \times H_0^1(\Omega) \tag{1.2}$$

generates a s.c. group on $H_0^1(\Omega) \times L_2(\Omega)$. Since $H_0^1(\Omega) \equiv \mathcal{D}(A^{1/2})$ [ ](we may assume without loss of generality that the fractional powers of A are well defined), it follows quickly that $\mathcal{A}$ generates a s.c. group also on the space

$$E = L_2(\Omega) \times [\mathcal{D}(A^{1/2})]' \equiv L_2(\Omega) \times H^{-1}(\Omega) \tag{1.3}$$

which we shall denote by $e^{\mathcal{A}t}$. We next introduce the Dirichlet map D (natural "harmonic" extension of boundary data into the interior) defined by $Du = y$ where $-A(\xi,\partial)y = 0$ in Ω and $y|_\Gamma = u$ in Γ. It is a well known result of elliptic theory [8] that

$$D \text{ is a continuous operator } H^s(\Gamma) \to H^{s+1/2}(\Omega), \ s \text{ real}. \tag{1.4}$$

We can now define the (unbounded) operator $\mathcal{B}^*: E \supset \mathcal{D}(\mathcal{B}^*) \to L_2(\Gamma)$ (with dense domain $\mathcal{D}(\mathcal{B}^*) \supset L_2(\Omega) \times H^{-1/2}(\Omega)$ by

$$\mathcal{B}^* v = D^* A^{*1/2} A^{-1/2} y_2, \quad v = [v_1, v_2] \in \mathcal{D}(\mathcal{B}^*) \tag{1.5}$$

where $(Du,y)_\Omega = (u,D^*y)_\Gamma$.

Our main results on the Riccati's feedback synthesis of the optimal control are

Theorem 1.3. (i) The unique optimal control u^o of problem (C.P.) can be expressed in feedback form as

$$u^0(t) = -\mathcal{B}^*\mathcal{P}(t) \begin{vmatrix} y^0(t) \\ \dot{y}^0(t) \end{vmatrix} \quad \text{a.e. in } t \in [0,T] \tag{1.6a}$$

or equivalently

$$u^0(t) = -D^*A^*P(t) \begin{vmatrix} y^0(t) \\ \dot{y}^0(t) \end{vmatrix}. \tag{1.6b}$$

Here

$$P(t)x = \int_t^T S^*(\tau-t)\,[1,0]\,\Phi(\tau,t)x\,d\tau, \qquad x \in E \tag{1.7}$$

where $S(t)$ is the sine operator corresponding to $-A$ and $\Phi(\tau,t)$ is the evolution operator of the optimal feedback system (see (2.9) - (2.10) below).

(ii) Moreover, $\mathcal{P}(t)$ is a self-adjoint positive definite operator and satisfies the following Differential Riccati equation:

$$\begin{aligned}\frac{d}{dt}(\mathcal{P}(t)x,y)_E = -(x_1,y_1)_\Omega &- (\mathcal{P}(t)x,\mathcal{A}y)_E \\ &- (\mathcal{P}(t)\mathcal{A}x,y)_E \\ &- (\mathcal{B}^*\mathcal{P}(t)x,\mathcal{B}^*\mathcal{P}(t)y)_\Gamma\end{aligned} \tag{R.D.E.}$$

for all $x,y \in \mathcal{D}(\mathcal{A})$ and a.e. in $t \in [0,T]$ with terminal condition $\mathcal{P}(T) = 0$. □

In order to give a pointwise meaning in t (not only a.e.) to the Riccati equation, we shall derive as a corollary the following integral version:

Corollary 1.4. The Riccati operator $\mathcal{P}(t)$ of Theorem 1.3 satisfies the following integral equation:

$$\begin{aligned}(\mathcal{P}(t)x,y)_E &= \int_t^T \left(\begin{bmatrix}1 & 0\\ 0 & 0\end{bmatrix} e^{\mathcal{A}(\tau-t)}x,\ e^{\mathcal{A}(\tau-t)}y\right)_E d\tau \\ &+ \int_t^T \left(\mathcal{B}^*\mathcal{P}(\tau)e^{\mathcal{A}(\tau-t)}x,\ \mathcal{B}^*\mathcal{P}(\tau)e^{\mathcal{A}(\tau-t)}y\right)_\Gamma d\tau\end{aligned} \tag{R.I.E.}$$

for all $x,\ y \in E$ and all $t \in [0,T]$.

The next section is devoted to a general outline of the proof of Theorem 1.3 and Corollary 1.4.

2. PROOF OF THEOREM 1.3. (outline)

As an abstract version of the mixed problem (1.1), we can take the input-solution formula (see [3])

$$y(t) = y(t,t_o = 0;\ y_o,y_1) = C(t)y_o + S(t)y_1 + (Lu)(t) \tag{2.1}$$

with $C(t)$ and $S(t)x = \int_0^t C(\tau)x d\tau$ cosine and sine operators generated by $-A$, where the operator L

$$(Lu)(t) = A \int_0^t S(t-\tau)Du(\tau)d\tau \tag{2.2}$$

is, by Theorem 1.1, bounded: $L_2(\Sigma) \to L_2(Q)$ (in fact even $\to C([0,T];\ L_2(\Omega))$ by Remark 1.1). Its dual L^*: $(Lu,v)_Q = (u,L^*v)_\Sigma$ is then

$$(L^*v)(t) = \int_t^T D^*A^*S^*(\tau-t)v(\tau)d\tau, \qquad 0 \le t \le T$$

bounded $L_2(Q) \to L_2(\Sigma)$. In order to treat the optimization problem, we introduce the Lagrangean

$$L(u,y,p) \equiv \frac{1}{2} \{|u|_\Sigma^2 + |y|_Q^2\} + (p,y - C(.)y_o - S(.)y_1 - Lu)_Q .$$

The optimality conditions $L_u(u^o,y^o,p^o) = L_y(u^o,y^o,p^o) = 0$ yield

$$p^o = -y^o; \qquad u^o = L^*p^o, \quad \text{thus } u^o = - L^*y^o \tag{2.4}$$

and thus, with the help of (3.1), (note that we have $[I + L^*L]^{-1}L^* = L^*[I + LL^*]^{-1}$ and $I - L[I + L^*L]^{-1}L^* = [I + LL^*]^{-1}$), we obtain

$$\begin{aligned} u^o &= - [I + L^*L]^{-1}L^* \{C(.)y^o + S(.)y_1\} \in L_2(\Sigma) \\ y^o &= [I + LL^*]^{-1} \{C(.)y_o + S(.)y_1\} \in L_2(Q) \end{aligned} \tag{2.5}$$

where the (selfadjoint) inverse operators are well defined and bounded on $L_2(\Sigma)$ and $L_2(Q)$, respectively. Notice that (2.5) provides the optimal solution (u^o,y^o) as $L_2(0,T)$ - trajectories with values in $L_2(\Gamma)$ and $L_2(\Omega)$ respectively, in terms of the initial data. Our goal, however, is to express the optimal control u^o in "feedback form", i.e. as an operator acting pointwise in time (or a.e. in t) on the "measured" solution $[y^o(t),\dot{y}^o(t)]$ (on line, or real time implementation) as in Theorem 1.3. To accomplish this, an evolution operator, which will describe the dynamics of the feedback system, will be introduced. Let s be an arbitrary time $0 \le s < T$. Henceforth we take s as the new initial time of our optimal control problem with corresponding initial datum $y_s = [y_{os},y_{1s}] \in E$; i.e. we consider the optimal control problem of the introduction over the time interval $[s,T]$ rather than over $[0,T]$. We shall denote the corresponding optimal solution by $y^o(t,s;y_s)$ and $u^o(t,s;y_s)$. In the new notation, the optimal solution on $[0,T]$, so far denoted by $y^o(t)$ and $u^o(t)$, will be $y^o(t,0;y_o)$ and $u^o(t,0;y_o)$ respectively. The same procedure leading to (3.5), once applied to the new problem, gives then

$$u^o(t,s;y_s) = -[I_s + L_s^* L_s]^{-1} L_s^* \{C(.-s)y_{os} + S(.-s)y_{1s}\} \tag{2.6a}$$

$$y^o(t,s;y_s) = [I_s + L_s L_s^*]^{-1} \{C(.-s)y_{os} + S(.-s)y_{1s}\} \tag{2.6b}$$

as elements of $L_2(s,T; L_2(\Gamma))$ and $L_2(s,T; L_2(\Omega))$, respectively

$$(L_s u)(t) \equiv A \int_s^t S(t-\tau)Du(\tau)d\tau \qquad s \leq t \leq T \tag{2.7}$$

$$(L_s^* v)(t) \begin{cases} (L^* v)(t) & s \leq t \leq T \\ 0 & 0 \leq t \leq s \end{cases} \quad \text{a.e.} \tag{2.8}$$

In order now to obtain the sought after evolution operator $\Phi(t,s)$ defined by

$$\begin{vmatrix} y^o(t,s;y_s) \\ \dot{y}^o(t,s;y_s) \end{vmatrix} = \Phi(t,s) \begin{vmatrix} y_{os} \\ y_{1s} \end{vmatrix} \qquad 0 \leq s \leq t \leq T \tag{2.9}$$

we can insert (2.6a) into the optimal dynamics

$$\begin{aligned} y^o(t,s;y_s) &= C(t-s)y_{os} + S(t-s)y_{1s} + \{L_s[u^o(.,s;y_s)]\}(t) \\ \dot{y}^o(t,s;y_s) &= AS(t-s)y_{os} + C(t-s)_{1s} + A \int_s^t C(t-\tau)Du^o(\tau,s;y_s)d\tau \end{aligned} \tag{2.10}$$

Alternatively, we can take the time derivative of (2.6b). In either case, we arrive at the explicit expression of $\Phi(t,s)$ given by $(0 \leq s \leq t \leq T)$

$$\Phi(t,s) = \begin{vmatrix} V_s C(.-s) & V_s S(.-s) \\ 3 & 4 \end{vmatrix} \equiv \begin{vmatrix} \Phi_1(t,s) \\ \Phi_2(t,s) \end{vmatrix} \tag{2.11}$$

where $V_s \equiv [I_s + L_s L_s^*]^{-1}$. The entries 3 and 4 are explicitly available, but we do not write them out as we shall not need them directly. Now, it is plain that for fixed $s, \Phi_1(.,s)$ is a bounded operator $E \to L_2(0,T; L_2(\Omega))$ and one can show [6] that likewise $\Phi_2(.,s)$ is a bounded operator $E \to L_2(0,T; H^{-1}(\Omega))$. Moreover, since obviously $||I_s + L_s L_s^*||_{\mathcal{L}(L_2(Q))} \geq 1$ for all s it follows that

$$||V_s||_{\mathcal{L}(L_2(Q)} \leq 1 \text{ for all } s \,. \tag{2.12}$$

All this leads to the following

Lemma 2.1. (i) For each fixed s,

$\Phi(.,s)$ is a linear bounded operator $E \to L_2(0,T,E)$.

Moreover

(ii) $||\Phi(.,s)||_{L(E \to L_2(0,T,E))} \leq M$, uniformly in s. □

Let us define now the operator $P(t)$ on E by

$$P(t)x \equiv \int_t^T e^{A^*(\tau-t)} \begin{bmatrix} 1 & 0 \\ 0 & 0 \end{bmatrix} \phi(\tau,t)x d\tau, \quad x \in E$$

$$\equiv \int_t^T e^{A^*(\tau-t)} \begin{vmatrix} \Phi_1(\tau,t)x \\ 0 \end{vmatrix} d\tau . \tag{2.13a}$$

For $y \in E$, compute

$$(P(t)x,y)_E = \int_t^T (\begin{vmatrix} \Phi_1(\tau,t)x \\ 0 \end{vmatrix} , e^{A(\tau-t)}y)_E d\tau$$

where of course

$$e^{At} = \begin{vmatrix} C(t) & S(t) \\ -AS(t) & C(t) \end{vmatrix} \tag{2.14}$$

to get that, in fact

$$P(t)x = \int_t^T \begin{vmatrix} C^*(\tau-t)\Phi_1(\tau,t)x \\ A^{1/2}A^{*1/2}S^*(\tau-t)\Phi_1(\tau,t)x \end{vmatrix} d\tau . \tag{2.13b}$$

Using the definition in (1.5), we then obtain

$$B^*P(t)x = D^*A^* \int_t^T S^*(\tau-t)\Phi_1(\tau,t)x d\tau \tag{2.15}$$

which, in view of (1.7) and the two component decomposition of $\Phi(t,s)$ in (2.11), yields

$$B^*P(t)x = D^*A^*P(t)x . \tag{2.16}$$

In order to express the optimal control u^o in a (pointwise, or a.e.) feedback form as claimed in Theorem 1.3, we compute via (2.5b) and (2.8)

$$- u^o(t,s;y_s) = \{L_s^* y^o(.,s,y_s)\}(t) = \{L_s^*\Phi_1(.,s)y_s\}(t), \quad s \leq t \leq T$$

$$= \int_t^T D^*A^*S^*(\tau-t)\Phi_1(\tau,s)y_s d\tau .$$

If we now take the initial time $s = t$ with initial data $y_s = y_t = [y^o(t),\dot{y}^o(t)]$ we obtain the desired pointwise relation (via (1.7))

$$-u^o(t) = D^*A^*P(t)\begin{vmatrix} y^o(t) \\ \dot{y}^o(t)\end{vmatrix} = \mathcal{B}^*\mathcal{P}(t)\begin{vmatrix} y^o(t) \\ \dot{y}^o(t)\end{vmatrix} . \tag{2.17}$$

This completes the proof of part (i) in Theorem 1.3. Some properties of $\mathcal{P}(t)$ are collected next.

Lemma 2.2. The operator $\mathcal{P}(t)$ defined by (2.13a) or (2.13b) satisfies the following properties:

(i) for each $t \in [0,T]$, $\mathcal{P}(t)$ is a bounded linear operator on E;

(ii) for each $t \in [0,T]$, the following identity, symmetric in x and y (both in E) holds

$$\begin{aligned}(\mathcal{P}(t)x,y)_E &= \int_t^T (\Phi_1(\tau,t)x,\Phi_1(\tau,t)y)_\Omega d\tau \\ &+ \int_t^T (\mathcal{B}^*\mathcal{P}(\tau)\Phi(\tau,t)x,\mathcal{B}^*\mathcal{P}(\tau)\Phi(\tau,t)y)_\Gamma d\tau\end{aligned} \tag{2.18}$$

Thus

1) $\mathcal{P}(t) = \mathcal{P}^*(t)$ and $\mathcal{P}(t)$ is self-adjoint

2) $\mathcal{P}(t)$ is positive definite

(iii) The minimal (optimal) value of the performance index J of the optimal control problem on $[s,t]$, $s < T$, that initiates at y_s at time s is

$$J^o(u^o(.,s;y_s),y^o(.,s;y_s)) = (\mathcal{P}(s)y_s,y_s)_E . \tag{2.19}$$

Hence, for any $x \in E$, the map $t \to (\mathcal{P}(t)x,x)_E$ is monotone decreasing.

Proof. (Sketch) (i) It follows immediately from Lemma 2.1 and the known property [2] that the map $t \to A^{*1/2}S^*(t)x$, $x \in L_2(\Omega)$ is well defined and continuous via the norm

$$|y|_{H^{-1}(\Omega)} = |A^{-1/2}y|_\Omega$$

(ii) We use the identity (see (2.14), (1.6a),

$$\Phi(\tau,t)x = e^{\mathcal{A}(\tau-t)}x + \int_t^\tau \begin{vmatrix} AS(\tau-\sigma)D\mathcal{B}^*\mathcal{P}(\sigma)\Phi(\sigma,t)x \\ AC(\tau-\sigma)D\mathcal{B}^*\mathcal{P}(\sigma)\Phi(\sigma,t)x\end{vmatrix} d\sigma . \tag{2.20}$$

Then, in the definition of $\mathcal{P}(t)$, we eliminate $e^{\mathcal{A}\cdot}$ from (2.19) while keeping Φ and proceed along the same lines as e.g. in [4, Proposition 3.3, (iii)]. Then, setting $x = y$, we obtain (ii1) and (ii2), and moreover also (iii) via (2.9) and (2.17). □

We now proceed to show that $\mathcal{P}(t)$ satisfies the Riccati D.E. as claimed in part

(ii) of Theorem 1.3. To this end, it is necessary - as a preliminary step - to give a meaning to the operator $\mathcal{B}^*\mathcal{P}(t)$ appearing there in the quadratic term. This point is the major difficulty that one encounters in deriving the Riccati equation (whether differential or integral) for the problem under study. A fundamental role in this direction is played by the following

Lemma 2.3. We have that

$\mathcal{B}^*\mathcal{P}(t)$ is a bounded linear operator $E \to L_2(\Sigma)$

i.e.

$$\int_0^T |\mathcal{B}^*\mathcal{P}(t)x|_\Gamma^2 dt \leq M \qquad x \in E, \ |x|_E \leq 1 \ . \ \square$$

In order to prove Lemma 2.3 we need a result, which we have proved in [5], which may be interpreted as a trace theorem for the mixed hyperbolic problem (1.1). It should be emphasized that our proof of it, given in [5], makes crucial use of Theorem 1.1., in particular of the boundedness of L^*: $L_2(Q) \to L_2(\Sigma)$. The result is

Lemma 2.4. [5, Theorem 2.1]. The operators J_1 and J_2

$$(J_1x)(t) \equiv D^*A^*S^*(t); \quad (J_2x)(t) \equiv D^*A^{*1/2}C^*(t) \tag{2.21}$$

are bounded linear operators from $L_2(\Omega) \to L_2(\Sigma)$.$\square$

Remark 2.1. The above result admits a trace theory interpretation, which we explain first when A is selfadjoint (and thus so are C(.) and S(.)). Then, D^*A is nothing but the operator of normal derivative on Γ: $D^*A = \frac{\partial}{\partial\nu}|_\Gamma$ [ ]. Thus, the terms in (2.21), rewritten as

$$D^*AS(t)x \quad \text{and} \quad DA^*C(t)A^{-1/2}y,$$

$x \in L_2(\Omega)$, $y \in [\mathcal{D}(A^{1/2}]' = H^{-1}(\Omega)$, are the normal traces of the fundamental solutions of equation (1.1) with $y_0 = A^{-1/2}y$, $y_1 = x$ and homogeneous Dirichlet B.C. ($u(t,\sigma) \equiv 0$). A similar interpretation holds in the non self-adjoint case, this time with $D^*A^* = \frac{\partial}{\partial\nu_{A^*}}|_\Gamma$ and with respect to the fundamental solutions of the homogeneous adjoint system to (1.1). Thus, the trace theory result of our Lemma 2.4 should be contrasted with standard version of the trace theorem in the case of the solution $y(t) = y(t,y_0,y_1)$ to the hyperbolic equation (1.1) with zero Dirichlet B.C., which gives instead $y(t) \in C([0,T]; H^1(\Omega))$, and hence

$$\frac{\partial}{\partial\nu} y(t,y_0,y_1) \in C([0,T]; H^{-1/2}(\Omega)), \ y_0,y_1 \in L_2(\Omega).\square$$

By taking the dual operator to J_1 and J_2: $(J_i x,u)_\Sigma = (x,J_i^* u)_\Omega$, we obtain that

Corollary 2.5. For $u \in L_2(\Sigma)$, the integrals

$$\int_0^t AS(\tau)Du(\tau)d\tau \text{ and } \int_0^t A^{1/2}C(\tau)Du(\tau)d\tau, \qquad 0 \leq t \leq T \tag{2.22}$$

are well defined as Bochner integrals, and hence they are absolutely continuous in $t \in [0,T]$. □

With Lemma 2.4. and Corollary 2.5 at hand, we can now proceed to prove Lemma 2.3.

Proof of Lemma 2.3. The identity (cf. [9])

$$S^*(\tau-t) = S^*(\tau)C^*(t) - C^*(\tau)S^*(t) \qquad t \in R \tag{2.23}$$

permits us to rewrite (2.15) as

$$\mathcal{B}^*\mathcal{P}(t)x = (U_1x)(t) + (U_2x)(t) \tag{2.24}$$

where we have to show that the operators

$$\begin{aligned} (U_1x)(t) &= D^*A^*C^*(t) \int_t^T S^*(\tau)\Phi_1(\tau,t)xd\tau \qquad &(a)\\ (U_2x)(t) &= -D^*A^*S^*(t) \int_t^T C^*(\tau)\Phi_1(\tau,t)xd\tau \qquad &(b)\end{aligned} \tag{2.25}$$

are linear bounded $E \to L_2(\Sigma)$. To this end, we shall work with the duals U_i^*: $(U_ix,g(.))_\Sigma = (x,U_i^*g(.))_E$, i.e. say for $i = 1$

$$\begin{aligned}\int_t^T ((U_1x)(t),g(t))_\Gamma dt &= \int_t^T \left(\int_t^T S^*(\tau)\Phi_1(\tau,t)xd\tau,\ C(t)ADg(t)\right)_\Gamma dt\\ &= \int_0^T \int_t^T (\Phi_1(\tau,t)x,\ S(\tau)C(t)ADg(t))_\Gamma d\tau dt\ .\end{aligned} \tag{2.26}$$

But we know that (Theorem 1.1)

$$\Phi_1(.,t) \text{ bounded linear } E \to L_2(t,T;\ L_2(\Omega))\ .$$

Hence taking the dual

$$\int_t^T (\Phi_1(\tau,t)x,f(\tau))_\Omega d\tau = (x,\Phi_{1t}^* f(.))_E$$

with $f \in L_2(t,T;\ L_2(\Omega))$, we have that

$$\Phi_{1t}^* \text{ bounded linear operator } L_2(t,T;\ L_2(\Omega)) \to E\ .$$

Notice that by Lemma 2.1

$$||\Phi^*_{1t}||_{L(L_2(t,T;L_2(\Omega)) \to E)} \leq M, \text{ uniformly in } t . \tag{2.27}$$

We can now rewrite (2.26) as

$$\int_0^T ((U_1x)(t), g(t))_\Gamma dt = (x, \int_0^T \Phi^*_{1t} \; S(.)C(t)ADg(t)dt)_E . \tag{2.28}$$

Thus

$$U_1^*g(.) = \int_0^T \Phi^*_{1t} \; \{S(.)C(t)ADg(t)\}dt \qquad \text{(a)}$$

and similarly (2.29)

$$U_2^*g(.) = - \int_0^T \Phi^*_{1t} \; \{C(.)S(t)ADg(t)\}dt . \qquad \text{(b)}$$

Hence, the sought after dual of $\mathcal{B}^*\mathcal{P}(t)$ is $U_1^* + U_2^*$. We claim that the operators

U_1^* and U_2^* are linear bounded $E \to L_2(\Sigma)$

as desired. In fact, by (2.27)

$$|U_2^*g(.)|_E \leq M \int_0^T \text{const}_T \; |S(t)ADg(t)|_\Omega dt < \infty$$

where the integral of the $L_2(\Omega)$-norm is finite by Corollary 2.5. Similarly for U_1^*, since $|A^{1/2}S(t)| \leq \text{const}_T$ (see above (2.20), whereby

$$|U_1^*g(.)|_E \leq M \int_0^T \text{const}_T \; |A^{1/2}C(t)Dg(t)|_\Omega dt < \infty .$$

Thus U_i^* map all of E into $L_2(\Sigma)$ and by the closed graph theorem are bounded. Finally, by duality, $\mathcal{B}^*\mathcal{P}(t)$ is a bounded linear operator $E \to L_2(\Sigma)$. Lemma 2.3 is proved. □

With the help of the fundamental Lemma 2.3, we can now proceed to derive the Differential Riccati equation. In fact, we can now assert that the quadratic term in (R.D.E.), i.e. $(\mathcal{B}^*\mathcal{P}(t)x, \mathcal{B}^*\mathcal{P}(t)y)_\Gamma$ is well defined a.e. in $t \in [0,T]$ for all $x, y \in E$. The following lemma gives most of the derivation of the Differential Riccati equation.

<u>Lemma 2.6.</u> (i) For $x \in \mathcal{D}(\mathcal{A})$, $y \in E$, we have

$$(\mathcal{B}^*\mathcal{P}(t)x, \mathcal{B}^*\mathcal{P}(t)y) = \int_t^T (e^{\mathcal{A}^*(\tau-t)} \begin{vmatrix} - \dfrac{\partial\Phi_1(\tau,t)x}{\partial t} \\ 0 \end{vmatrix} - \mathcal{P}(t)\mathcal{A}x, y)_E d\tau . \tag{2.30}$$

Proof. We first define the operator $\mathcal{B}$: $L_2(\Gamma) \supset \mathcal{D}(\mathcal{B}) \to E$ by

$$(\mathcal{B}u,v)_E = (u,\mathcal{B}^*v)_\Gamma \quad u \in \mathcal{D}(\mathcal{B}), \quad v \in \mathcal{D}(\mathcal{B}^*) \tag{2.31}$$

with $\mathcal{B}^*$ given by (1.5), and then compute via (2.15) - (2.16)

$$(\mathcal{B}\mathcal{B}^*\mathcal{P}(t)x,y)_E = ([\mathcal{B}\mathcal{B}^*\mathcal{P}(t)x]_1,y_1)_\Omega + ([\mathcal{B}\mathcal{B}^*\mathcal{P}(t)x]_2,y_2)_{H^{-1}(\Omega)}$$

$$= (\mathcal{B}^*\mathcal{P}(t)x,\mathcal{B}^*y)_\Gamma$$

(by (1.5))

$$= (\mathcal{B}^*\mathcal{P}(t)x,D^*A^{*1/2}A^{-1/2}y_2)_\Gamma$$

$$= (A^{1/2}D\mathcal{B}^*\mathcal{P}(t)x,A^{-1/2}y_2)_\Omega$$

$$= (AD\mathcal{B}^*\mathcal{P}(t)x,y_2)_{H^{-1}(\Omega)} \quad .$$

Thus

$$([\mathcal{B}\mathcal{B}^*\mathcal{P}(t)x]_1,y_1)_\Omega = 0 \qquad y_1 \in L_2(\Omega) \tag{2.31a}$$

$$[\mathcal{B}\mathcal{B}^*\mathcal{P}(t)x]_2 = AD\mathcal{B}^*\mathcal{P}(t)x \in H^{-1}(\Omega) \ . \tag{2.31b}$$

Next from (2.6b) and (2.9) we get

$$\Phi_1(\tau,t)x = [I_t + L_tL_t^*]^{-1} \{C(.-t)x_1 + S(.-t)x_2\} \tag{2.32a}$$

i.e.

$$\Phi_1(\tau,t)x + L_tL_t^*\Phi_1(\tau,t)x = C(.-t)x_1 + S(.-t)x_2 \ . \tag{2.32b}$$

Thus, by (2.32a)

$$\int_t^T (\begin{vmatrix}\phi_1(\tau,t)\mathcal{B}\mathcal{B}^*\mathcal{P}(t)x \\ 0\end{vmatrix}, e^{A(\tau-t)}y)_E \, d\tau$$

$$= \int_t^T ([I_t + L_tL_t^*]^{-1} \{C(.-t)[\mathcal{B}\mathcal{B}^*\mathcal{P}(t)x]_1 + S(.-t)[\mathcal{B}\mathcal{B}^*\mathcal{P}(t)x]_2\}, [e^{A(\tau-t)}y]_1)_\Omega dt$$

$$= \int_t^T (\text{zero} + [I_t + L_tL_t^*]^{-1}AS(.-t)D\mathcal{B}^*\mathcal{P}(t)x, \ [e^{A(\tau-t)}y]_1)_\Omega d\tau \tag{2.33}$$

where in the last step we have used (2.31).

Now, by selfadjointness of $P(t)$ (Lemma 2.2 (ii1)) and by (2.13a), we get

$$(\mathcal{B}^*P(t)x,\ \mathcal{B}^*P(t)y)_\Gamma = (P(t)\mathcal{B}\mathcal{B}^*P(t)x,y)_E$$
$$= \int_t^T (e^{A^*(\tau-t)} \begin{vmatrix} \Phi_1(\tau,t)\mathcal{B}\mathcal{B}^*P(t)x \\ 0 \end{vmatrix}, y)_E d\tau \quad . \tag{2.34}$$

Now from (2.7) - (2.8), (2.3) we have

$$\int_0^T (\frac{\partial}{\partial t}\{L_tL_t^*\Phi_1(\tau,t)x\}(\tau), [e^{A(\tau-t)}y]_1 d\tau$$
$$= \int_t^T (\frac{\partial}{\partial t}\int_t^\tau AS(\tau-\sigma)D[L^*\Phi_1(.,\tau)x](\sigma)d\sigma,\ [e^{A(\tau-t)}y]_1)_\Omega d\tau$$
$$= \int_t^T (-AS(\tau-t)D[L^*\Phi_1(.,\tau)x](t) + L_tL_t^*\frac{\partial\Phi_1}{\partial t}(.,t)x,\ [e^{A(\tau-t)}y]_1)_\Omega d\tau \tag{2.35}$$
$$= \int_t^T (-AS(\tau-t)D\mathcal{B}^*P(t)x + L_tL_t^*\frac{\partial\Phi_1}{\partial t}(.,t)x,\ [e^{A(\tau-t)}y]_1)_\Omega d\tau$$

where in the last step we have used (2.15). Hence differentiating (2.32b) in t and using (2.35) yields

$$\int_t^T (\frac{\partial\Phi_1}{\partial t}(\tau,t)x,\ [e^{A(\tau-t)}y]_1)_\Omega d\tau$$
$$= \int_t^T ([I_t + L_tL_t^*]^{-1}\{S(.-t)Ax_1 - C(.-t)x_2\} + [I_t + L_tL_t^*]^{-1}\{-AS(.-t)D\mathcal{B}^*P(t)x\},$$
$$[e^{A(\tau-t)}y]_1)_\Omega d\tau \ . \tag{2.36}$$

Thus, by (2.33) and (2.36)

$$\int_t^T (\begin{vmatrix} \Phi_1(\tau-t)\mathcal{B}\mathcal{B}^*P(t)x \\ 0 \end{vmatrix},\ e^{A(\tau-t)}y)_E d\tau$$
$$= \int_t^T ([I_t + L_tL_t^*]^{-1}\{S(.-t)Ax_1 - C(.-t)x_2\} - \frac{\partial\Phi_1}{\partial t}(\tau-t)x,\ [e^{A(\tau-t)}y]_1)_\Omega d\tau$$
$$= -\int_t^T (\begin{vmatrix} \Phi_1(\tau,t)Ax + \frac{\partial\Phi_1}{\partial t}(\tau,t)x \\ 0 \end{vmatrix},\ e^{A(\tau-t)}y)_E d\tau \tag{2.37}$$

by (2.32a) in the last step. Thus (2.34) and (2.37) yield the desired conclusion via (2.13a). □

Derivation of the Differential Riccati equation is now immediate: from (2.13a) we compute

$$(\mathcal{P}(t)x,y)_E = \int_t^T (\begin{vmatrix} \Phi_1(\tau,t)x \\ 0 \end{vmatrix}, e^{A(\tau-t)}y)_E d\tau$$

and hence for $y \in \mathcal{D}(A)$

$$\frac{d}{dt}(\mathcal{P}(t)x,y)_E = -(\begin{vmatrix} \Phi_1(t,t)x \\ 0 \end{vmatrix}, y)_E + \int_t^T (\begin{vmatrix} \frac{\partial \Phi_1(\tau,t)x}{\partial t} \\ 0 \end{vmatrix}, e^{A(\tau-t)}y)_E d\tau$$

$$- (\mathcal{P}(t)x,y)_E \text{ .}$$

Since $\Phi_1(t,t)x \equiv x_1$ ( ), invoking Lemma 2.6 (ii) gives the sought after D.R.E.

$$\frac{d}{dt}(\mathcal{P}(t)x,y)_E = -(x_1,y_1)_\Omega - (\mathcal{P}(t)x,Ay)_E$$

$$- (\mathcal{P}(t)Ax,y)_E - (\mathcal{B}^*\mathcal{P}(t)x, \mathcal{B}^*\mathcal{P}(t)y)_\Gamma$$

Theorem 1.3 is thus fully proved. □

We now sketch a proof of Corollary 1.4. We need the following

Lemma 2.7. We have that

$\mathcal{B}^*\mathcal{P}(t)e^{At}$ is a bounded linear operator $E \to L_2(\Sigma)$. □

Proof. The proof is similar to that of the fundamental Lemma 2.3, and thus will only be sketched. By (2.15) and (2.14) we have explicitly for $x \in E$:

$$\mathcal{B}^*\mathcal{P}(t)e^{At}x = D^*A^* \int_t^T S^*(\tau-t)\Phi_1(\tau,t) \begin{vmatrix} C(t)x_1 + S(t)x_2 \\ AS(t)x_1 + C(t)x_2 \end{vmatrix} d\tau \text{ .}$$

The idea, as in Lemma 2.3, is to consider the operator $\mathcal{O}$: $L_2(\Sigma) \to E$, which is dual to $\mathcal{B}^*\mathcal{P}(t)e^{At}$ and show that this dual is bounded. The dual operator $\mathcal{O}$ is now (compare with (2.29))

$$\mathcal{O}g(.) = \int_0^T e^{A^*t}\Phi^*_{1t}\{S(.)C(t)ADg(t) - C(.)S(t)ADg(t)\}dt$$

where Φ^*_{1t} is defined above (2.27), and $\mathcal{O}$ is bounded $L_2(\Sigma) \to E$ as in the proof of Lemma 2.3. □

It is now standard how to complete the proof of Corollary 1.4, i.e. how to go from the D.R.E. to the (second) Riccati Integral Equation R.I.E.

REFERENCES

[1] Curtain, R., A. Pritchard: An abstract theory for unbounded control action for distributed parameter systems, SIAM J. Control & Opt. 15 (1977), 566-611. Also their book: Infinite dimensional linear systems theory, Lecture Notes in Control 8, Springer-Verlag 1978.

[2] Fattorini, H.O.: Ordinary Differential equations in linear topological spaces, I and II, J. Diff. Eqs. 5 (1968), 72-105, and 6 (1969), 537-565.

[3] Lasiecka, I., R. Triggiani: A cosine operator approach to modelling $L_2(0,T;L_2(\Gamma))$-boundary input hyperbolic equations, Applied Math. & Optim. 7 (1981), 35-93.

[4] Lasiecka, I., R. Triggiani: Dirichlet boundary control problem for parabolic equations with quadratic cost: analyticity and Riccati's feedback sythesis, SIAM J. Control & Optimiz. 21 (1983)

[5] Lasiecka, I., R. Triggiani, Regularity of hyperbolic equations under $L_2(0,T;L_2(\Gamma))$-Dirichlet boundary terms, submitted

[6] Lasiecka, I., R. Triggiani: The quadratic cost problem for $L_2(0,T;L_2(\Gamma))$-boundary input hyperbolic equations, in progress.

[7] Lions, J.L.: Optimal Control of systems governed by partial differential equations, Springer-Verlag 1971.

[8] Necas, J.: Les methodes directes en theories des equations elliptiques, Masson et cie, Eds.,1967.

[9] Travis, C.C., G.F. Webb: Second order differential equations in Banach spaces, in Nonlinear Equations in Abstract spaces, edited by V. Lakshmikantham, Academic Press 1978, 331-361.

[10] Vinter, R., T. Johnson: Optimal control of nonsymmetric hyperbolic systems in n-variables on the half-space, SIAM J. Control 15 (1977), 129-143.

ON THE IDENTIFIABILITY OF PARAMETERS IN DISTRIBUTED SYSTEMS

S. Nakagiri

Department of Applied Mathematics
Faculty of Engineering
Kobe University
Rokkodai, Nada, Kobe 657, Japan

1. INTRODUCTION

We will explain the identifiability problem. To identify unknown parameters in a system, we usually use the following model reference method:

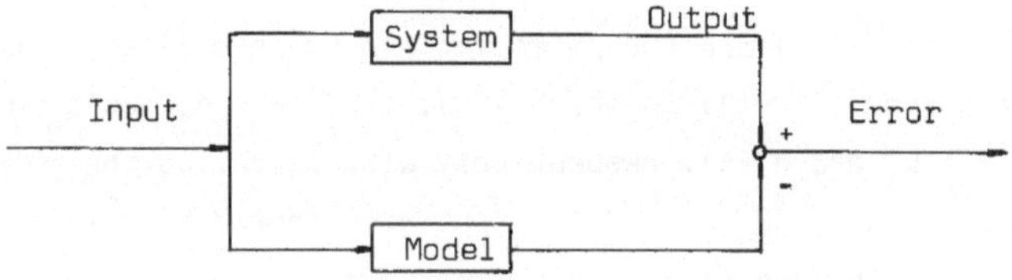

In this case, one problem is whether the model parameters coincide with those of the system or not when the error between the system and model outputs becomes zero. This is the socalled identifiability problem, one of the inverse problems. The origin of the problem is old and is of considerable importance in the field of control engineering (see [3,8,9,11]) and the references cited therein). However we think that the full development of the problem has not been achieved until now.

Recently this problem has attracted much interest and there have appeared a number of papers which deal the problem in distributed systems [4,9,10-13,15-17]. The purpose of this paper is to give a short survey of those recent developments in the identifiability theory.

2. IDENTIFIABILITY OF ONE DIMENSIONAL PARABOLIC DISTRIBUTED SYSTEMS

Consider the parabolic distributed system described by the following partial differential equation

$$\frac{\partial u}{\partial t} = \frac{\partial}{\partial x}\left(a(x)\frac{\partial u}{\partial x}\right) + b(x)u, \quad x \in (0,1),\ t > 0, \tag{1}$$

with the boundary and initial conditions

$$\alpha_0 u(0,t) - (1-\alpha_0)\frac{\partial u}{\partial x}(0,t) = h_0(t), \quad t > 0,$$
$$\alpha_1 u(1,t) + (1-\alpha_1)\frac{\partial u}{\partial x}(1,t) = h_1(t), \quad t > 0, \tag{2}$$

$$u(x,0) = u_0(x), \quad x \in (0,1). \tag{3}$$

Here we assume that $a(x) > 0$, $x \in [0,1]$, $a \in C^1[0,1]$, $b \in C[0,1]$, $0 \leq \alpha_0, \alpha_1 \leq 1$, $h_0, h_1 \in C(R^+)$ and $u_0 \in L_2[0,1]$.

The output y of the measurement system is given by

$$y(x_p,t) = Cu(x,t), \quad x_p \in \Omega_p \subset [0,1], \quad t > 0, \tag{4}$$

where C is a measurement operator and Ω_p denotes a set in which the output y is defined.

$a, b, \alpha_0, \alpha_1, u_0$ in (1) - (3) are the parameters to be identified by the measurement (4). The system (1) - (3) in which $a(x)$, $b(x)$, α_0, α_1 and u_0 are replaced by $a^m(x) > 0$, $b^m(x)$, α_0^m, α_1^m and $u_0^m(x)$, respectively will be called the model system. The corresponding model state and the model output are denoted by $u^m(x,t)$, $y^m(x_p,t)$, respectively. All quantities subscripted by m are known.

<u>Definition 1.</u> The parameters a, b, α_0, α_1 and/or u_0 are said to be identifiable if $a(x) = a^m(x)$, $x \in [0,1]$, $b(x) = b^m(x)$, $x \in [0,1]$, $\alpha_0 = \alpha_0^m$, $\alpha_1 = \alpha_1^m$ and/or $u_0(x) = u_0^m(x)$ a.e. $x \in [0,1]$ follow from the relation

$$\varepsilon(x_p,t) = y(x_p,t) - y^m(x_p,t) = 0, \quad x_p \in \Omega_p, \ t > 0. \tag{5}$$

Let A denote a realization in $L_2[0,1]$ of Sturm-Liouville's operator $\frac{\partial}{\partial x}(a(x)\frac{\partial}{\partial x}) + b(x)$ with the homogeneous boundary conditions corresponding to (2). The realization of the model is denoted by A^m.

As is well known there exist two sets of eigenvalues and eigenfunctions $\{\lambda_n, \psi_n : n = 1,2,\ldots\}$ of A and $\{\lambda_n^m, \psi_n^m : n = 1,2,\ldots\}$ of A^m. Note that λ_n and λ_n^m are simple. Both $\{\psi_n\}$ and $\{\psi_n^m\}$ constitute complete orthonormal system in $L_2[0,1]$, however, the set $\{\lambda_n, \psi_n\}$ is known and $\{\lambda_n^m, \psi_n^m\}$ is unknown.

<u>Theorem 1.</u> Let $\Omega_p = [0,1]$ and $h_0 = h_1 \equiv 0$ in (2). Then all parameters a, b, α_0, α_1 and u_0 are identifiable if and only if u_0^m satisfies

$$\langle u_0^m, \psi_n^m \rangle = \int_0^1 u_0^m(x)\psi_n^m(x)dx \neq 0, \quad n = 1,2,\ldots \,. \tag{6}$$

<u>Proof:</u> Since $\Omega_p = [0,1]$, the zero output error implies

$$\sum_{n=1}^{\infty} e^{\lambda_n t} \langle u_o, \psi_n \rangle \psi_n(x) = \sum_{n=1}^{\infty} e^{\lambda_n^m t} \langle u_o^m, \psi_n^m \rangle \psi_n^m(x) \quad \text{for all } x \in [0,1] \text{ and } t > 0. \tag{7}$$

By (6), (7) and the completeness of eigenfunctions, we have $\{\lambda_n, \psi_n\} = \{\lambda_n^m, \psi_n^m\}$ so that $A = A^m$. From this the identifiability of parameters a, b, α_o, α_1 follows. Using (7) again, we obtain the identifiability of u_o. The proof of necessity is not difficult.

Applying the same method as above, we have the following result.

Theorem 2. Let $\Omega_p = [0,1]$, $u_o = u_o^m \equiv 0$ and let h_o, h_1 satisfy

(i) $h_o \equiv 0$ and $h_1 \not\equiv 0$ or

(ii) $h_1 \equiv 0$ and $h_o \not\equiv 0$.

Then the parameters a, b, α_o and α_1 are identifiable.

In Theorem 1 and Theorem 2 we assume that $\Omega_p = [0,1]$. But this assumption is not practical in application. Usually the measurement is possible at a finite number of points in the domain. So, we study the identifiability in the case of pointwise measurement.

We first consider one point measurement, i.e., the case where the output y is given by

$$y(t) = Cu(x,t) = u(x_p,t), \quad x_p \in [0,1], \ t > 0. \tag{8}$$

In what follows we assume that a, α_o, α_1 and u_o are unknown but it is known that $a(x) \equiv 1$ for all $x \in [0,1]$ (normal type), $b \in C^1[0,1]$ and $0 \le \alpha_o, \alpha_1 < 1$ (not Dirichlet type).

Differently from the above theorems it is assumed that the eigenfunctions ψ_n, ψ_n^m are normalized so that

$$\psi_n(0) = \psi_o^m(0) = 1 \quad (\text{since } \alpha_o, \alpha_1, \alpha_o^m, \alpha_1^m \ne 1).$$

The following theorem is an improvement of Pierce's result [13].

Theorem 3. Let $x_p = 0$ or 1 in (8), $u_o = u_o^m \equiv 0$ and let h_o, h_1 satisfy

(i) $h_o \equiv 0$ and $h_1 \not\equiv 0$ or

(ii) $h_1 \equiv 0$ and $h_o \not\equiv 0$.

Then the parameters b, α_o and α_1 are identifiable.

Proof: Let $x_p = 0$ and (ii) be satisfied. Then the zero output error implies that

$$\int_0^t \Big(\sum_{n=1}^{\infty} r_n e^{\lambda_n(t-s)} \psi_n(0) - \sum_{n=1}^{\infty} r_n^m e^{\lambda_n^m(t-s)} \psi_n^m(0) \Big) h_o(s)ds = 0, \ t > 0 \tag{9}$$

where $r_n = 1/||\psi_n||^2$, $r_n^m = 1/||\psi_n^m||^2$. Then by Titchmarsh's theorem on convolutions, we have from (9) that

$$\sum_{n=1}^{\infty} r_n e^{\lambda_n t} = \sum_{n=1}^{\infty} r_n^m e^{\lambda_n^m t}, \quad t > 0. \tag{10}$$

Notice that the series in (10) are locally integrable on R^+. By the uniqueness of Dirichlet series it follows from (10) that

$$\lambda_n = \lambda_n^m, \quad ||\psi_n|| = ||\psi_n^m||, \quad n = 1,2,\dots .$$

Since the spectral function for A with $a \equiv 1$ is given by

$$\sigma(\lambda) = \sum_{\lambda_n \le \lambda} 1/||\psi_n||^2,$$

both spectral functions for A and A^m coincide. Hence by the Gel'fand-Levitan theory [7] we have $A = A^m$, which means the identifiability of parameters b, α_0 and α_1.

Let the boundary inputs h_0, h_1 be zero. Replacing the measurement (8) by the boundary measurement

$$y(t) = \{u(0,t),u(1,t)\}, \quad t > 0, \tag{11}$$

we obtain the following identifiability result including the initial value.

Theorem 4. Let $h_0 = h_1 \equiv 0$ and let the measurement be given by (11). Then the parameters b, α_0, α_1 and u_0 are identifiable if and only if u_0^m satisfies (6).

The above theorem can be proved by the Gel'fand-Levitan theory as in [13]. In [17] a proof was given without using the Gel'fand-Levitan theory directly.

If the condition (6) is not satisfied, parameters can not be determined uniquely. In this case the problem of determining the degree of the freedom of parameters is considered.

Define a set $E^m \subset \sigma(A^m)$ by $E^m = \{\lambda_n^m : \langle u_0^m, \psi_n^m \rangle = 0\}$. If $\# E^m$ is finite, this problem can be reduced to solve the second order ordinary differential equation.

Let $\# E^m = N$ and $E^m = \{\lambda_{n_1}^m, \dots, \lambda_{n_N}^m\}$. We denote by ., I, Λ^m and $\Theta^m = \Theta^m(x)$ the inner product in R^N, the unit matrix in R^N_N, the diagonal matrix

$$\begin{pmatrix} \lambda_{n_1}^m & & 0 \\ & \ddots & \\ 0 & & \lambda_{n_N}^m \end{pmatrix}$$ and the N-vector function ${}^t(\psi_{n_1}^m(x), \dots, \psi_{n_N}^m(x))$,

respectively. The next theorem was proved by Suzuki [16].

Theorem 5. Let the assumption in Theorem 4 be satisfied and let $\# E^m = N < \infty$. If the output error by the measurement (11) vanishes, then the parameters b, α_o and α_1 satisfy

$$b = b^m + 2\frac{d}{dx}(G^m.\Theta^m)$$

$$\alpha_o = \alpha_o^m + (G^m.\Theta^m)(0) - [\alpha_o^m + (G^m.\Theta^m)(0)]$$

$$\alpha_1 = \alpha_1^m - (G^m.\Theta^m)(1) - [\alpha_1^m - (G^m.\Theta^m)(1)]$$

for some $G^m \in C^2([0,1];\ R^N)$ which is a solution of

$$\frac{d^2}{dx^2}G^m = \{(2\frac{d}{dx}(G^m.\Theta^m) + b^m)I - \Lambda^m\}G^m, \tag{12}$$

where [ ] denotes the Gauss symbol. Moreover the correspondence between (b, α_o, α_1) and G^m is one to one.

In Theorem 5 the initial value u_o can also be represented in terms of b^m, α_o^m, α_1^m, u_o^m and G^m, but since this is too complicated we will not give it here (see [16]).

The Dirichlet boundary condition is removed in Theorem 3 - 5. In this case Seidman [15] proved the following theorem by using the Borg theory [2].

Theorem 6. Let $h_o = h_1 \equiv 0$, $\alpha_o = \alpha_1 = \alpha_o^m = \alpha_1^m = 1$ and let the measurement be given by

$$y(t) = \frac{\partial u}{\partial x}(0,t), \quad t > 0.$$

If b is symmetric with respect to $\frac{1}{2}$ and u_o^m satisfies (6), then b is identifiable.

We return to the system (1) - (3). If $b \equiv 0$, $\alpha_o = \alpha_1 = 0$ (Neumann type), the first eigenvalue $\lambda_1 = 0$ and the corresponding eigenfunction $\psi_1(x) \equiv 1$. In [10] Murayama has shown the following result under the assumption that $a \in C^3[0,1]$.

Theorem 7. Let $b = b^m \equiv 0$, $\alpha_o = \alpha_1 = \alpha_o^m = \alpha_1^m = 0$, $h_o = h_1 \equiv 0$ and let the measurement be given by (11). If u_o^m satisfies

$$\langle u_o^m, \psi_n^m \rangle \neq 0, \quad n = 2,3,\ldots \quad \text{and}$$

$$\int_0^1 dx/\sqrt{a(x)} = \int_0^1 dx/\sqrt{a^m(x)},$$

then a and u_o are identifiable.

This theorem follows from the Liouville transformation and the Gel'fand-Levitan theory.

Concluding this section we note that all theorems can be extended to hyperbolic distributed systems [11].

3. IDENTIFIABILITY OF ABSTRACT SYSTEMS

In this section we study the identifiability within the framework of linear operator theory. The dynamics of a system and its model are described by the following evolution equations in a Hilbert space H:

$$\frac{du(t)}{dt} = Au(t), \qquad \frac{du^m(t)}{dt} = A^m u^m(t). \tag{13}$$

The observation space Y is also a Hilbert space and the measurements are given by

$$y(t) = Bu(t) \in Y, \qquad y^m(t) = Bu^m(t) \in Y, \qquad t > 0. \tag{14}$$

Here we assume that A and A^m generate strongly continuous semi-groups T(t) and $T^m(t)$, respectively, B is a bounded linear operator from $Z \subset H$ into Y with some Banach space Z. For the equations (13) the initial values

$$u(0) = u_o, \qquad u^m(0) = u_o^m \tag{15}$$

are considered to be inputs. Now the identifiability problem for the abstract systems (13), (15) can be stated as follows:

Under what condition do $A = A^m$ and/or $u_o = u_o^m$ follow from the zero output error

$$\varepsilon(t) = y(t) - y^m(t) = 0 \quad \text{in} \quad Y, \quad t > 0 \ ? \tag{16}$$

As usually we denote by (B,A) and (B,A^m) the sets of dynamical equations in (13), (14).

Definition 2. The pair (B,A) is said to be initially observable if

$$\text{Range } T(t) \subset Z \quad \text{for all} \quad t > 0 \quad \text{and} \quad \bigcap_{t>0} \text{Ker } BT(t) = \{0\}.$$

We also assume that A and A^m are self-adjoint and their resolvents are compact. Then the spectrum of A is a point spectrum and there exists a set of eigenvalues and eigenfunctions $\{\lambda_n, \psi_{nj}: j = 1,\dots,m_n,\ n = 1,2,\dots\}$ $(\lambda_1 > \lambda_2 > \dots)$ of A (cf. [1], [18]).

Let P_n be the n-th eigenprojector corresponding to λ_n, i.e.,

$$P_n u = \sum_{j=1}^{m_n} \langle u, \psi_{nj} \rangle \psi_{nj}, \qquad u \in H,$$

where <,> is the inner product of H. The semi-group T(t) generated by A is analytic and is given by

$$T(t)u = \sum_{n=1}^{\infty} e^{\lambda_n t} P_n u, \qquad u \in H. \tag{17}$$

We denote by $\{\lambda_n^m, \psi_{nj}: j = 1,\ldots,M_n,\ n = 1,2,\ldots\}$ $(\lambda_1^m > \lambda_2^m > \ldots)$ and P_n^m the set of eigenvalues and eigenfunctions of A^m and the n-th eigenprojector, respectively.

Definition 3. The eigenvalues $\{\lambda_n\}$ of A are said to be identifiable if (16) implies $\lambda_n = \lambda_n^m$ for all $n = 1,2,\ldots$.

Definition 4. The initial value u_o (resp. u_o^m) is said to be strongly exciting if $P_n u_o \neq 0$ (resp. $P_n^m u_o^m \neq 0$) in H for all $n = 1,2,\ldots$.

Theorem 8. Let $Z = H$. If u_o and u_o^m satisfy

$$BP_n u_o \neq 0, \quad BP_n^m u_o^m \neq 0 \quad \text{in } Y, \quad n = 1,2,\ldots ,$$

then the eigenvalues $\{\lambda_n\}$ of A are identifiable.

Theorem 8 requires some information of unknown functions. The condition $BP_n u_o \neq 0$ is true, however, the coincidence of all eigenvalues does not follow if the condition is not satisfied. The next corollary is immediate and is more useful than Theorem 8.

Corollary 9. Let $Z = H$. If u_o and u_o^m are strongly exciting and (B,A), (B,A^m) are initially observable, then the eigenvalues $\{\lambda_n\}$ of A are identifiable.

We next consider the identifiability of constant parameters in operators. In the systems (13), (15), it is assumed that

$$u_o = u_o^m \text{ (known!)} \quad \text{and } A = aA_o + b, \quad A^m = a^m A_o + b^m. \tag{18}$$

Here a, $a^m > 0$ and b, b^m are real constants and A_o is an a priori known operator.

Definition 5. The constant parameters a, b in A are said to be identifiable if (16) implies $a = a^m$ and $b = b^m$.

See [11] for another definition of the identifiability of a, b without using the model.

Since A is self-adjoint with compact resolvent, A_o is also. We write the set of eigenvalues and eigenfunctions of A_o and the n-th eigenprojector by $\{\mu_n, \Phi_{nj}\}$ $(\mu_1 > \mu_2 > \ldots)$ and P_n^o, respectively. By (18) we have $\{\lambda_n, \psi_{nj}\} = \{a\mu_n + b, \Phi_{nj}\}$ and $\{\lambda_n^m, \psi_{nj}\} = \{a^m\mu_n + b^m, \Phi_{nj}\}$. Then it is verified by (17) that

$$u(t),\ u^m(t) \in \bigcap_{n=0}^{\infty} D(A_o^n) \quad \text{for all } t > 0.$$

Let k be a natural number and $Z = D(A_o^k)$. We know that Z is a Banach space with the graph norm $||\cdot||_{D(A_o^k)}$:

$$||x||_{D(A_o^k)} = ||x|| + ||A_o^k x|| \quad \text{for } x \in D(A_o^k).$$

Theorem 10. Let $Z = D(A_o^k)$. Then the constant parameters a, b in A are identifiable if and only if there exist two distinct n_1 and n_2 such that $BP_{n_1}^o u_o \neq 0$ and $BP_{n_2}^o u_o \neq 0$ in Y.

Corollary 11. Let $Z = D(A_o^k)$. If there exist n_1, n_2 $(n_1 \neq n_2)$ such that $P_{n_1}^o u_o \neq 0$ and $P_{n_2}^o u_o \neq 0$ and (B, A_o) is initially observable, then the parameters a, b in A are identifiable.

Theorem 10 can be applied to the identifiability of a, b in the operator $a\Delta + b$ (Δ is the Laplace operator) defined on a bounded domain $\Omega \subset R^N$ from the observation by pointwise measurement $y(t) = u(x_p, t)$, $x_p \in \bar{\Omega}$. For related results on the constant parameter identifiability, we refer to [4,9,11,12].

Finally we consider the identifiability of the operator A itself. To solve it we assume that B = I, the identity operator on H (H = Y). Since the number of initial values has a close relation to the identifiability of A, we give the following definition.

Definition 6. The operator A and the set of initial values $\{u_{0,1}, \ldots, u_{0,k}\}$ are said to be identifiable if

$$A = A^m \quad \text{and} \quad u_{0,1} = u_{0,1}^m, \ldots, u_{0,k} = u_{0,k}^m \quad \text{in } H$$

follow from

$$\varepsilon_1(t) = u_1(t) - u_1^m(t) = 0, \ldots, \varepsilon_k(t) = u_k(t) - u_k^m(t) = 0 \quad \text{in } H,\ t > 0,$$

where $u_i(t)$ and $u_i^m(t)$ $(i = 1, \ldots, k)$ are the (mild) solutions of the system and the model corresponding to the initial values $u_{0,i}$ and $u_{0,i}^m$, respectively.

Theorem 12. If the set of model initial values $\{u_{0,1}^m, \ldots, u_{0,k}^m\}$ satisfies

$$\text{rank} \begin{pmatrix} \langle u_{0,1}^m, \Psi_{n1} \rangle & \cdots\cdots & \langle u_{0,1}^m, \Psi_{nM_n} \rangle \\ \langle u_{0,2}^m, \Psi_{n1} \rangle & \cdots\cdots & \langle u_{0,2}^m, \Psi_{nM_n} \rangle \\ \vdots & & \vdots \\ \langle u_{0,k}^m, \Psi_{n1} \rangle & \cdots\cdots & \langle u_{0,k}^m, \Psi_{nM_n} \rangle \end{pmatrix} = M_n, \quad n = 1,2,\ldots \tag{19}$$

then the operator A and the set $\{u_{0,1}, \ldots, u_{0,k}\}$ are identifiable.

The condition (19) can be identified with the rank condition for controllability or observability in the following sense (see [5,6,14,18]).

The set $E^M = \{u_{0,1}^m, \ldots, u_{0,k}^m\}$ satisfies (19) if and only if

(i) (E^m, A^m) is initially observable, where the set E^m is identified with the operator $\tilde{E}^m : H \to R^k$ given by

$$\tilde{E}^m u = (\langle u_{0,1}^m, u\rangle, \ldots, \langle u_{0,k}^m, u\rangle) \in R^k,$$

or

(ii) (E^m, A^m) is approximately controllable, where the set E^m is identified with the operator $\underset{\sim}{E}^m : R^k \to H$ given by

$$\underset{\sim}{E}^m(x_1, \ldots, x_k) = \sum_{i=1}^{k} x_i u_{o,i}^m \in H.$$

As an application of Theorem 12, we can treat the identifiability of coefficients of higher order elliptic partial differential operators and that of a potential in the Schrödinger operator [11].

Acknowledgements: I wish to express my gratitude to Professor R. Curtain for her helpful comments and suggestions on section 3.

REFERENCES

[1] Balakrishnan, A.V.: Applied Functional Analysis (Second edition), Springer-Verlag 1981.

[2] Borg, G.: Eine Umkehrung der Sturm-Liouvilleschen Eigenwertaufgabe, Acta Math. 78 (1946), 1-96.

[3] Chavant, G.: Analyse fonctionelle et identification de coefficient répartis dans les équation aux derivées partielles, Thesis, Paris 1971.

[4] Courdesses, M., M.G. Polis, M. Amouroux: On the identifiability of parameters in a class of parabolic distributed systems, IEEE Trans. Automatic Control 1981.

[5] Curtain, R.F., A.J. Pritchard: Infinite Dimensional Linear System Theory, Lecture Notes in Control and Information Sciences 8, Springer-Verlag 1978.

[6] Fattorini, H.O.: On complete controllability of linear systems, J.Diff.Eq. 3 (1967), 391-402.

[7] Gel'fand, M., B.M. Levitan: On the determination of a differential equation from its spectral function, A.M.S. Transl.,(2) 1 (1955), 253-304.

[8] Goodson, R.E., M.G. Polis, Parameter identification in distributed systems, Proc. IEEE, 64 (1976), 45-61.

[9] Kitamura S., S. Nakagiri: Identifiability of spatially-varying and constant parameters in distributed systems of parabolic type, SIAM J. Control & Optimiz. 15 (1977), 785-802.

[10] Murayama, R.: The Gel'fand-Levitan theory and certain inverse problem for the parabolic equation, J. Fac. Sci. Univ. Tokyo 28 (1981), 317-330.

[11] Nakagiri, S.: Identifiability of linear systems in Hilbert spaces, to appear in SIAM J. Control & Optimization.

[12] Nakagiri, S., S. Kitamura, H. Murakami: Mathematical treatment of the constant parameter identifiability of distributed systems of parabolic type, Math. Semi. Notes 5 (1977), 97-105.

[13] Pierce, A.: Unique identification of eigenvalues and coefficients in a parabolic problem, SIAM J. Control & Optimization 17 (1979), 494-499.

[14] Sakawa, Y.: Controllability for partial differential equations of parabolic type, SIAM J. Control 12 (1974), 389-400.

[15] Seidman, T.I.: Ill-posed problems arising in boundary control and observation for diffusion equations, in: Anger ed., Inverse and Improperly Posed Problems in Differential Equations, Academie-Verlag 1979, 233-247.

[16] Suzuki, T.: Uniqueness and nonuniqueness in an inverse problem for the parabolic equation, to appear in J. Diff. Eq.

[17] Suzuki, T., R. Murayama: A uniqueness theorem in an identification problem for coefficients of parabolic equations, Proc. Japan Acad. 56 (1980), 259-263.

[18] Triggiani, R.: Extensions of rank conditions for controllability and observability to Banach spaces and unbounded operators, SIAM J. Control & Optimization 14 (1976), 313-338.

THE POLE AND ZERO STRUCTURE OF A CLASS OF LINEAR SYSTEMS

L. Pandolfi *)

Istituto di Matematica
Politecnico di Torino
Corso Duca degli Abruzzi 24
I-10100 Torino, Italy

1. INTRODUCTION

The study of the structure of the transfer function and of the system matrix of linear finite dimensional systems is one of the oldest topic in system theory. At the beginning, scalar transfer functions were studied. A scalar rational function is identified by its zeros and its poles (and by a multiplicative constant). Hence it was natural to derive the properties of the system by looking at the zeros and the poles of its transfer function. The study of the pole-zero structure of a system proved in fact to be useful not only for the applications (system design, and sensitivity analysis for example), but also to study the structure of a linear system (for example, the zeros are invariant under the action of feedbacks and output injections, and plays a special role in the invariant description of the action of special groups of transformation on linear systems). References on the above subjects can be found in [3],[7],[8]).

A complete analysis of the zeros and the poles of multi-input multi-output systems appeared in the book by Rosenbrock ([13]). Due to the importance of this topics, and because of the results obtained for finite dimensional systems, several authors investigated the zeros of infinite dimensional systems. We can quote for example the papers [5], [1], [2], [10] in which the authors study the pole zero structure of transfer functions of a given class. In [9], [4] (and references therein) the authors study the zeros of the system matrix of a system with delay. A recent paper of Pojolanen ([12]) consider the limiting positions of the poles of the transfer function function, under high gain feedback. In [11] some results by Rosenbrock were generalized for systems with delays. In this paper, we present an improved version of the results in [11], and we show that they hold for a wider class of distributed control systems.

---

*) This paper has been written according to the programmes of the GNAFA-CNR.

2. THE CLASS OF CONTROL SYSTEMS

In this paper we consider the class of those control systems that can be described, after Laplace transformation, by the relationships

$$\Delta(z)x(z) = B(z)u(z) + f(z) \tag{1.a}$$

$$y(z) = C(z)x(z) + D(z)u(z). \tag{1.b}$$

We observe that z will consistently be used to denote complex numbers, and that g(z) means the Laplace transform of g(t), when g(t) is a real function.

In Eq. (1.a,b), x, $f \in \mathbb{R}^n$, $u \in \mathbb{R}^m$, $y \in \mathbb{R}^p$. $\Delta(z)$, $B(z)$, $C(z)$, $D(z)$ are matrices of suitable dimensions. The elements of the above matrices are meromorphic functions. We assume that Eq. (1.a) can be solved for x(z), i.e.

Assumption: $\det(\Delta(z))$ is not identically zero.

The following systems belongs to the class of systems just defined:

1. Linear finite dimensional systems

$$K\dot{x} = Ax + Bu$$
$$y = Cx + Du$$

(We must assume that there exists a number s such that (sK+A) is invertible).

2. Systems with delays

$$\dot{x} = \int_{-h}^{0} dA(s)x(t+s) + \int_{-h}^{0} dB(s)u(t+s)$$

$$y(t) = \int_{-h}^{0} dC(s)x(t+s) + \int_{-h}^{0} dD(s)u(t+s)$$

(Also systems of neutral type, or systems described by difference equations).

3. Integral equations

$$x(t) = \int_{0}^{t} A(t-s)x(s) + \int_{0}^{t} B(t-s)u(s)ds$$

$$y(t) = \int_{0}^{t} C(t-s)x(s) + \int_{0}^{t} D(t-s)u(s)ds.$$

4. Integro-differential equations

$$\dot{x} = \int_{0}^{t} A(t-s)x(s)ds + \int_{0}^{t} B(t-s)u(s)ds$$

y(t) as in ex. 3.

Other systems which belongs to the given class can be obtained for example by combining the above ones, or by assuming that input derivatives act on the system (i.e. that $u(t) = \sum_{0i}^{\mu} H_i v^{(i)}(t)$).

The system described by the Eq. (1.a), (1.b) will be referred as system (S). The following matrix S(z) is the "system matrix" of the system (S):

$$S(z) = \begin{bmatrix} \Delta(z) & -B(z) \\ C(z) & D(z) \end{bmatrix},$$

while the transfer function is the matrix $T(z) = D(z) + C(z)\Delta^{-1}(z)B(z)$. We observe that we can define T(z) because $\det(\Delta(z)) \neq 0$.

Remark 2.1. The matrices S(z) and T(z) are matrices of meromorphic functions. We do not make any other assumption. Hence, the class of transfer functions described in [1] is a special case of the one just introduced.

3. LOCAL STRUCTURE OF A MEROMORPHIC MATRIX

The main tool for the study of the zeros of a rational matrix is its Smith-Mac Millan form. Such kind of normal form exists also for matrices of meromorphic functions (see [5], [14]). However, we shall use a local version, which was introduced, for the case of square holomorphic matrices, in [6] (see also [10]). Let H(z) be a meromorphic function defined in a neighboorhood of a complex number z_o (for simplicity we assume $z_o = 0$). It is possible to find matrices F(z), G(z) which are holomorphic near z_o, and which have holomorphic inverses, such that F(z)H(z)G(z) = $= M_H(z)$, where

$$\begin{aligned} M_H(z) &= [M'(z),0] && m > p, \\ &= M'(z) && m = p, \\ &= \begin{bmatrix} M'(z) \\ 0 \end{bmatrix} && m < p, \end{aligned}$$

$M'(z) = \mathrm{diag}(z^{s_1}M_1(z),\dots,z^{s_{k-1}}M_{k-1}(z),z^{s_k}M_k(z))$.

The blocks $M_i(z)$ are $d_i x d_i$ holomorphic matrices, and the exponents s_i are entire numbers. The matrix $M_i(z)$ is invertible, unless it is identically zero. We assume that the zero block, if present, has been written in the last position. If $M_k(z) = 0$, we assume by definition that $s_k = +\infty$. Moreover, we can order the blocks in such a way that $s_i < s_{i+1}$. We shall call the matrix $M_H(z)$ a local Smith-Mac Millan form of H(z) (near the given point z_o). The matrix $M_H(z)$ is not unique. However, the numbers d_i, s_i are uniquely determined by the matrix H(z). The matrix

$$\begin{pmatrix} s_1, s_2, \dots, s_{k-1}, s_k \\ d_1, d_2, \dots, d_{k-1}, d_k \end{pmatrix} \tag{2}$$

is the structure matrix of the point z_o. We say that this point is a pole of H(z) if $s_1 < 0$. Its multiplicity is the number

$$m_H(z_o) = -\sum_{s_i<0} s_i d_i.$$

We say that z_o is a zero if $s_k > 0$, and that it is a non-trivial zero if at least one of the number s_i is positive and finite. The multiplicity of the zero z_o is the number

$$\overline{m}_H(z_o) = \sum_{0<s_i<\infty} s_i d_i.$$

The pole-structure matrix of the point z_o, is the submatrix of the matrix (2) which is obtained by taking the columns which have negative elements in the first position. The submatrix of (2) formed by the columns which have positive numbers in the first positions is the zero-structure matrix of the point z_o.

Now we present three results that have been proved in [11]. They will used in the following.

Let Ξ be a set of holomorphic functions $\xi(z)$, such that:

1. The vectors $\xi(0)$ are independent.
2. There exist holomorphic functions $\eta(z)$ such that near z_o, we have:

$$H(z)\xi_i(z) = z^{r_i}\eta_i(z), \quad r_i > 0.$$

Theorem 3.1. If z_o (say $z_o = 0$) is a zero of $H(z)$, of multiplicity r, we can find a set Ξ of functions, such that properties 1, 2 hold, and such that $r = \Sigma r_i$.

Theorem 3.2. Let $H(z)$ be a pxm meromorphic matrix such that:

i) $\max_z \operatorname{rank} H(z) = m$
ii) $p \geq m$.

If we can find a set of holomorphic functions Ξ such that properties 1, 2 hold, then $H(z)$ has a non-trivial zero at the point z_o (say $z_o \equiv 0$), whose multiplicity is at least Σr_i.

Let now $\tilde{\Xi}$ be a set of holomorphic functions $\xi_i(z)$ defined near $z_o(=0)$ and such that

1. The vectors $\xi_i(0)$ are independent.
2. We have $H(z)\xi_i(z) = z^{-r_i}\xi_i(z)$, $r_i > 0$, for $z \neq 0$.
3. The functions $\eta_i(z)$ are holomorphic functions, defined in a neighboorhood of the origin, and the vectors $\eta_i(0)$ are independent.

Let $m(\Xi)$ be the sum of the exponents r_i. Then

Theorem 3.3. If we can find a set $\tilde{\Xi}$ of functions such that properties 1, 2, 3 hold, then $H(z)$ has a pole of multiplicity

$$\max_{\Xi} m(\tilde{\Xi}). \tag{3}$$

Conversely, if z_0 (=0) is a pole of H(z) of multiplicity r, then we can find a set $\tilde{\Xi}$ of functions such that

$$r = m(\tilde{\Xi}).$$

We said already that the aim of this paper is to study the relations between the structure of S(z) and that of T(z) at a given point z_0, that we assume to be zero without restriction. It is well known [13] that these relations hold for canonical lumped parameter systems. Hence we must give a suitable definition of canonical systems, which can be applied to the elements of the larger class of systems under study. For this, we write $\Delta(z)$ in its local Smith-Mac Millan form, and we write it in the block form

$$\begin{bmatrix} \Delta_1(z) & 0 \\ 0 & \Delta_2(z) \end{bmatrix}. \tag{4}$$

Where $\Delta_1(z)$ is the square matrix which contains the blocks with nonpositive exponents. Of course, det $\Delta_2(z)$ is not identically zero. We write B(z) and C(z) in the corresponding block form

$$\begin{bmatrix} B_1(z) \\ B_2(z) \end{bmatrix}, \qquad [C_1(z),\ C_2(z)].$$

<u>Definition 1.</u> We say that the system (S) is locally controllable at a given point z_0, when B(z) is a holomorphic matrix near z_0, and the matrix $B_2(z_0)$ is surjective.

<u>Definition 2.</u> We say that the system (S) is locally observable at a given point z_0, when the matrix C(z) is holomorphic in a neighboorhood of z_0, and when Ker $C_2(z_0) = 0$.

Of course, we shall say that a system is locally canonical at z_0, when it is both locally controllable and locally observable at z_0.

We observe explicitly that when (S) is a delayed system, then the matrices $\Delta(z)$, B(z), C(z) are matrices of entire functions, and the above definitions coincide with the definitions of spectral controllability and spectral observability ([11],[14]).

In the next sections we shall consider the properties of the poles and the zeros which appear at a given point z_0, where the system is canonical.

4. THE ZERO-STRUCTURE OF THE MATRICES S(z) AND T(z)

In the paper [11] we proved that the matrices S(z) and T(z) have the same zeros, with the same multiplicities, when S(z) is the system matrix of a canonical system with delays. In this section, we shall improve that result. In fact, the following statement holds:

Theorem 4.1. Let z_o be a complex number such that D(z) is bounded near z_o. We assume that the system (S) is canonical at z_o. Then, the zero-structure of the matrix S(z) is equal to the zero-structure of T(z).

Proof. We shall present a proof of this result under the assumption that $p > m$, and that S(z) and T(z) have full rank a.e. on the complex plane. These assumptions are restrictive, but can be removed as in [11], sect. 4. For simplicity of notations, we assume that $z_o = 0$.

First of all, we consider the zero-structure matrix of the transfer function. Let it be

$$\begin{pmatrix} d_r,\dots,d_k \\ s_r,\dots,s_k \end{pmatrix}$$

($s_k < +\infty$, since T(z) has rank m for a.e. z). Hence the local Smith-Mac Millan form of T(z) near zero has (k-r+1) blocks with positive exponents. Let us consider the i-th block. Theorem 3.1. asserts that we can find functions $u_j(z)$ such that

$$T(z)u_j(z) = z^{s_i}\phi_j(z), \qquad 1 \le j \le d_i$$

and that the vectors $u_j(0)$ are independent.

Let us define

$$x_j(z) = \Delta^{-1}(z)B(z)u_j(z), \qquad \xi_j(z) = (x_j^*(z),u_j^*(z))^*.$$

It is easy to see that the vectors $x_j(z)$ are bounded near zero. In fact, let us write $\Delta(z)$ in the form (4), so that

$$x_j(z) = (x_j^{1*}(z),x_j^{2*}(z))^*.$$

The function $x_j^1(z)$ is bounded. Hence,

$$y_j(z) - C_1(z)x_j^1(z) = C_2(z)\Delta_2^{-1}(z)B_2(z)u_j(z)$$

is bounded. We assumed that the system is observable. Hence, Ker $C_2(0) = \{0\}$, so that $x_j^2(z) = \Delta_2^{-1}(z)B_2(z)u_j(z)$ is bounded near the point $z_o = 0$.

A simple calculation shows that

$$S(z)\xi_j(z) = z^{s_i}\psi_j(z), \qquad \psi_j(0) \neq 0.$$

Moreover, the vectors $\xi_j(0)$ are independent, since the u-components of these vectors are independent. Hence, if we think to write the matrix S(z) in Smith-Mac Millan form near $z_o = 0$, we see that its zero-structure matrix has the submatrix

$$\begin{pmatrix} d'_r,\dots,d'_k \\ s_r,\dots,s_k \end{pmatrix} \qquad d'_i \geq d_i . \tag{5}$$

In order to prove the theorem, we must show that $d'_i = d_i$, and that the matrix (5) is in fact the zero-structure matrix of S(z).

We observe that

$$\overline{m}_S(0) \geq \Sigma d'_i s_i \geq \Sigma d_i s_i = \overline{m}_T(0).$$

The first inequality is strict only if the matrix (5) is not the zero-structure of S(z). The second inequality is strict if $d'_i > d_i$, at least for one index i. Hence, the proof of the theorem will be completed if we can show that $\overline{m}_S(0) \leq \overline{m}_T(0)$. We use again Theorem 3.1. Let $\overline{m} = \overline{m}_S(0)$. We can find functions $\phi_i(z)$ such that the vectors $\phi_i(0)$ are independent, and

$$S(z)\phi_i(z) = z^{s_i}\psi_i(z), \quad \psi_i(0) \neq 0, \ \Sigma s_i = \overline{m}.$$

We write $\Delta(z)$ in the diagonal form $\mathrm{diag}(\Delta'(z),\Delta''(z))$, where $\Delta'(z)$ is the matrix of the blocks of the Smith-Mac Millan form of $\Delta(z)$ which correspond to negative exponents. Accordingly with this form of $\Delta(z)$ we write $B(z) = [B'^*(z),B''^*(z)]^*$, $C(z) = [C'(z),C''(z)]$, $\phi_i(z) = (\phi_i'^*(z),\phi_i''^*(z),u_i^*(z))^*$. Of course, $\phi'_i(z)$ has a zero of order greater then s_i for z = 0, since $S(z)\phi_i(z)$ has a zero of order s_i, so that

$$\begin{bmatrix} \Delta''(z) & B''(z) \\ C''(z) & D(z) \end{bmatrix} \begin{bmatrix} \phi''_i(z) \\ u''_i(z) \end{bmatrix} = z^{\overline{s}_i}\eta_i(z), \qquad \eta_i(0) \neq 0, \quad \overline{s}_i \geq s_i .$$

The elements of the matrix on the left side of the above equality are matrices of holomorphic functions defined in a neighboorhood of zero. Hence we can apply Lemmas 4.4, 4.5 in [11]. In this way we find that the vectors $u_i(0)$ are independent, and that $T(z)u_i(z)$ has a zero, for z = 0, of order at least $\overline{s}_i$. Hence, by Theorem 3.2 the multiplicity of the zero of T(z) is no less then the multiplicity of the zero of S(z). This completes the proof.

Remark. The above theorem says nothing about those points which are poles of the matrices B(z), C(z). In fact, we could think to exclude, for the definitions of observability and controllability, the assumption that the matrices B(z) and C(z) be regular. However, the following example shows that the above theorem does not hold in this more general situation.

Let us consider the matrix of rational function

$$S(z) = \begin{bmatrix} (1/z) & 0 & (1/z^2) \\ 0 & z & 1 \\ 1 & 1 & 0 \end{bmatrix} .$$

The matrix S(z) is the system matrix of the control system

$$\left.\begin{aligned} \int_0^t x_1(s)ds &= \int_0^t (t-s)u(s)ds \\ \dot{x}_2(t) &= u(t) \\ y(t) &= x_1(t) + x_2(t) \quad . \end{aligned}\right\}$$

It is easy to see that

$$\begin{bmatrix} 1 & 0 & 0 \\ 0 & 0 & 1 \\ (-z^2) & 1 & 0 \end{bmatrix} \begin{bmatrix} (\sqrt{z}) & 0 & (\sqrt{z}^2) \\ 0 & z & 1 \\ 1 & 1 & 0 \end{bmatrix} \begin{bmatrix} 0 & 1 & (-\sqrt{2}) \\ 0 & 1 & \sqrt{2} \\ 1 & -z & z/2 \end{bmatrix} = \mathrm{diag}\ (\sqrt{z}^2, 2, z),$$

so that the system matrix has a simple zero for $z = 0$. The transfer function has no zero for $z = 0$, since $T(z) = 2/z$.

Now we observe that

$$\begin{bmatrix} 1 & 0 & 0 \\ 0 & 1 & -z \\ 0 & 0 & 1 \end{bmatrix} \begin{bmatrix} z & 1 & 1 \\ 0 & z & 1 \\ 1 & 0 & \sqrt{z} \end{bmatrix} \begin{bmatrix} 1 & 1 & -1 \\ 1 & 0 & -1 \\ -z & -z & z+1 \end{bmatrix} = \begin{bmatrix} 1 & 0 & 0 \\ 0 & -z & 0 \\ 0 & 0 & \sqrt{z} \end{bmatrix}.$$

Hence, the system matrix

$$S(z) = \begin{bmatrix} z & 1 & 1 \\ 0 & z & 1 \\ 1 & 0 & \sqrt{z} \end{bmatrix}$$

has a unique zero, which is simple, for $z = 0$. The transfer function which corresponds to $S(z)$ is $T(z) = (2z-1)/z^2$, which has a zero for $z = \sqrt{2}$. Hence, the assumption that the matrix $D(z)$ be regular in z_o must be retained in Theorem 4.1.

5. THE POLE-STRUCTURE OF THE TRANSFER FUNCTION

In this section, we consider the pole-structure of the matrix $T(z)$. In the finite dimensional case, the poles of the transfer function depend on the zeros of $\Delta(z)$. A similar result holds also in this case. In fact:

Theorem 5.1. Let z_o be a given complex number, and assume that the matrices $B(z)$, $C(z)$, $D(z)$ are holomorphic at z_o. Then, the system (S) is canonical at z_o if and only if the pole-structure of $T(z)$ at z_o coincide with the zero-structure of $\Delta(z)$ at z_o.

Proof. Again, we assume that $z_o = 0$. We write $\Delta(z)$ in the form (4) and, correspondingly, we write $B(z)$, $C(z)$ in block form. Then, we know that system (S) is canonical if and only if $\mathrm{Ker}\ C_2(0) = \{0\}$, $\mathrm{Ker}\ B_2^*(0) = \{0\}$. Let

$$\begin{pmatrix} s_r, \dots, s_k \\ d_r, \dots, d_k \end{pmatrix} \qquad (6)$$

be the zero-structure matrix of $\Delta(z)$. Of course, $s_k < \infty$. Let us assume first that the system (S) is canonical. We show that a matrix of the form

$$\begin{pmatrix} s_r, \dots, s_k \\ d'_r, \dots, d'_k \end{pmatrix} \qquad d'_i \geq d_i$$

is a submatrix of the pole-structure matrix of $T(z)$ at $z_o = 0$. In fact, let us consider the j-th block of $\Delta(z)$, $j \geq r$, which is a $d_j x d_j$-matrix. Let $u_\nu(z)$, $1 \leq \nu \leq d_j$ be vectors such that $B(0)u_\nu(0)$ are independent vectors, and such that the non zero components of $B_2(z)u_\nu(z)$ correspond to the entries of the j-th block of $\Delta(z)$. Functions $u_\nu(z)$ with these properties exist, since $B_2(0)$ has full rank (of course, we do not require that $B_1(z)u_\nu(z)$ be zero). We consider now the functions

$$\begin{pmatrix} \Delta_1^{-1}(z)B_1(z) \\ \Delta_2^{-1}(z)B_2(z) \end{pmatrix} u_\nu(z) .$$

The components of these functions which do not correspond to the j-th block of $\Delta(z)$ vanish for $z \to 0$. The components which correspond to the entries of the j-th block have a pole of order s_j. Hence,

$$T(z)u_\nu(z) = z^{-s_j}\psi_\nu(z).$$

The vectors $\psi_\nu(0)$ are not zero, since Ker $C_2(0) = \{0\}$. Hence the local Smith-Mac Millan form of $T(z)$ has a block with exponent $-s_j$, of dimension at least $d_j x d_j$. This completes the first part of the proof. A consequence of the above argument is that $\overline{m}_\Delta(0) \geq m_T(0)$. In this relation, the equality can hold if and only if the matrix (6) is the pole-structure matrix of $T(z)$. Hence, the proof of the first part of the theorem will be completed if we show that $m_T(0) \leq \overline{m}_\Delta(0)$.

Let us observe that

$$T(z) = C_1(z)\Delta_1^{-1}(z)B_1(z) + C_2(z)\Delta_2^{-1}(z)B_2(z) + D(z),$$

and that $D(z) + C_1(z)\Delta_1^{-1}(z)B_1(z)$ is bounded. Hence, it is a consequence of Theorem 3.3 that the multiplicity of the pole of $T(z)$ is equal to the multiplicity of the pole of the function

$$C_2(z)\Delta_2^{-1}(z)B_2(z) = \{C_2(z)[\mathrm{adj}\ \Delta_2(z)]B_2\}/(\det \Delta_2(z)).$$

Hence, the multiplicity of the pole of $T(z)$ is at most the order of the zero of $(\det \Delta_2(z))$, i.e. $\overline{m}_\Delta(0)$. So, we have proved that if system (S) is canonical, then the zero-structure of $\Delta(z)$ coincide with the pole-structure of $T(z)$.

Now we prove the converse part of the theorem. We assume that the system (S) is not canonical. In particular, we assume that (S) is controllable, but not observable. This is not restrictive. In fact, if the system (S) is observable, it cannot be controllable, and the system identified by the system matrix $S^*(z)$ has the required properties (and, of course, transposition does not change the pole-zero structure of a matrix). If the system is neither observable nor controllable, we can add columns to the matrix $B(z)$, till we obtain a controllable system. By Theorem 3.3 this operation does not reduce the order of the pole of $T(z)$. So we must consider the order of the pole of the matrix $C_2(z)\Delta_2^{-1}(z)B_2(z)$, that we write in the form

$$C_2(z)\Delta_2^{-1}(z)B_2(z) = [C_r(z),\dots,C_k(z)]\begin{bmatrix} z^{-s_r}I & & \\ & \ddots & \\ & & z^{-s_k}I \end{bmatrix}\begin{bmatrix} B_r(z) \\ \vdots \\ B_k(z) \end{bmatrix} .$$

(We note that it is not restrictive to assume that $M_i(z)$ are the identity matrices of suitable dimensions).

The matrix $[B_r(z),\dots,B_k(z)]$ is of full rank. Hence it can be reduced, after right multiplication, to the form $[I,0]$, and we must consider the order of the pole of the matrix

$$[C_r(z)z^{-s_r},\dots,C_k(z)z^{-s_k}].$$

We assumed that the system (S) is not canonical. Hence at least one of the columns of the matrix $C(z)$ will be linearly dependent on the other columns, when $z = 0$. As a consequence, a right multiplication by an invertible matrix, reduce one of the blocks of the above transfer function to the form $[C_i'(z),\phi(z)]z^{-s_i}$, and $\phi(0) = 0$. By using Theorem 3.3 we can deduce that the order of the pole of the transfer function is strictly less then $\Sigma s_i d_i$, i.e. that $m_T(0) < \bar{m}_\Delta(0)$. Hence, if the system is not canonical, the pole of $T(z)$ and the corresponding zero of $\Delta(z)$ cannot have the same structure.

Remark. The following example shows that, like in the previous section, we cannot remove the assumptions that the matrices $B(z)$, $C(z)$, $D(z)$ be regular near z_0. Let $S(z)$ be the matrix

$$S(z) = \begin{bmatrix} 0 & 0 & \sqrt{z^2} \\ 0 & 0 & 0 \\ 1 & 0 & 0 \end{bmatrix} .$$

It seems impossible to find any reasonable definition of controllability and observability such that the system described by the matrix $S(z)$ be canonical. However, the transfer function is the function

$$T(z) = 1/z^3$$

which has a pole of order greater then the zero of det $\Delta(z)$.

6. CONCLUSIONS

In this paper, we proved two results for a class of distributed parameter systems, which relate the zeros of the transfer functions with the zeros of the system matrix, and the poles of the transfer function with the zeros of the matrix $\Delta(z)$. The class of systems that we studied contains linear autonomous finite dimensional systems. When specialized to this subclass, our results give two known theorems by Rosenbrock ([13], Ch. 3). However, we note explicitly that many systems which are important for applications (for example, systems of partial differential equations), are not covered by our study.

REFERENCES

[1] Callier, F.M., V. Hon Lam Cheng, C.A. Desoer: Dynamic Interpretation of Poles and Transmission Zeros for Distributed Parameters Multivariable Systems, IEEE Trans. Cyrcuit Systems, IEEE-CAS 28 (1981), 300-306.

[2] Callier, F.M., C.A. Desoer, Stabilization, Tracking and Disturbance Rejection in Multivariable Convolution Systems, Ann. Soc. Sci. Bruxelles 94 (1980), 7-51.

[3] Fallside, F.: Control Systems design by pole-zero assignment, Academic Press, London, 1977.

[4] Frost, M.C., C. Storey: Equivalence of a Matrix over R(s,z) with its Smith Form, Int. J. Cont. 28 (1978), 665-671.

[5] Hautus, H.L.J.: The Formal Laplace Transform for Smooth Linear systems, in "Mathematical Systems Theory", Marchesini G., Mitter S.K., Ed. Springer Verlag, Berlin, 1976.

[6] Kappel, F., H.K. Wimmer: An Elementary Divisor Theory for Autonomous Linear Functional Differential Equations, J. Diff. Equations 21 (1976), 134-147.

[7] MacFarlane, A.G.J., N. Karcanias: Poles and Zeros of Linear Multivariable Systems; A survey of the Algebraic, Geometric and Complex Variable Theory, Int. J. Control 24 (1976), 33-74.

[8] Morse, A.S.: Structural Invariants of Linear Multivariable Systems, SIAM J. Control 11 (1973), 446-460.

[9] Olbrot, A.W., S.H. Zak: On Zeros of Retarded Systems, Arch. Autom. I Tel. 25 (1980), 445-451.

[10] Pandolfi, L.: On the Zeros of the Transfer Functions of Delayed Systems, Systems and Contr. Letters 1 (1981), 204-210.

[11] Pandolfi, L.: The Transmission Zeros of Systems with Delays, Int. J. Control, to appear.

[12] Pohjolanen, S.: Computation of Transmission Zeros of Distributed Parameter Systems, Int. J. Contr. 33 (1981), 199-212.

[13] Rosenbrock, H.H.: State Space and Multivariable Theory, Nelson, London, 1970.

[14] Wimmer, H.K.: Exponential Solutions of Systems of Linear Differential Equations of Infinite Order, J. Differential Equations 33 (1979), 39-44.

OPTIMAL CONTROL OF ROTATION OF A FLEXIBLE ARM

Y. Sakawa*, R. Ito**, N. Fujii*

* Faculty of Engineering Science, Osaka University, Toyonaka, Osaka 560, Japan
** Mitsubishi Heavy Industries, Ltd, Takasago Machinery Works, Hyogo-ken, Japan

1. INTRODUCTION

When we rotate a flexible arm or rod in a horizontal plane about an axis through the arm's fixed end, transverse vibration may occur. The problem considered in this paper is to control motor torque in such a way that at the end of rotation there is no vibration of the arm and the flexible arm must be completely at rest. We calculate the optimal control of the motor torque such that the flexible arm rotates in the above mentioned manner. Such a problem occurs when we control a manipulator or a robot arm.

In this paper, we first derive a partial differential equation and boundary conditions which govern the transverse vibration of the flexible arm. By using the Galerkin approximations, a set of ordinary differential equations is obtained. On the basis of this set of ordinary nonlinear differential equations, the optimal control is calculated by employing an iterative algorithm [1]. Several satisfactory numerical results will be presented.

2. EQUATIONS OF MOTION

The equations of motion of the driving motor are given by

$$\dot{\theta}(t) = \omega(t),$$

$$\dot{\omega}(t) = u(t) = \tau(t)/J, \tag{1}$$

where θ is the angle of rotation, ω is the angular velocity, τ is the torque generated by the motor, and J is the moment of inertia of the motor as well as the arm. Exactly speaking, J is not constant because of the vibration. However, since the displacement due to the vibration is not so large, we assume that J is constant. Since $|\tau|$ cannot exceed some maximum value, we impose the constraint

$$|u(t)| \leq u_m, \tag{2}$$

where u_m is the maximum value of control u(t).

The partial differential equation governing the vibration of the arm can be derived by considering the equations of motion for the differential segment of the arm shown in Fig. 1 and Fig. 2 with respect to the rotating coordinate (r,w). Let $\rho(r)$ be the mass per unit length, and let f(t,r), M(t,r), and S(t,r) be the axial tensile force, the bending moment, and the shearing force, respectively, acting at the position r and at time t.

The differential segment can be regarded as a rigid body, and the forces and the bending moments acting on the differential segment are shown in Fig. 2. Since the coordinate system (r,w) rotates with the angular velocity ω, the equations of motion in r and w directions are respectively given by [2]

$$\rho\ddot{r}\, dr = \frac{\partial f}{\partial r} dr + \rho dr(2\omega\dot{w} + \dot{\omega}w + \omega^2 r), \tag{3}$$

$$\rho\ddot{w}\, dr = \frac{\partial S}{\partial r} dr + \rho dr(-2\omega\dot{r} - \dot{\omega}r + \omega^2 w). \tag{4}$$

The second terms in the right sides of (3) and (4) appear due to the rotation of the coordinate system and include the Coriolis force and the centrifugal force.

Balance of moments acting on the differential segment gives the relation

$$\frac{\partial M}{\partial r} dr + S\, dr - f\frac{\partial w}{\partial r} dr = 0,$$

from which we obtain

$$S = -\frac{\partial M}{\partial r} + f\frac{\partial w}{\partial r}. \tag{5}$$

At the fixed end r = 0, the arm is clamped. At the other end r = L, the arm is assumed to be free. Therefore, the transverse displacement w(t,r) satisfies the boudary condition

$$\begin{aligned} &w(t,0) = w'(t,0) = 0, \\ &w''(t,L) = w'''(t,L) = 0, \end{aligned} \tag{6}$$

where a prime denotes the derivative with respect to r, e.g., $w'(t,0) = \partial w(t,0)/\partial r$.

We assume that there is no longitudinal vibration of the arm. Therefore, $\dot{r} = \ddot{r} = 0$. Using this relation in (3) yields

$$\frac{\partial f}{\partial r} = -\rho(2\omega\dot{w} + \dot{\omega}w + \omega^2 r). \tag{7}$$

Since the axial tensile force f(t,r) vanishes at the free end r = L, from (7) we obtain

$$f(t,r) = \int_r^L \rho(2\omega\dot{w} + \dot{\omega}w + \omega^2 r)dr. \tag{8}$$

In view of (6), we assume that $|\dot{w}/\omega r| << 1$ and $|\dot{\omega}w/\omega^2 r| << 1$. Consequently,

$$f(t,r) = \omega^2(t) \int_r^L \rho(r)rdr. \tag{9}$$

From (4) it follows that

$$\ddot{w} = \frac{1}{\rho}\frac{\partial S}{\partial r} + \omega^2 w - ru. \tag{10}$$

Let E be the Young's modulus of the arm material, and let I(r) be the moment of inertia of the cross-section. Then we see that [3]

$$M = EI(r)w''(t,r). \tag{11}$$

Using the relations (5) and (11), from (10) we obtain

$$\frac{\partial^2 w}{\partial t^2} = -\frac{1}{\rho}\frac{\partial^2}{\partial r^2}(EI(r)\frac{\partial^2 w}{\partial r^2}) + \frac{1}{\rho}\frac{\partial}{\partial r}(f\frac{\partial w}{\partial r}) + \omega^2 w - ru. \tag{12}$$

In the special case where the arm is uniform, (12) can be rewritten as

$$\frac{\partial^2 w}{\partial t^2} + \alpha\frac{\partial^4 w}{\partial r^4} - \frac{\omega^2}{2}\frac{\partial}{\partial r}[(L^2 - r^2)\frac{\partial w}{\partial r}] - \omega^2 w = -ru, \tag{13}$$

where $\alpha = EI/\rho$.

3. THE GALERKIN APPROXIMATIONS

To obtain a finite-dimensional model for (12), let us assume that

$$w(t,r) \cong \sum_{i=1}^{N} y_i(t)\phi_i(r), \tag{14}$$

where $\phi_i(r)$ are the basis functions for the Galerkin approximations. In our case, the eigenfunctions of the following eigenvalue problem

$$d^4\phi(r)/dr^4 = \lambda\phi(r),$$
$$\phi(0) = \phi'(0) = \phi''(L) = \phi'''(L) = 0, \tag{15}$$

are taken as the basis functions. It can be easily seen that (15) has a non-zero solution if and only if $\lambda = (\beta/L)^4$, where β is a positive solution of

$$\cosh\beta\cos\beta + 1 = 0. \tag{16}$$

Let β_i be the solutions of (16) such that $0 < \beta_1 < \beta_2 < \ldots$. Then the eigenvalues of (15) are given by $\lambda_i = \mu_i^4$, where

$$\mu_i = \beta_i/L, \quad i = 1,2,\ldots . \tag{17}$$

It can be easily seen that the corresponding normalized eigenfunctions are given by

$$\phi_i(r) = [\cosh \mu_i r - \cos \mu_i r - \gamma_i(\sinh \mu_i r - \sin \mu_i r)]/\sqrt{L}, \tag{18}$$

where

$$\gamma_i = (\cosh \mu_i L + \cos \mu_i L)/(\sinh \mu_i L + \sin \mu_i L).$$

It is clear that the function $w(t,r)$ given by (14) satisfies the boundary condition (6), and that

$$(\phi_i,\phi_j) = \int_0^L \phi_i(r)\phi_j(r)dr = 0 \quad (i \neq j),$$
$$\|\phi_i\| = (\phi_i,\phi_i)^{1/2} = 1. \tag{19}$$

We substitute (14) into (12), and determine $y_i(t)$ in such a way that the error of the Galerkin approximations (14) is orthogonal in $L_2(0,L)$ to each $\phi_i \in L_2(0,L)$, $i = 1,\ldots,N$. Computing along this line, we obtain the following set of ordinary differential equations

$$\ddot{y}_i = -\sum_{j=1}^{N} (a_{ij} + \omega^2 b_{ij})y_j + \omega^2 y_i + d_i u, \quad (i = 1,2,\ldots,N) \tag{20}$$

where

$$a_{ij} = \int_0^L EI(r)\left(\frac{\phi_i(r)}{\rho(r)}\right)''\phi_j''(r)dr,$$
$$b_{ij} = \int_0^L \left(\frac{\phi_i(r)}{\rho(r)}\right)'\phi_j'(r)\left[\int_r^L \rho(x)x\,dx\right]dr, \tag{21}$$
$$d_i = -\int_0^L r\phi_i(r)dr.$$

Let us introduce new variables $z_i = \dot{y}_i$. Then (20) can be written as

$$\dot{y}_i = z_i,$$
$$\dot{z}_i = -\sum_{j=1}^{N} (a_{ij} + \omega^2 b_{ij})y_j + \omega^2 y_i + d_i u. \tag{22}$$

Thus, the whole system of state equations consists of (1) and (22). If the arm is uniform, then corresponding to (13), (22) can be written as

$$\dot{y}_i = z_i \tag{23...}$$

$$\dot{z}_i = (\omega^2 - \alpha\mu_i^4)y_i - \omega^2 \sum_{j=1}^{N} b_{ij}y_j + d_i u, \quad (i = 1,\dots,N) \qquad (\dots 23)$$

where

$$b_{ij} = \frac{1}{2}\int_0^L (L^2 - r^2)\phi_i'(r)\phi_j'(r)dr. \qquad (24)$$

Let us introduce (2N+2)-dimensional state vector x defined by

$$x = (y_1,\dots,y_N,z_1,\dots,z_N,\theta,\omega)^T.$$

Then (22) or (23) together with (1), and the initial conditions are expressed as

$$\dot{x}(t) = f(x(t), u(t)), \quad x(0) = 0, \qquad (25)$$

where f is the (2N+2)-dimensional vector function, and it is assumed that initially the system is at complete rest. The function f is nonlinear in x, however it is linear in u.

4. OPTIMAL CONTROL PROBLEM AND ALGORITHM

Let us introduce a function

$$J_o(x) = [\sum_{i=1}^{N} (y_i^2 + c_o z_i^2) + c_1(\theta - \theta_d)^2 + c_2\omega^2]/2, \qquad (26)$$

where θ_d is a desired angle of rotation, and c_o, c_1, and c_2 are some positive constants, respectively. Let T be the time of rotation which is given properly corresponding to θ_d, and let x(t) be the solution of (25). Now our problem is to find such an optimal control u(t) that minimizes the functional $J_o(x(T))$ under the constraint (2). This problem is equivalent to minimizing

$$\int_0^T L(x(t), u(t))dt \qquad (27)$$

under (2), where L(x,u) is given by

$$L(x(t), u(t)) = \frac{dJ_o(x(t))}{dt} = \sum_{i=1}^{2N+2} \frac{\partial J_o(x)}{\partial x_i} f_i(x(t), u(t)).$$

When f(x,u) is defined by (1) and (23), L(x,u) is expressed as

$$L(x,u) = \sum_{i=1}^{N} [y_i z_i + c_o z_i((\omega^2 - \alpha\mu_i^4)y_i + d_i u - \omega^2 \sum_{j=1}^{N} b_{ij}y_j)] + c_1(\theta - \theta_d)\omega + c_2\omega u. \qquad (28)$$

Now let us introduce the Hamiltonian function

$$H(x,u,\lambda) = L(x,u) + \lambda f(x,u),$$

where $\lambda = (\lambda_1,\ldots,\lambda_{2N+2})$. The class Ω of admissible controls is defined as the set of all measurable functions $u(t)$ satisfying (2).

To solve the optimal control problem, we employ the following algorithm [1] :

Step 0. Select a nominal control $u^0 \in \Omega$. Let $x^0(t)$ be the corresponding nominal trajectory. Set $i = 1$.

Step 1. Compute $\lambda^{i-1}(t)$ by solving the differential equation

$$d\lambda^{i-1}(t)/dt = - H_x(x^{i-1}(t),u^{i-1}(t),\lambda^{i-1}(t)), \quad \lambda^{i-1}(T) = 0, \tag{29}$$

where

$$H_x = (\partial H/\partial x_1,\ldots,\partial H/\partial x_{2N+2}).$$

Step 2. Define the function

$$K(x,u,\lambda;\bar{u},C) = H(x,u,\lambda) + C(u - \bar{u})^2, \tag{30}$$

where $C \geq 0$. Select a constant C^i appropriately. Determine $x^i(t)$ and $u^i(t)$ which satisfy both

$$\begin{aligned} &K(x^i(t),u^i(t),\lambda^{i-1}(t);u^{i-1}(t),C^i) \\ &= H(x^i(t),u^i(t),\lambda^{i-1}(t)) + C^i(u^i(t) - u^{i-1}(t))^2 \\ &= \min_{|u|\leq u_m} K(x^i(t),u,\lambda^{i-1}(t);u^{i-1}(t),C^i), \end{aligned} \tag{31}$$

and the differential equation

$$dx^i(t)/dt = f(x^i(t),\ u^i(t)), \quad x^i(0) = 0. \tag{32}$$

This is possible by integrating (32) from $t = 0$ to $t = T$ step by step while seeking $u^i(t)$ that minimizes K.

Step 3. Calculate

$$J(u^i) = \int_0^T L(x^i(t),\ u^i(t))dt. \tag{33}$$

If $J(u^i) > J(u^{i-1})$, make C^i larger and go to Step 2. Otherwise, set $i := i+1$ and go to Step 1.

Stop the computation if the sequence $\{C^i\}$ is bounded and the sequence $\{u^i(t)\}$ of the controls converges.

In this algorithm we minimize K instead of H. Since the function K contains a quadratic penalty term $C^i(u - u^{i-1})^2$ for the possible large change of control, instability of the algorithm at the early stage of computation can be avoided by taking large C^i.

In [1] it is proved that there is a constant $M > 0$ independent of i such that the inequality

$$J(u^i) - J(u^{i-1}) \leq -(2C^i - M) \int_0^T [u^i(t) - u^{i-1}(t)]^2 dt \tag{34}$$

holds for any i. If we choose C^i such that

$$C^i > M/2, \quad i = 1,2,\dots , \tag{35}$$

then the sequence $\{J(u^i)\}$ of the cost functional decreases monotonically and converges. In [1] the following proposition is also proved: Suppose that the sequence $\{C^i\}$ is bounded. If the sequence $\{u^i(t)\}$ converges in the sense that

$$\lim_{i\to\infty} u^i(t) = \hat{u}(t) \quad \text{almost everywhere on } [0,T], \tag{36}$$

then the control $\hat{u}(t)$ and the corresponding solution $\hat{x}(t)$ of (25) satisfy the necessary conditions for optimality, i.e.,

$$H(\hat{x}(t),\hat{u}(t),\hat{\lambda}(t)) = \min_{|u|\leq u_m} H(\hat{x}(t),u,\hat{\lambda}(t)), \tag{37}$$

where $\hat{\lambda}(t)$ is the solution of

$$d\hat{\lambda}(t)/dt = - H_x(\hat{x}(t),\hat{u}(t),\hat{\lambda}(t)), \quad \hat{\lambda}(T) = 0. \tag{38}$$

5. NUMERICAL RESULTS

In this section, we present some results of numerical computation using the algorithm explained in the preceding section. We considered a hollow iron rod as shown in Fig. 3, where $D = 2$ cm, $d = 1.6$ cm, with the length $L = 3.0$ m. The parameters are as follows: $E = 1.96 \times 10^{11}$ (N/m^2), $\rho = 0.89$ (kg/m), $I = 4.64 \times 10^{-9}$ (m^4), $\alpha = EI/\rho = 1.03 \times 10^3$ (m^4/s^2), $\theta_d = 120^o$, $T = 1.6$ s, $u_m = 4$(s^{-2}). For this particular numerical example, it was recognized numerically that

$$\max_t |x_{i+1}(t)|/\max_t |x_i(t)| \approx 0.1$$

for each i. Therefore we set $N = 3$. This shows that the first several modes are important compared to the remainders. It was also recognized numerically that

$$|w(t,r)| \leq |w(t,L)| \quad \text{for any } t.$$

Fig. 4 shows the initial nominal solution, where w_L denotes $w(t,L)$, the displacement at the free end. Fig. 5 shows the optimal solution, which was obtained after 70 iterations. Fig. 6 shows $w(t,r)$ as a function of r at several instants during the rotation.

Fig. 7 shows the values of cost functional $J(u^i)$, the parameter values C^i, and the variations of control functions $\int_0^T |u^i(t) - u^{i-1}(t)| dt$, respectively, versus iteration number. It is clear from Fig. 7 that at the end of iteration $J(u^i) \cong 0$, that the sequence $\{C^i\}$ is bounded, and that the sequence of the control functions $\{u^i(t)\}$ converges.

6. CONCLUDING REMARKS

The optimal control of rotation of a flexible arm has been discussed. Controlling the motor following the computed optimal control is an open loop control scheme. Therefore, at the end of the optimal control, it is necessary to switch the control scheme to the closed-loop control for attaining the desired state exactly. This dual-mode control scheme is recommended from the practical point of view.

Sometimes the flexible arm carries a mass concentrated near the end point $r = L$. This case can be treated as a problem with variable density function $\rho(r)$, and the system parameters can be calculated by using (21).

REFERENCES

[1] Sakawa, Y, Y. Shindo: On global convergence of an algorithm for optimal control, IEEE Trans. on Automatic Control, AC-25 (1980), 1149-1153.

[2] Goto, K: Mechanics, Gakujutsu-Tosho Shuppansha, Tokyo, 1975 (in Japanese).

[3] Dym, C.L., I.H. Shames: Solid Mechanics: A Variational Approach, McGraw-Hill, 1973.

LIST OF FIGURE CAPTIONS

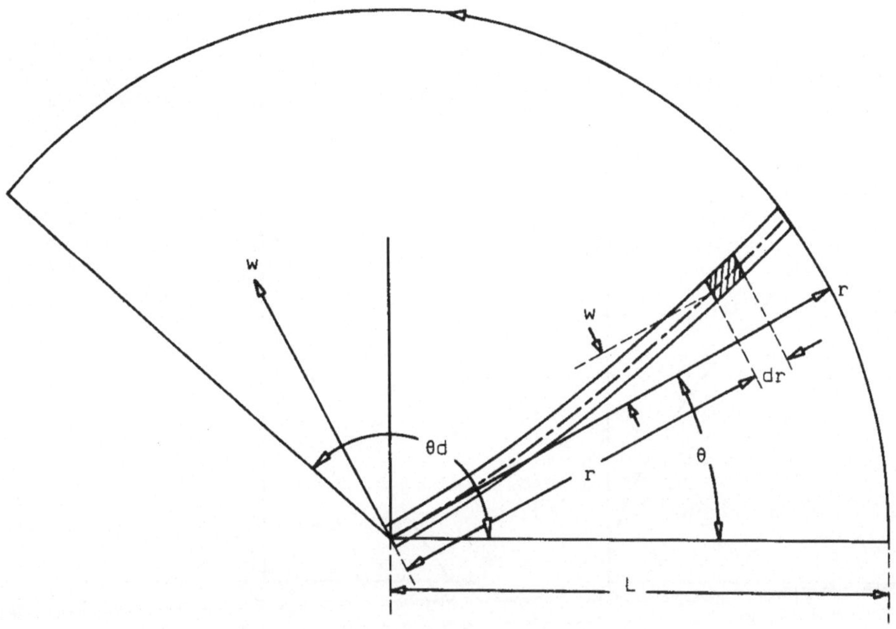

Fig. 1

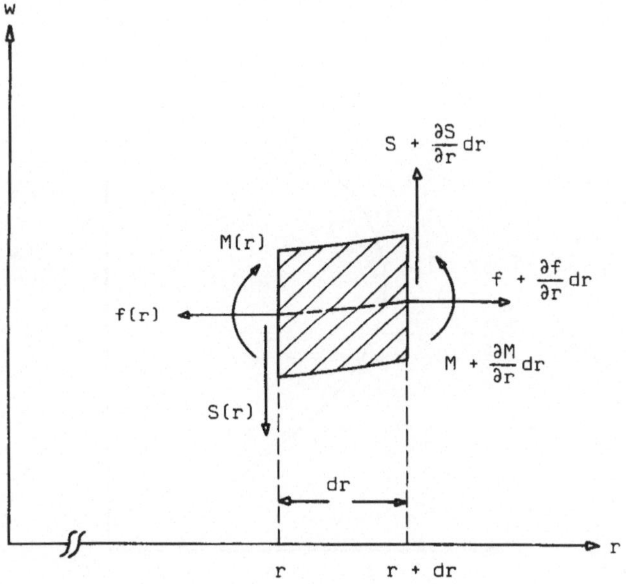

Fig. 2

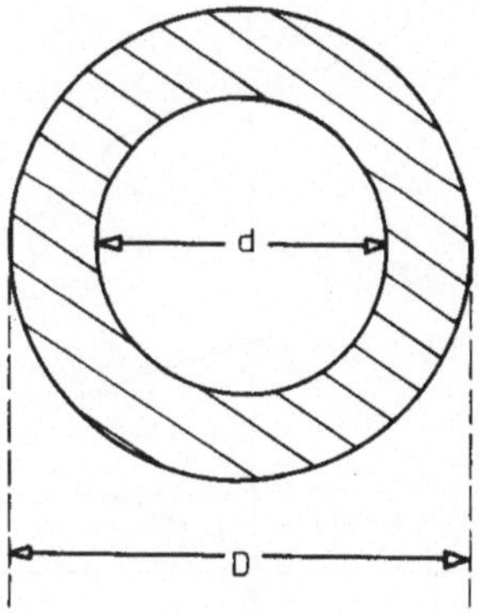

Fig. 3

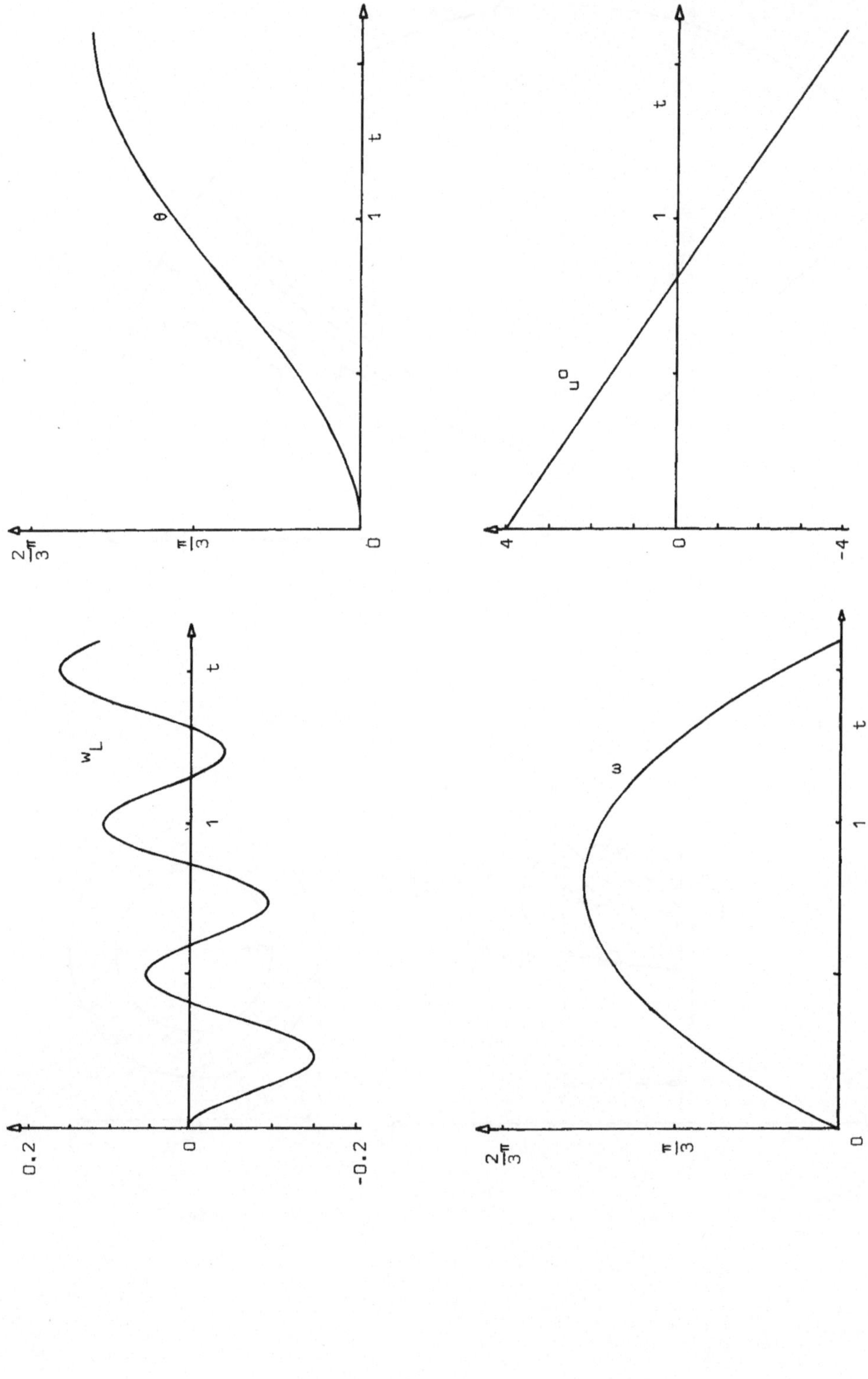
0.2
0
-0.2
w_L
1
t
$\frac{2}{3}\pi$
$\frac{\pi}{3}$
0
ω
1
t
$\frac{2}{3}\pi$
$\frac{\pi}{3}$
0
θ
1
t
4
0
-4
u^o
1
t

Fig. 4

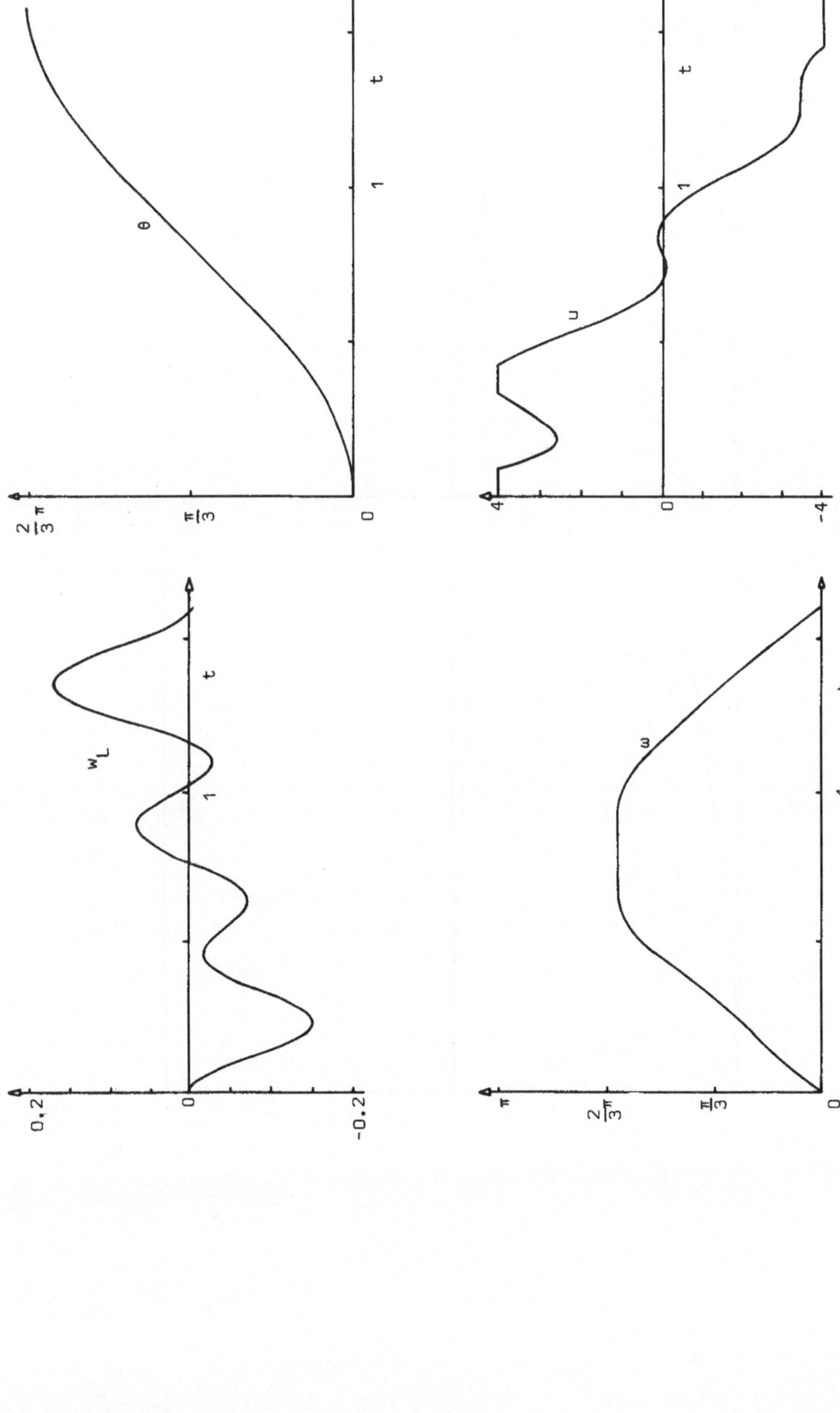

Fig. 5

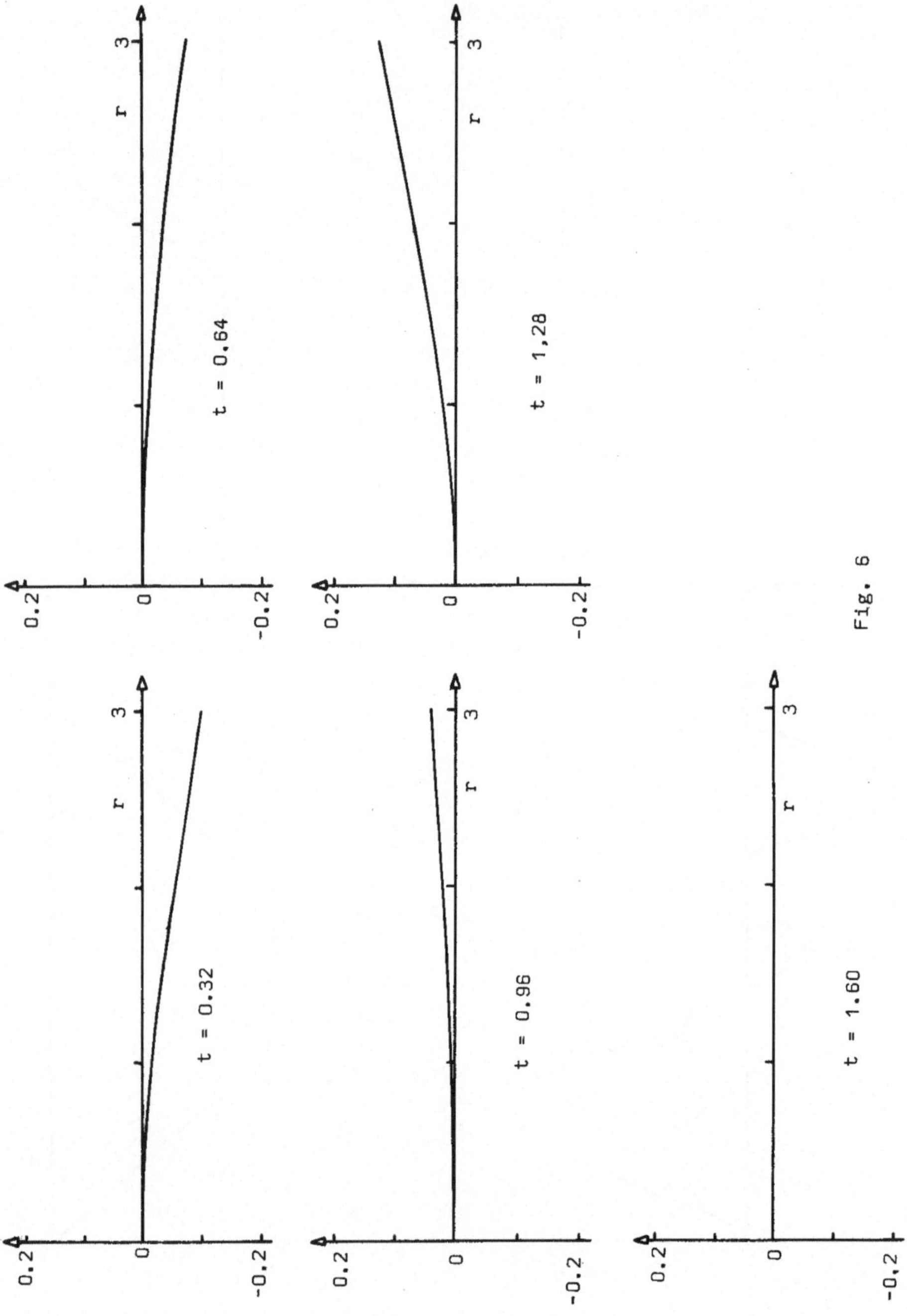

Fig. 6

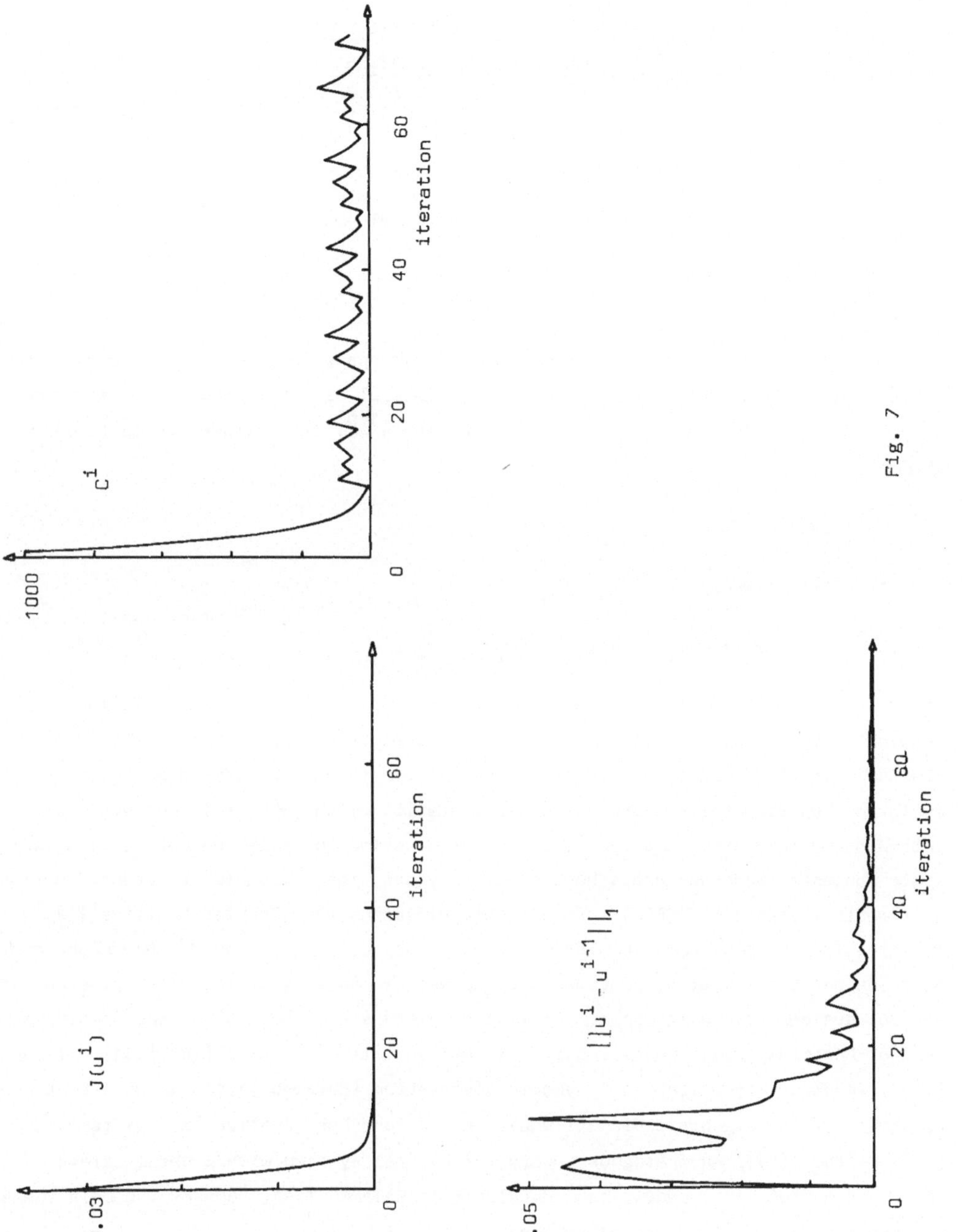
1000
0
20
40
60
iteration
C^i
0.03
0
20
40
60
iteration
J(u^i)
0.05
0
20
40
60
iteration
||u^i - u^(i-1)||_1

Fig. 7

NEUTRAL FUNCTIONAL DIFFERENTIAL EQUATIONS AND SEMIGROUPS OF OPERATORS

D. Salamon

Forschungsschwerpunkt Dynamische Systeme
Universität Bremen
D-2800 Bremen 33, West Germany

1. INTRODUCTION

The object of this paper is to develop a state space approach for linear neutral functional differential equations (NFDE) with general delays in the state- and input/output-variables in the state spaces $\mathbb{R}^n \times L^p$ and $W^{1,p}$. We consider the controlled NFDE

$$\frac{d}{dt}(x(t) - Mx_t - \Gamma u_t) = Lx_t + Bu_t \tag{1}$$

and the observed NFDE

$$\dot{x}(t) = L^T x_t + M^T \dot{x}_t, \quad y(t) = B^T x_t + \Gamma^T \dot{x}_t , \tag{2}$$

which is obtained from (1) by transposition of matrices. The duality relations between these two systems involve two different state concepts (Salamon [15]) and play an essential role in our theory. A particular outcome of this duality is an evolution equation for a state space repesentation of the NFDE (1). Moreover, we introduce certain structural operators which describe the relation between the two state concepts. Such operators have previously been introduced for retarded functional differential equations (RFDE) in a slightly different way (Bernier-Manitius [1], Manitius [9], Delfour-Manitius [6]). In this class of systems the structural operator approach has turned out of provide a very powerful and elegant tool for the study of various systems theoretic properties like completeness & small solutions (Manitius [9], Delfour-Manitius [6]), controllability & observability (Manitius [10],[11], Salamon [12]), feedback stabilization & dynamic observation (Salamon [13]), as well as the treatment of the algebraic Riccati equation (Delfour-Lee-Manitius [5], Vinter-Kwong [16], Delfour [4]). An analogous theory for neutral systems with general delays in input and output has recently been developed in Salamon [14]. Some of the main ideas of this approach will be presented here.

A preliminary section is devoted to the discussion of the two state concepts for NFDEs and their duality (Section 2). Semigroups and structural operators for the description of the free motions of neutral systems will be introduced in Section 3.

In this section we will also prove some basic relations between the various operators. Finally, we consider systems with delays in control and observation (Section 4) and indicate some consequences of our results concerning controllability and observability properties of neutral systems (Section 5).

Notation

We will always assume that $x(t)$, $\mathscr{x}(t) \in \mathbb{R}^n$ and $u(t)$, $y(t) \in \mathbb{R}^m$ and we define $x_t(\tau) = x(t+\tau)$, $u_t(\tau) = u(t+\tau)$ for $-h \leq \tau \leq 0$ where $0 < h < \infty$. Correspondingly L, M, B, Γ are bounded, linear maps from $C = C([-h,0];\mathbb{R}^n)$ respectively $C([-h,0];\mathbb{R}^m)$ into $\mathbb{R}^n$. These can be represented by matrix functions $\eta(\tau)$, $\mu(\tau)$, $\beta(\tau)$, $\gamma(\tau)$ of bounded variation in the following way

$$L\phi = \int_{-h}^{0} d\eta(\tau)\phi(\tau), \quad M\phi = \int_{-h}^{0} d\mu(\tau)\phi(\tau), \quad \phi \in C([-h,0];\mathbb{R}^n).$$

$$B\xi = \int_{-h}^{0} d\beta(\tau)\xi(\tau), \quad \Gamma\xi = \int_{-h}^{0} d\gamma(\tau)\xi(\tau), \quad \xi \in C([-h,0];\mathbb{R}^m).$$

The linear maps L^T, M^T, B^T, Γ^T from C into $\mathbb{R}^n$ respectively $\mathbb{R}^m$ are represented by the transposed matrix functions in an obvious manner. Without loss of generality we can assume that these matrix functions are normalized, i.e. vanish for $\tau \geq 0$, are constant for $\tau \leq -h$ and left continuous for $-h < \tau < 0$. Moreover, we will always assume that

$$-1 \notin \sigma(\lim_{\tau \uparrow 0} \mu(\tau)) \tag{3}$$

in order to guarantee existence and uniqueness for the solutions to (1) and (2).

Furthermore, we make use of the abbreviations $L^p = L^p([-h,0];\mathbb{R}^n)$ and $M^p = \mathbb{R}^n \times L^p$. The Sobolev space $W^{1,p} = W^{1,p}([-h,0];\mathbb{R}^n)$ will be identified with a dense subspace of M^p via the continuous embedding

$$\iota: W^{1,p} \to M^p, \qquad \iota\phi = (\phi(0) - M\phi, \phi).$$

Analogously, we define $\iota^T: W^{1,q} \to M^q$ by $\iota^T\psi = (\psi(0) - M^T\psi, \psi)$. The adjoint ι^{T*} of this map is a continuous embedding of $M^p = M^{q*}$ into the dual space $W^{-1,p}$ of $W^{1,q}$ $(1/p + 1/q = 1)$.

The usual duality pairing between M^q and M^p will be denoted by

$$\langle\psi,\phi\rangle = \psi^{0^T}\phi^0 + \int_{-h}^{0} \psi^{1^T}(\tau)\phi^1(\tau)d\tau$$

for $\phi = (\phi^0,\phi^1) \in M^p$ and $\psi = (\psi^0,\psi^1) \in M^q$. Finally, we introduce the map $\pi: \mathbb{R}^n \times L^p \times L^p \to W^{-1,p}$ which associates with every triple

$f = (f^0, f^1, f^2) \in \mathbb{R}^n \times L^p \times L^p$ the following bounded linear functional $\pi f \in W^{-1,p}$ on $W^{1,q}$

$$\langle \psi, \pi f \rangle_{W^{1,q}, W^{-1,p}} = \psi^T(0) f^0 + \int_{-h}^{0} \psi^T(\tau) f^1(\tau) d\tau + \int_{-h}^{0} \dot{\psi}^T(\tau) f^2(\tau) d\tau .$$

The map $\pi^T : \mathbb{R}^n \times L^q \times L^q \to W^{-1,q}$ is defined analogously.

2. STATE SPACE DESCRIPTION AND DUALITY

The 'classical' way of introducing the state of a functional differential equation (FDE) is to specify an initial state of suitable length which describes the past history of the solution. An alternative (dual) state concept can be obtained by defining the initial state of the FDE to be an additional forcing term of suitable length which determines the future behaviour of the solution (Miller). These two notions of the state are dual to each other in the sense that the evolution of the state in the sense of Miller is described by the adjoint semigroup of the one which corresponds to the transposed equation in terms of the 'classical' state concept (Burns, Herdman). A modified version of these ideas applies to neutral systems in the state spaces M^p and $W^{1,p}$ (Salamon [14],[15]).

Let us begin with the discussion of the 'classical' state concept for the observed NFDE

$$\Omega^T \qquad \begin{aligned} \dot{x}(t) &= L^T x_t + M^T \dot{x}_t , \\ y(t) &= B^T x_t + \Gamma^T \dot{x}_t . \end{aligned}$$

This system admits a unique solution $x \in W^{1,q}_{loc}([-h,\infty), \mathbb{R}^n)$ for every initial condition

$$x(\tau) = \psi(\tau), \qquad -h \leq \tau \leq 0 , \tag{4}$$

where $\psi \in W^{1,q}$ (Henry [7]). Correspondingly, the state of system Ω^T at time $t \geq 0$ will be defined to be the solution segment $x_t \in W^{1,q}$.

In the case $\Gamma = 0$ the output does not depend on the derivative of the solution and hence the above system can be extended to the product space M^q. For this sake we rewrite system Ω^T as follows

$$\Sigma^T \qquad \begin{aligned} \dot{z}(t) &= L^T x_t , \\ x(t) &= z(t) + M^T x_t , \\ y(t) &= B^T x_t . \end{aligned}$$

It has been shown by Burns, Herdman, and Stech [2] that system Σ^T admits a unique solution pair $z \in W^{1,q}_{loc}([0,\infty);\mathbb{R}^n)$, $x \in L^q_{loc}([-h,\infty);\mathbb{R}^n)$ for every initial condition

$$z(0) = \psi^0, \quad x(\tau) = \psi^1(\tau), \qquad -h \leq \tau < 0, \tag{5}$$

where $\psi = (\psi^0,\psi^1) \in M^q$. Correspondingly, the state of system Σ^T at time $t \geq 0$ will be defined to be the pair $(z(t),x_t) \in M^q$. Note that the embedding $\iota^T: W^{1,q} \to M^q$ maps every state to system Ω^T into the corresponding state of system Σ^T (in the case $\Gamma = 0$).

We have seen that an *extension* of the NFDE (2) to the product space M^q is only possible if $\Gamma = 0$. The opposite situation occurs in the case of the controlled NFDE (2). This time a *restriction* to absolutely continuous solutions is only possible if $\Gamma = 0$ (no derivatives in the input). Correspondingly we obtain the systems

$$\Sigma \qquad \begin{aligned} \dot{w}(t) &= Lx_t + Bu_t, \\ x(t) &= w(t) + Mx_t + \Gamma u_t, \\ w(0) &= \phi^0, \ x(\tau) = \phi^1(\tau), \quad -h \leq \tau < 0, \qquad (6.1) \\ u(\tau) &= \xi(\tau), \quad -h \leq \tau < 0, \qquad (6.2) \end{aligned}$$

($\phi \in M^p$, $\xi \in L^p([-h,0];\mathbb{R}^m)$) and

$$\Omega \qquad \begin{aligned} \dot{x}(t) &= Lx_t + M\dot{x}_t + Bu_t, \\ x(\tau) &= \phi(\tau), \quad -h \leq \tau \leq 0, \qquad (7.1) \\ u(\tau) &= \xi(\tau), \quad -h \leq \tau < 0, \qquad (7.2) \end{aligned}$$

($\phi \in W^{1,p}$, $\xi \in L^p([-h,0];\mathbb{R}^m)$). The fact that the future behaviour of the solution depends also on the past values of the control function $u(t)$ indicates that the input segment $u_t \in L^p([-h,0];\mathbb{R}^m)$ should be included in the state of the systems Σ and Ω. This difficulty can be overcome if we introduce the announced dual state concept. For this sake we replace the action of the initial functions ϕ and ξ on the right hand side of the equation by extra forcing terms.

Then system Σ transforms into

$$\tilde{\Sigma}\quad \begin{aligned} \dot{w}(t) &= \int_{-t}^{0} d\eta(\tau)x(t+\tau) + \int_{-t}^{0} d\beta(\tau)u(t+\tau) + f^1(-t),\\ x(t) &= w(t) + \int_{-t}^{0} d\mu(\tau)x(t+\tau) + \int_{-t}^{0} d\gamma(\tau)u(t+\tau) + f^2(-t),\\ w(0) &= f^0, \end{aligned}$$

where the triple $f = (f^0, f^1, f^2) \in \mathbb{R}^n \times L^p \times L^p$ is given by

$$f^0 = \phi^0, \tag{8.1}$$

$$f^1(\sigma) = \int_{-h}^{\sigma} d\eta(\tau)\phi^1(\tau-\sigma) + \int_{-h}^{\sigma} d\beta(\tau)\xi(\tau-\sigma), \qquad -h \le \sigma \le 0, \tag{8.2}$$

$$f^2(\sigma) = \int_{-h}^{\sigma} d\mu(\tau)\phi^1(\tau-\sigma) + \int_{-h}^{\sigma} d\gamma(\tau)\xi(\tau-\sigma), \qquad -h \le \sigma \le 0. \tag{8.3}$$

The initial state of system $\tilde{\Sigma}$ is defined to be the bounded linear functional $\pi f \in W^{-1,p}$ on $W^{1,q}$. This definition is motivated from the fact that the solution $x(t)$ of $\tilde{\Sigma}$ vanishes for $t \ge 0$ iff $\pi f = 0$ (see Lemma 2.1 below). Correspondingly the state of $\tilde{\Sigma}$ at time $t \ge 0$ is given by $\pi(w(t), w^t, x^t) \in W^{-1,p}$ where $w^t, x^t \in L^p$ are the forcing terms of $\tilde{\Sigma}$ after a time shift. These are of the form

$$w^t(\sigma) = \int_{\sigma-t}^{\sigma} d\eta(\tau)x(t+\tau-\sigma) + \int_{\sigma-t}^{\sigma} d\beta(\tau)u(t+\tau-\sigma) + f^1(\sigma-t), \tag{9.1}$$

$$x^t(\sigma) = \int_{\sigma-t}^{\sigma} d\mu(\tau)x(t+\tau-\sigma) + \int_{\sigma-t}^{\sigma} d\gamma(\tau)u(t+\tau-\sigma) + f^2(\sigma-t). \tag{9.2}$$

Motivated from the one-to-one correspondence between this state at time $t \ge 0$ and the future behaviour $x(t+s)$, $s \ge 0$, of the solution one might regard the bounded, linear functional $\pi(w(t), w^t, x^t)$ - defined by (9) and (8) - as the 'real' state of system Σ.

Analogously, system Ω can be transformed into

$$\tilde{\Omega}\quad \begin{aligned} \dot{x}(t) &= \int_{-t}^{0} d\eta(\tau)x(t+\tau) + \int_{-t}^{0} d\mu(\tau)\dot{x}(t+\tau) + \int_{-t}^{0} d\beta(\tau)u(t+\tau) + f^1(-t),\\ x(0) &= f^0, \end{aligned}$$

where the initial state $f = (f^0, f^1) \in M^p$ is of the form

$$f^0 = \phi(0), \tag{10.1}$$

$$f^1(\sigma) = \int_{-h}^{\sigma} d\eta(\tau)\phi(\tau-\sigma) + \int_{-h}^{\sigma} d\mu(\tau)\dot{\phi}(\tau-\sigma) + \int_{-h}^{\sigma} d\beta(\tau)\xi(\tau-\sigma). \tag{10.2}$$

The state $(x(t),x^t) \in M^p$ of $\tilde{\Omega}$ at time $t \geq 0$ is given by

$$x^t(\sigma) = \int_{\sigma-t}^{\sigma} d\eta(\tau)x(t+\tau-\sigma) + \int_{\sigma-t}^{\sigma} d\mu(\tau)\dot{x}(t+\tau-\sigma) + \int_{\sigma-t}^{\sigma} d\beta(\tau)u(t+\tau-\sigma)$$
$$+ f^1(\sigma-t), \quad -h \leq \sigma \leq 0. \tag{11}$$

The next lemma has been proved in Salamon [15]. It shows that the embedding $\iota^{T*}: M^p \rightarrow W^{-1,p}$ maps every state of $\tilde{\Omega}$ into the corresponding state of $\tilde{\Sigma}$.

<u>Lemma 2.1.</u> Let $\Gamma = 0$ and let $f \in M^p$, $f \in \mathbb{R}^n \times L^p \times L^p$ as well as $u \in L^p_{loc}([0,\infty);\mathbb{R}^m)$ be given. Moreover let $x(t)$ be the unique solution of $\tilde{\Omega}$ and $w(t)$, $x(t)$ the unique solution pair of $\tilde{\Sigma}$. Then $x(t) = x(t)$ for all $t \geq 0$ if and only if $\pi f = \iota^{T*}f$.

Now we are in the position to formulate the basic duality result for neutral systems which has been proved in Salamon [15].

<u>Theorem 2.2.</u> Let $u \in L^p_{loc}([0,\infty);\mathbb{R}^n)$ be given.

(i) Let $f \in \mathbb{R}^n \times L^p \times L^p$ and $\psi \in W^{1,q}$. Moreover, let $\pi(w(t),w^t,x^t) \in W^{-1,p}$ be the state of $\tilde{\Sigma}$ - defined by (9) - and let $x(t)$ be the unique solution of Ω^T, (4) with output $y(t)$. Then

$$\langle\psi,\pi(w(t),w^t,x^t)\rangle = \langle x_t,\pi f\rangle + \int_0^t y^T(t-s)u(s)ds, \quad t \geq 0.$$

(ii) Let $f \in M^p$ and $\psi \in M^q$. Moreover, let $(x(t),x^t) \in M^p$ be the state of $\tilde{\Omega}$ - defined by (11) - and let $z(t)$, $x(t)$ be the unique solution pair of Σ^T, (5) with output $y(t)$. Then

$$\langle\psi,(x(t),x^t)\rangle = \langle(z(t),x^t),f\rangle + \int_0^t y^T(t-s)u(s)ds, \quad t \geq 0.$$

<u>3. SEMIGROUPS AND STRUCTURAL OPERATORS</u>

Throughout this section we restrict our discussion to the free motions of (1) and (2). This means that we have to deal with the following four systems

Σ	$\dot{w}(t) = Lx_t$ $x(t) = w(t) + Mx_t$	$\dot{z}(t) = L^Tx_t$ $x(t) = z(t) + M^Tx_t$	Σ^T
Ω	$\dot{x}(t) = Lx_t + M\dot{x}_t$	$\dot{x}(t) = L^Tx_t + M^T\dot{x}_t$	Ω^T

The systems on the left hand side correspond to the NFDE (1) and those on the right hand side to the transposed NFDE (2). On each side the system below represents the restriction of the upper system to absolutely continuous solutions. A diagonal relation is given by the above duality result (Theorem 2.2).

The evolution of these four systems in terms of the 'classical' state concept (initial functions) can be described by the following four strongly continuous semigroups

$$S(t): M^p \to M^p, \qquad S^T(t): M^q \to M^q,$$

$$\mathcal{S}(t): W^{1,p} \to W^{1,p}, \qquad \mathcal{S}^T(t): W^{1,q} \to W^{1,q}.$$

The semigroup $S(t)$ on M^p has recently been introduced by Burns, Herdman and Stech [2] and associates with every $\phi \in M^p$ the corresponding state $S(t)\phi = (w(t),x_t)$ of Σ, (6.1) at time $t \geq 0$. The semigroup $\mathcal{S}(t): W^{1,p} \to W^{1,p}$ maps every $\phi \in W^{1,p}$ into the corresponding solution segment $\mathcal{S}(t)\phi = x_t \in W^{1,p}$ of Ω, (7.1). The infinitesimal generators of $S(t)$ and $\mathcal{S}(t)$ are given by

$$\text{dom } A = \{\phi \in M^p | \phi^1 \in W^{1,p}, \phi^0 = \phi^1(0) - M\phi^1\},$$

$$A\phi = (L\phi^1, \dot{\phi}^1),$$

and

$$\text{dom } \mathcal{A} = \{\phi \in W^{1,p} | \dot{\phi} \in W^{1,p}, \dot{\phi}(0) = L\phi + M\dot{\phi}\},$$

$$\mathcal{A}\phi = \dot{\phi} .$$

The (transposed) semigroups $S^T(t)$ and $\mathcal{S}^T(t)$ are defined analogously. They are not the adjoint operators of $S(t)$ and $\mathcal{S}(t)$. However, Theorem 2.2 allows us to give an interpretation of the adjoint semigroups $S^{T*}(t)$ and $\mathcal{S}^{T*}(t)$ in terms of the dual state concept for the original system equation.

Corollary 3.1.

(i) Let $f \in \mathbb{R}^n \times L^p \times L^p$ be given and let $\pi(w(t),w^t,x^t)$ be the corresponding state of $\tilde{\Sigma}$ with input $u(t) \equiv 0$. Then we have $\pi(w(t),w^t,x^t) = \mathcal{S}^{T*}(t)\pi f$.

(ii) Let $f \in M^p$ be given and let $(x(t),x^t) \in M^p$ be the corresponding state of $\tilde{\Omega}$ with input $u(t) \equiv 0$. Then we have $(x(t),x^t) = S^{T*}(t)f$.

Proof. If $x(t)$ is the solution of Ω^T corresponding to the initial state $x_o = \psi \in W^{1,q}$, then by Theorem 2.2,

$$<\psi,\pi(w(t),w^t,x^t)> = <x_t,\pi f> = <\mathcal{S}^T(t)\psi,\pi f>.$$

This proves (i). (ii) follows analogously.

Q.E.D.

Remarks 3.2.

(i) It can be proved straight forward that the infinitesimal generator A^{T*} of $S^{T*}(t)$ is of the following form. Given $f, g \in M^p$, we have $f \in \text{dom } A^{T*}$ and $A^{T*} f = g$ if and only if the following equations hold

$$- \eta(-h) f^0 = [I + \mu(-h)] g^0 + \int_{-h}^{0} g^1(\tau) d\tau,$$

$$f^1(\sigma) - \eta(\sigma) f^0 = [I + \mu(\sigma)] g^0 + \int_{\sigma}^{0} g^1(\tau) d\tau, \qquad -h \le \sigma \le 0.$$

(ii) The domain of the infinitesimal generator $\mathcal{A}^{T*}$ of $\mathcal{S}^{T*}(t)$ is given by dom $\mathcal{A}^{T*}$ = ran ι^{T*} (see e.g. Salamon [14, Lemma II.3.2]).

(iii) By definition, the semigroup $\mathcal{S}(t)$ represents the restriction of $S(t)$ to the domain of its generator which is given by dom A = ran ι. The same holds for the semigroups $\mathcal{S}^T(t)$ and $S^T(t)$ and, by duality, for the adjoint semigroups $S^{T*}(t)$ and $\mathcal{S}^{T*}(t)$ with interchanged roles. These facts can be expressed by the formulas

$$\iota \mathcal{S}(t) = S(t)\iota, \qquad \iota^T \mathcal{S}^T(t) = S^T(t)\iota^T,$$
$$\iota^{T*} S^{T*}(t) = \mathcal{S}^{T*}(t)\iota^{T*}, \qquad \iota^* S^*(t) = \mathcal{S}^*(t)\iota^*. \tag{12}$$

THE STRUCTURAL OPERATORS

We have seen in Section 2 that the solution segment of system Σ (respectively Ω) at time h can be derived from the initial function in two steps. These two operations can be expressed by socalled 'structural operators' F and G (respectively $\mathcal{F}$ and $\mathcal{G}$). Roughly speaking, the operator F maps the initial function into the corresponding forcing term of the equation and the operator G maps this forcing term into the corresponding solution segment at time h.

More precisely, the four operators

$$F: M^p \to W^{-1,p}, \qquad G: W^{-1,p} \to M^p,$$

$$\mathcal{F}: W^{1,p} \to M^p, \qquad \mathcal{G}: M^p \to W^{1,p}$$

are defined as follows.

Let $\phi \in M^p$ and $\xi = 0$, then $F\phi = \pi f \in W^{-1,p}$ where the triple $f = (f^0, f^1, f^2)$ is defined by (8).

Let $f \in \mathbb{R}^n \times L^p \times L^p$, then $G\pi f = (w(h), x_h) \in M^p$ where $w(t), x(t)$ is the solution pair of $\tilde{\Sigma}$, $u(t) \equiv 0$.

Let $\phi \in W^{1,p}$ and $\xi = 0$, then $F\phi = f \in M^p$ is defined by (10).

Let $f = (f^0, f^1) \in M^p$, then $Gf = x_h \in \tilde{W}^{1,p}$ where $x(t)$ is the solution of $\tilde{\Omega}$, $u(t) \equiv 0$.

<u>Lemma 3.3.</u> The above operators $G: M^p \to W^{1,p}$ and $\mathrm{G}: W^{-1,p} \to M^p$ are well defined, bounded, linear and bijective.

<u>Proof.</u> If follows from the existence, uniqueness and continuous dependence of the solutions to $\tilde{\Omega}$ (Salamon [14, Section II.1]) that the operator $G: M^p \to W^{1,p}$ is well defined, bounded and linear. It is easy to see that this operator is always bijective.

In order to prove the assertions of the lemma for the operator G, we introduce the operator $\tilde{\mathrm{G}}: \mathbb{R}^n \times L^p \times L^p \to M^p$ which associates with every forcing term $f = (f^0, f^1, f^2)$ the corresponding solution segment $\tilde{\mathrm{G}}f = (w(h), x_h) \in M^p$ of system $\tilde{\Sigma}$. This operator is obviously bounded and linear. Moreover, it follows from Lemma 2.1 that $\ker \tilde{\mathrm{G}} = \ker \pi$. Hence $\tilde{\mathrm{G}}$ induces an injective operator $\tilde{\tilde{\mathrm{G}}}$ from $\mathbb{R}^n \times L^p \times L^p/\ker \pi$ into M^p. Note that the map $[f] \to \pi f$ from $\mathbb{R}^n \times L^p \times L^p/\ker \pi$ onto $W^{-1,p}$ is an isomorphism. We conclude that there exists a unique bounded, linear, one-to-one map G from $W^{-1,p}$ into M^p such that $\mathrm{G}\pi f = \tilde{\tilde{\mathrm{G}}}[f] = \tilde{\mathrm{G}}f$ for every triple $f \in \mathbb{R}^n \times L^p \times L^p$. Again it is easy to see that the operator G is onto.

Q.E.D.

As a consequence of Lemma 2.1 and Corollary 3.1 we obtain the following important relations between the structural operators and semigroups.

<u>Theorem 3.4.</u>

(i) $\mathrm{S}(h) = \mathrm{GF}$, $S(h) = GF$, $S^{T*}(h) = \mathrm{FG}$, $\mathrm{S}^{T*}(h) = FG$.

(ii) $\mathrm{FS}(t) = S^{T*}(t)\mathrm{F}$, $FS(t) = \mathrm{S}^{T*}(t)F$,

$\mathrm{S}(t)\mathrm{G} = \mathrm{G}S^{T*}(t)$, $S(t)G = G\mathrm{S}^{T*}(t)$.

(iii) $\mathrm{F}\iota = \iota^{T*}F$, $\mathrm{G}\iota^{T*} = \iota G$.

<u>Proof.</u> (i) follows directly from the definition of the operators F, G, F and G.

Now let $f \in M^p$ be given and let $x(t)$, $t \geq 0$, be the unique solution of $\tilde{\Omega}$ corresponding to the input $u(t) \equiv 0$. Moreover, let $x^t \in L^p$ be defined by (11). Then

$$\dot{x}(t+s) = \int_{-s}^{0} d\eta(\tau)x(t+s+\tau) + \int_{-s}^{0} d\mu(\tau)\dot{x}(t+s+\tau) + x^t(-s), \quad s \geq 0,$$

and hence $G(x(t), x^t) = x_{t+h}$. By Corollary 3.1, this implies

$$G\mathrm{S}^{T*}(t)f = G(x(t), x^t) = x_{t+h} = S(t)x_h = S(t)Gf.$$

Now it follows from (i) that the following equation holds for every $\phi \in W^{1,p}$

$$GFS(t)\phi = S(t+h)\phi = S(t)GF\phi = G\mathrm{S}^{T*}(t)F\phi.$$

This proves the equations on the right hand side of (ii) since $\mathcal{G}$ is injective. The remaining assertions in (ii) follow analogously.

In order to prove (iii), let $\mathcal{f} \in M^p$ and $f \in \mathbb{R}^n \times \tilde{L}^p \times L^p$ satisfy $\iota^{T^*}\mathcal{f} = \pi f$. Moreover, let $x(t)$, $t \geq 0$, be the unique solution of $\tilde{\Omega}$ and $w(t)$, $x(t)$, $t \geq 0$, the unique solution pair of Σ corresponding to the input $u(t) \equiv 0$. Then $x(t) = x(t)$ for every $t \geq 0$ (Lemma 2.1) and hence

$$\iota \mathcal{G}\mathcal{f} = \iota x_h = (x(h) - Mx_h, x_h) = (w(h), x_h) = G\pi f = G\iota^{T^*}\mathcal{f}.$$

By (i) and (12), this implies that the following equation holds for every $\phi \in W^{1,p}$

$$GF\iota\phi = S(h)\iota\phi = \iota \mathcal{S}(h)\phi = \iota \mathcal{G}\mathcal{F}\phi = G\iota^{T^*}\mathcal{F}\phi.$$

Hence (iii) follows from the injectivity of G (Lemma 3.3).

Q.E.D.

Corollary 3.5.

(i) $S^T(h) = \mathcal{G}^*\mathcal{F}^*$, $\mathcal{S}^T(h) = G^*F^*$, $\mathcal{S}^*(h) = \mathcal{F}^*\mathcal{G}^*$, $S^*(h) = F^*G^*$.

(ii) $\mathcal{F}^*S^T(t) = \mathcal{S}^*(t)\mathcal{F}^*$, $F^*\mathcal{S}^T(t) = S^*(t)F^*$,

$S^T(t)\mathcal{G}^* = \mathcal{G}^*\mathcal{S}^*(t)$, $\mathcal{S}^T(t)G^* = G^*S^*(t)$

(iii) $\mathcal{F}^*\iota^T = \iota^*F^*$, $\mathcal{G}^*\iota^* = \iota^T G^*$.

The relations of Theorem 3.4 may be illustrated by the following commuting diagramm

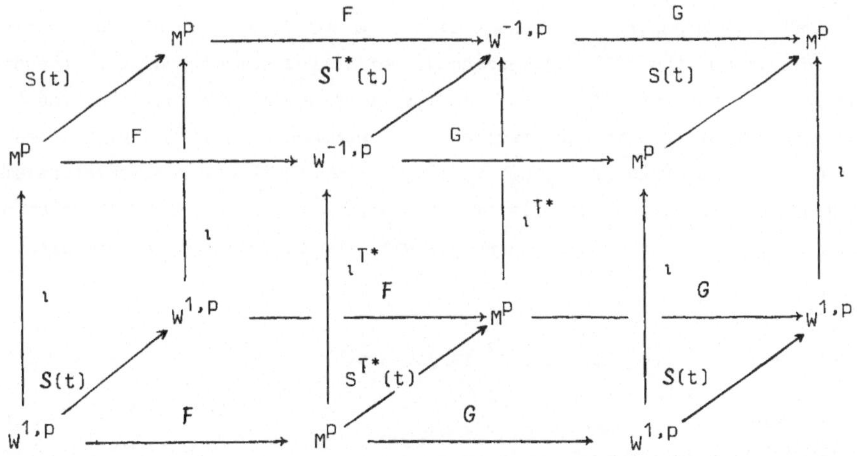

Taking the adjoint operators we obtain the commuting diagramm below (Corollary 3.5)

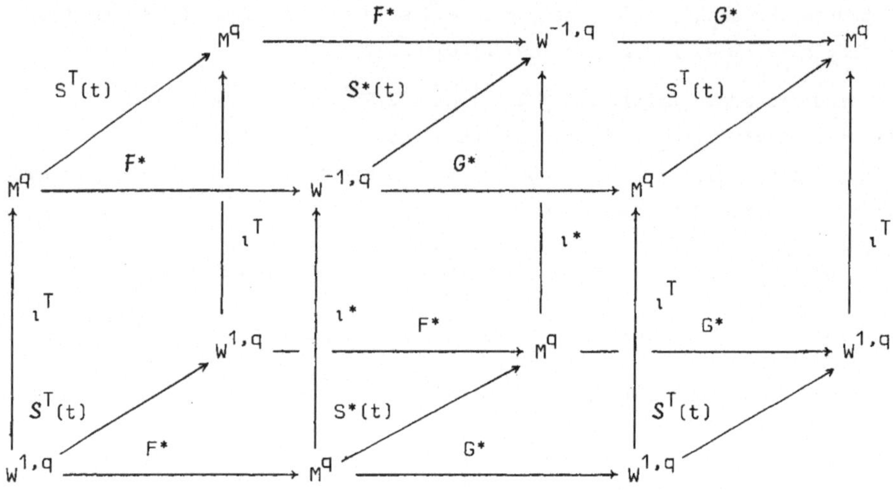

These relations are obtained by just dualizing Theorem 3.4. However, it is important not only to make use of these equations in a purely formal way but to understand their meaning. More precisely, we will see that the equations of Corollary 3.5 play the same role for the description of the systems Σ^T and Ω^T as those of Theorem 3.4 do for the systems Σ and Ω. For this sake we have to show that the operators $F^*: W^{1,q} \to M^q$ and $G^*: M^q \to W^{1,q}$ (respectively $\mathcal{F}^*: M^q \to W^{-1,q}$ and $\mathcal{G}^*: W^{-1,q} \to M^q$) are the structural operators of system Ω^T (respectively Σ^T). For the operators F^* and $\mathcal{F}^*$ this can be estabilished in a straight forward way.

For proving the desired properties of the operators G^* and $\mathcal{G}^*$ one has to do a bit more. First give a concrete representation of $\mathcal{G}$ and G in terms of the fundamental matrix solution of the NFDE (1). Secondly, derive the corresponding representation of the adjoint operators G^* and $\mathcal{G}^*$. Finally, make use of the fact that the transposed of the fundamental solution is precisely the fundamental solution of the transposed equation (2) in order to prove that G^* and $\mathcal{G}^*$ are the structural operators of the systems Ω^T and Σ^T. Further details of these arguments can be found in Salamon [14, Section II.2]). Here we content ourselves with the statement of the results.

<u>Lemma 3.6.</u>

(i) Let $\psi \in W^{1,q}$. Then $F^*\psi = g \in M^q$ is given by

$$g^0 = \psi(0), \tag{13.1}$$

$$g^1(\sigma) = \int_{-h}^{\sigma} d\eta^T(\tau)\psi(\tau-\sigma) + \int_{-h}^{\sigma} d\mu^T(\tau)\dot{\psi}(\tau-\sigma), \quad -h \le \sigma \le 0. \tag{13.2}$$

(ii) Let $\psi \in M^q$. Then $\mathcal{F}^*\psi = \pi^T g \in W^{-1,q}$ where the triple $g \in \mathbb{R}^n \times L^q \times L^q$ is given by

$$g^0 = \psi^0, \tag{14.1}$$

$$g^1(\sigma) = \int_{-h}^{\sigma} d\eta^T(\tau)\psi^1(\tau-\sigma), \quad -h \leq \sigma \leq 0, \tag{14.2}$$

$$g^2(\sigma) = \int_{-h}^{\sigma} d\mu^T(\tau)\psi^1(\tau-\sigma), \quad -h \leq \sigma \leq 0. \tag{14.3}$$

(iii) Let $g \in M^q$. Then $G^*g = x_h \in W^{1,q}$ where $x(t)$, $t \geq 0$, is the unique solution of

$$\tilde{\Omega}^T \quad \begin{aligned} \dot{x}(t) &= \int_{-t}^{0} d\eta^T(\tau)x(t+\tau) + \int_{-t}^{0} d\mu^T(\tau)\dot{x}(t+\tau) + g^1(-t), \\ x(0) &= g^0. \end{aligned}$$

(iv) Let $g \in \mathbb{R}^n \times L^q \times L^q$. Then $G^*\pi^T g = (z(h), x_h) \in M^q$ where $z(t)$, $x(t)$, $t \geq 0$, is the unique solution pair of

$$\tilde{\Sigma}^T \quad \begin{aligned} \dot{z}(t) &= \int_{-t}^{0} d\eta^T(\tau)x(t+\tau) + g^1(-t), \quad z(0) = g^0, \\ x(t) &= z(t) + \int_{-t}^{0} d\mu^T(\tau)x(t+\tau) + g^2(-t), \quad t \geq 0. \end{aligned}$$

Remarks 3.7.

(i) For retarded systems ($M = 0$) there has been defined a structural operator $\bar{F}: M^p \to M^p$ which maps $\phi \in M^p$ into the pair $\bar{F}\phi = (f^0, f^1) \in M^p$ defined by (8.1) and (8.2) with $\xi = 0$ (Delfour-Manitius [6]). This operator makes the following diagramm commute

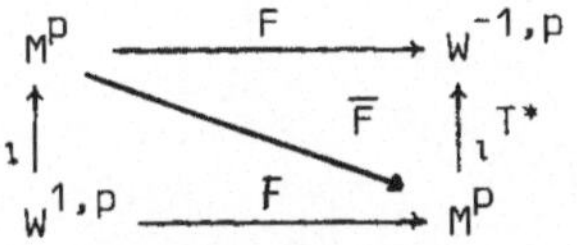

$$\mathcal{F} = \bar{F}\iota, \quad F = \iota^{T*}\bar{F}.$$

In general there does not exist such an operator $\bar{F}$ since ran F will not be contained in ran ι^{T*}. Therefore it is necessary to deal with *two* structural operators F and $\mathcal{F}$ for neutral systems. The adjoint operators $\mathcal{F}^*$ and F^* correspond to the transposed equation with *interchanged roles* (Lemma 3.6).

(ii) The operator $\bar{G}: M^p \to M^p$ which was introduced by Manitius [9] for retarded systems is given by $\bar{G} = G\iota^{T*} = \iota\mathcal{G}$.

(iii) The operator family $S(t)$ defines a C_0-group if and only if $F: M^p \to W^{-1,p}$ (or equivalently $\mathcal{F}: W^{1,p} \to M^p$) is bijective. A sufficient condition is that

$$\det A_{-1} \neq 0, \quad A_{-1} = \lim_{\tau \downarrow -h} \mu(\tau) - \mu(-h). \tag{15}$$

If $\mu(\tau)$ is absolutely continuous with L^q-derivative on some interval $(-h, \varepsilon-h]$, $\varepsilon > 0$, then (15) is also necessary for F to be a Banach isomorphism (Salamon [14, Prop. III. 1.15]; see also Burns-Herdman-Stech [2]).

4. DELAYS IN CONTROL AND OBSERVATION

Throughout this section we consider the control system Σ, Ω, Σ^T and Ω^T introduced in Section 2.

As a consequence of Theorem 2.2 we will first derive an evolution equation for the state space description of the systems $\tilde{\Sigma}$ and $\tilde{\Omega}$ in the case $\Gamma = 0$. For this sake let us introduce the output operator $\mathcal{B}^T : W^{1,q} \to \mathbb{R}^m$ by defining

$$\mathcal{B}^T\psi = \int_{-h}^{0} d\beta^T(\tau)\psi(\tau), \qquad \psi \in W^{1,q}.$$

This operator has the following properties.

Remarks 4.1.

(i) For every $T > 0$ there exists some constant $b_T > 0$ such that the following equation holds for every $\psi \in W^{1,q}$

$$||\mathcal{B}^T S^T(.)\psi||_{L^q([0,T];\mathbb{R}^m)} \leq b_T ||\iota^T\psi||_{M^q} .$$

This follows from the fact that the output function $y(.)$ of system Σ^T depends continuously on the initial state.

(ii) For every $u(.) \in L^p([0,T];\mathbb{R}^m)$ we have

$$\int_0^T S^{T^*}(T-s)\mathcal{B}^{T^*}u(s)ds \in \operatorname{ran} \iota^{T^*}$$

and

$$||\iota^{T^{*-1}} \int_0^T S^{T^*}(T-s)\mathcal{B}^{T^*}u(s)ds||_{M^p} \leq b_T||u(.)||_{T,p} .$$

This follows from (i) duality (see e.g. Salamon [14, Section I.3]). (ii) can also be obtained as a consequence of the following result.

Corollary 4.2. Let $u(.) \in L^p_{loc}([0,\infty);\mathbb{R}^m)$ be given and let $\Gamma = 0$.

(i) Let $f \in \mathbb{R}^n \times L^p \times L^p$ and let $\pi(w(t),w^t,x^t) \in W^{-1,p}$ be the corresponding state of $\tilde{\Sigma}$ - defined by (9). Then

$$\pi(w(t),w^t,x^t) = S^{T^*}(t)\pi f + \int_0^t S^{T^*}(t-s)\mathcal{B}^{T^*}u(s)ds.$$

(ii) Let $f \in M^p$ and let $(x(t),x^t) \in M^p$ be the corresponding state of $\tilde{\Omega}$ - defined by (11). Then

$$(x(t),x^t) = S^{T^*}(t)f + \iota^{T^{*-1}} \int_0^t S^{T^*}(t-s)\mathcal{B}^{T^*}u(s)ds.$$

Proof. Let $x(t)$ be the unique solution of Ω^T with initial function $x_0 = \psi \in W^{1,q}$ and output $y(t) = \mathcal{B}^T S^T(t)\psi$, $t \geq 0$. Then, by Theorem 2.2,

$$<\psi, \pi(w(t), w^t, x^t)>_{W^{1,q}, W^{-1,p}}$$

$$= <S^T(t)\psi, \pi f>_{W^{1,q}, W^{-1,p}} + \int_0^t <\mathcal{B}^T S^T(t-s)\psi, u(s)>_{\mathbb{R}^m} ds$$

$$= <\psi, S^{T*}(t)\pi f + \int_0^t S^{T*}(t-s)\mathcal{B}^{T*}u(s)ds>_{W^{1,q}, W^{-1,p}} .$$

This proves (i). (ii) follows from (i) and Lemma 2.1.

Q.E.D.

Evolution equations of the above type play a central role in infinite dimensional linear systems theory (see e.g. Curtain-Pritchard [3]) as well as for nonlinear perturbation of linear systems via fixed point methods (see e.g. Ichikawa-Pritchard [8]). They are of particular importance for the treatment of the linear quadratic control problem and for the derivation of approximation results.

SOME FURTHER STRUCTURAL OPERATORS

In order to relate the above evolution equation to the original state concept, we introduce some further structural operators E and D from $L^p([-h,0];\mathbb{R}^m)$ into $W^{-1,p}$ (respectively $\mathcal{E}$ and $\mathcal{D}$ from $L^p([-h,0],\mathbb{R}^m)$ into M^p). These operators describe the action of the input segments ξ and u_h on the right hand side of $\tilde{\Sigma}$ (respectively $\tilde{\Omega}$). More precisely, we define

$$E\xi = \pi f \in W^{-1,p}, \quad f^0 = 0,$$

$$f^1(\sigma) = \int_{-h}^{\sigma} d\beta(\tau)\xi(\tau-\sigma), \quad f^2(\sigma) = \int_{-h}^{\sigma} d\gamma(\tau)\xi(\tau-\sigma).$$

$$D\xi = \pi f \in W^{-1,p}, \quad f^0 = 0,$$

$$f^1(\sigma) = \int_{\sigma}^{0} d\beta(\tau)\xi(\tau-\sigma-h), \quad f^2(\sigma) = \int_{\sigma}^{0} d\gamma(\tau)\xi(\tau-\sigma-h).$$

$$[\mathcal{E}\xi]^0 = 0, \quad [\mathcal{E}\xi]^1(\sigma) = \int_{-h}^{\sigma} d\beta(\tau)\xi(\tau-\sigma).$$

$$[\mathcal{D}\xi]^0 = 0, \quad [\mathcal{D}\xi]^1(\sigma) = \int_{\sigma}^{0} d\beta(\tau)\xi(\tau-\sigma-h)$$

for every $\xi \in L^p([-h,0];\mathbb{R}^m)$.

Remarks 4.3.

(i) Operators of the type $\mathcal{E}$ have previously been introduced by Vinter-Kwong [16] and Delfour [4] for the study of retarded systems.

(ii) If w(t), x(t) is any solution of Σ, (6), then the corresponding state of system $\tilde{\Sigma}$, (8), at time $t \geq 0$ is given by $\pi(w(t),w^t,x^t) = F(w(t),x_t) + Eu_t \in W^{-1,p}$.

Analogously, for any solution $x(t)$ of Ω, (7), the corresponding state of system $\tilde{\Omega}$, (10), at time $t \geq 0$ is given by $(x(t),x^t) = \mathcal{F}x_t + \mathcal{E}u_t \in M^p$.

These facts are of particular importance in connection with Corollary 4.2.

(iii) The adjoint operators E^*, D^*: $W^{1,q} \to L^q([-h,0];\mathbb{R}^m)$ and $\mathcal{E}^*$, $\mathcal{D}^*$: $M^q \to L^q([-h,0],\mathbb{R}^m)$ are given by the following explicit formulas

$$[E^*\psi](\sigma) = \int_{-h}^{\sigma} d\beta^T(\tau)\psi(\tau-\sigma) + \int_{-h}^{\sigma} d\gamma^T(\tau)\dot{\psi}(\tau-\sigma), \quad \psi \in W^{1,q},$$

$$[D^*\psi](\sigma) = \int_{\sigma}^{0} d\beta^T(\tau)\psi(\tau-\sigma-h) + \int_{\sigma}^{0} d\gamma^T(\tau)\dot{\psi}(\tau-\sigma-h), \quad \psi \in W^{1,q},$$

$$[\mathcal{E}^*\psi](\sigma) = \int_{-h}^{\sigma} d\beta^T(\tau)\psi^1(\tau-\sigma), \quad \psi \in M^q,$$

$$[\mathcal{D}^*\psi](\sigma) = \int_{\sigma}^{0} d\beta^T(\tau)\psi^1(\tau-\sigma-h), \quad \psi \in M^q.$$

(iv) An operator of the type $\mathcal{D}^*$ has previously been introduced by Manitius [10],[11] for the study of controllability properties of RFDEs with undelayed input variables.

The next result is a direct consequence of the definition of the structural operators and of Remark 4.3 (iii).

Proposition 4.4.

(i) Given $\phi \in M^p$ and ξ, $u_h \in L^p([-h,0];\mathbb{R}^m)$, then the corresponding solution w(t), x(t) of Σ, (6) can be described by

$$(w(h),x_h) = G(F\phi + E\xi + Du_h).$$

(ii) Given $\phi \in W^{1,p}$ and ξ, $u_h \in L^p([-h,0];\mathbb{R}^m)$, then the corresponding solution $x(t)$ of Ω, (7) can be described by

$$x_h = \mathcal{G}(\mathcal{F}\phi + \mathcal{E}\xi + \mathcal{D}u_h).$$

(iii) Given $\psi \in W^{1,q}$, then the corresponding output $y(t)$ of Ω^T, (4) can be described by

$$y(t) = (E^*\psi + D^*G^*F^*\psi)(-t), \quad 0 \leq t \leq h.$$

(iv) Given $\psi \in M^q$, then the corresponding output y(t) of Σ^T, (5) can be described by

$$y(t) = (E^*\psi + \mathcal{D}^*G^*F^*\psi)(-t), \quad 0 \leq t \leq h.$$

5. F-CONTROLLABILITY AND OBSERVABILITY

In this section we show how the structural operator approach leads to a new controllability concept for NFDEs and to a duality relation between controllability and observability. Such results have been open problems, even in the retarded case.

Let us first introduce the reachable subspaces

$$R = \{(w(t),x_t,u_t) \in M^p \times L^p([-h,0];\mathbb{R}^m) \mid t \geq 0;\ w(.),x(.) \text{ satisfy } \Sigma,\ (6) \text{ for } \phi = 0,\ \xi = 0 \text{ and some input } u(.) \in L^p([0,t];\mathbb{R}^m)\},$$

$$\mathcal{R} = \{(x_t,u_t) \in W^{1,p} \times L^p([-h,0];\mathbb{R}^m) \mid t \geq 0;\ x(.) \text{ satisfies } \Omega,\ (7) \text{ for } \phi = 0,\ \xi = 0 \text{ and some input } u(.) \text{ in } L^p([0,t];\mathbb{R}^m)\}$$

of Σ and Ω as well as the unobservable subspaces

$$N^T = \{\psi \in M^q \mid \text{the output } y(t) \text{ of } \Sigma^T,\ (5) \text{ vanishes for } t \geq 0\}$$

$$\mathcal{N}^T = \{\psi \in W^{1,q} \mid \text{the output } y(t) \text{ of } \Omega^T,\ (4) \text{ vanishes for } t \geq 0\}$$

of Σ^T and Ω^T. These subspaces are related by means of the structural operators.

Lemma 5.1.

(i) Let $\psi \in W^{1,q}$, $g \in M^q$, $d \in L^q([-h,0];\mathbb{R}^m)$ be given. Then

$$(F^*\psi,E^*\psi) \perp R \Leftrightarrow \psi \in \mathcal{N}^T,$$

$$(g,d) \perp R \Leftrightarrow G^*g \in \mathcal{N}^T,\quad d = -D^*G^*g.$$

(ii) Let $\psi \in M^q$, $g \in R^n \times L^q \times L^q$, $d \in L^q([-h,0];\mathbb{R}^m)$ be given. Then

$$(F^*\psi,E^*\psi) \perp \mathcal{R} \Leftrightarrow \psi \in N^T,$$

$$(\pi^T g,d) \perp \mathcal{R} \Leftrightarrow G^*\pi^T g \in N^T,\quad d = -\mathcal{D}^*G^*\pi^T g.$$

Proof. First note that $(F^*\psi,E^*\psi) \perp R$ if and only if ψ is orthogonal to $[F\ E]R = \{F\phi + E\xi \mid (\phi,\xi) \in R\}$. By Remark 4.3 (ii), this is the reachable subspace of system $\tilde{\Sigma}$. Hence it follows from Theorem 2.2 that $\psi \perp [F\ E]R$ if and only if the corresponding output $y(.)$ of system Ω^T, (4) satisfies

$$\int_0^t y^T(t-s)u(s)ds = 0 \quad \forall\, t \geq 0 \quad \forall\, u \in L^p([0,t];\mathbb{R}^m).$$

This means that $\psi \in N^T$.

Secondly, note that

$$R = \{(G[F\phi+E\xi],0) \mid (\phi,\xi) \in R\} + \{(GD\zeta,\zeta) \mid \zeta \in L^p([-h,0];\mathbb{R}^m)\}$$

(Proposition 4.4). Hence $(g,d) \perp R$ if and only if $g \perp G[F\ E]R$ and

$$\langle g, GD\zeta\rangle_{M^q,M^p} + \langle d,\zeta\rangle_{L^q([-h,0];\mathbb{R}^m),L^p([-h,0];\mathbb{R}^m)} = 0$$

for every $\zeta \in L^p([-h,0];\mathbb{R}^m)$. This is equivalent to $(F^*G^*g, E^*G^*g) \perp R$, $d + D^*G^*g = 0$, and hence to $G^*g \in N^T$, $d = -D^*G^*g$. This proves (i). The proof of (ii) is strictly analogous.

Q.E.D.

Recall that $[F\ E]R \subset W^{-1,p}$ and $[\mathcal{F}\ \mathcal{E}]\mathcal{R} \subset M^p$ are the reachable subspaces of the systems $\tilde{\Sigma}$ and $\tilde{\Omega}$ (Remark 4.3 (ii)) and that the dual state concept may be considered as the 'real' state of the NFDE (1) (Section 2). Consequently, one might regard $[F\ E]R$ and $[\mathcal{F}\ \mathcal{E}]\mathcal{R}$ as the 'real' reachable subspaces of the systems Σ and Ω. This motivates the following definition of F-controllability. Such a notion has previously been introduced by Manitius [11] for RFDEs with undelayed input variables.

Definition 5.2.

(i) System Σ is said to be (approximately) F-controllable if $\mathrm{cl}([F\ E]R) = \mathrm{cl}(\mathrm{ran}\ [F\ E])$.

(ii) System Ω is said to be (approximately) F-controllable if $\mathrm{cl}([\mathcal{F}\ \mathcal{E}]\mathcal{R}) = \mathrm{cl}(\mathrm{ran}\ [\mathcal{F}\ \mathcal{E}])$.

(iii) System Ω^T is said to be observable if $\mathcal{N}^T \subset \ker F^*$ or equivalently

$$y(t) = 0 \quad \forall\, t \geq 0 \Rightarrow x(t) = 0 \quad \forall\, t \geq 0.$$

(iv) System Σ^T is said to be observable if $N^T \subset \ker \mathcal{F}^*$ or equivalently

$$y(t) = 0 \quad \forall\, t \geq 0 \Rightarrow x(t) = 0 \quad \forall\, t \geq 0.$$

The following duality result is a direct consequence of Lemma 5.1 and Definition 5.2.

Corollary 5.3.

(i) System Σ is F-controllable if and only if system Ω^T is observable.

(ii) System Ω is F-controllable if and only if system Σ^T is observable.

Remarks 5.4.

(i) Every observable system Ω^T has the property

$$\left.\begin{array}{l} x(t) = 0 \quad \forall\, t \geq h \\ y(t) = 0 \quad \forall\, t \geq 0 \end{array}\right\} \Rightarrow x(t) = 0 \quad \forall t \geq 0 \tag{16}$$

(*observability of nontrivial small solutions*). By Proposition 4.4, this is equivalent to

$$\ker F^*G^*F^* \cap \ker E^*G^*F^* \cap \ker (D^*G^*F^* + E^*) \subset \ker F^*. \tag{17}$$

This condition together with spectral observability is also necessary for observability of Ω^T (Salamon [14, Theorem IV.3.5]).

(ii) In this case $\Gamma = 0$ condition (16) is equivalent to the corresponding property

$$\left.\begin{array}{l} x(t) = 0 \quad \forall\, t \geq h \\ y(t) = 0 \quad \forall\, t \geq 0 \end{array}\right\} \Rightarrow x(t) = 0 \quad \forall\, t \geq 0 \tag{18}$$

of system Σ. This follows from the fact that - for every solution $x(t)$ of Σ which vanishes for $t \geq T$ and has a zero output - the function

$$x(t) = -\int_t^T x(s)ds\ , \quad t \geq -h,$$

defines a solution of Ω^T with zero output. *We conclude that - in the case* $\Gamma = 0$ - *system* Ω^T *is observable if and only if system* Σ^T *is.*

(iii) Again by Proposition 4.4, (18) is equivalent to

$$\ker F^*G^*F^* \cap \ker E^*G^*F^* \cap \ker (\mathcal{D}^*G^*F^* + \mathcal{E}^*) \subset \ker F^*. \tag{19}$$

In the case of RFDEs ($M = 0$, $\Gamma = 0$) with undelayed input variables ($B\xi = B_0\xi(0)$) we obtain $\mathcal{E}^* = 0$ and hence (19) is precisely the necessary condition for F-controllability which was obtained by Manitius [11].

(iv) Let L, M, B and Γ be given by

$$\begin{array}{ll} L\phi = A_0\phi(0) + A_1\phi(-h), & M\phi = A_{-1}\phi(-h), \\ B\xi = B_0\xi(0) + B_1\xi(-h), & \Gamma\xi = B_{-0}\xi(0) + B_{-1}\xi(-h). \end{array} \tag{20}$$

Then (17) is equivalent to

$$\max_{\lambda\in\mathbb{C}} \operatorname{rank} \begin{bmatrix} A_0-\lambda I & A_1+\lambda A_{-1} & B_0+\lambda B_{-0} & B_1+\lambda B_{-1} \\ A_1+\lambda A_{-1} & 0 & B_1+\lambda B_{-1} & 0 \end{bmatrix}$$

$$= n + \max_{\lambda\in\mathbb{C}} \operatorname{rank} [A_1+\lambda A_{-1} \quad B_1+\lambda B_{-1}]$$

(Salamon [14, Theorem IV. 3.7]).

(v) It follows from Lemma 5.1 that R is dense in the product space $M^p \times L^p([-h,0];\mathbb{R}^m)$ (i.e. Σ is *approximately controllable*) if and only if $N^T = \{0\}$ or equivalently the solutions of Ω^T satisfy

$$y(t) = 0 \quad \forall\, t \geq 0 \Rightarrow x(t) = 0 \quad \forall\, t \geq -h$$

(i.e. Ω^T is *strictly observable*).

This time we obtain the necessary condition

$$\left.\begin{array}{l} x(t) = 0 \quad \forall\, t \geq 0 \\ \\ y(t) = 0 \quad \forall\, t \geq 0 \end{array}\right\} \Rightarrow x(t) = 0 \quad \forall\, t \geq -h \tag{22}$$

(*observabiltiy of small solutions*) or equivalently

$$\ker F^* \cap \ker E^* = \{0\} \tag{23}$$

(Proposition 4.4). This condition is clearly stronger than (17). Hence it follows easily from (i) that (23) together with spectral observability is equivalent to $N^T = \{0\}$.

(vi) Again - in the case $\Gamma = 0$ - we obtain that (22) is equivalent to the corresponding property of system Σ^T (compare (ii)). *Hence Ω^T is strictly observable if and only if Σ^T is.*

(vii) If L, M, B and Γ are given by (20), then (23) is equivalent to

$$\max_{\lambda \in \mathbb{C}} \operatorname{rank} [A_1 + \lambda A_{-1} \;\; B_1 + \lambda B_{-1}] = n \tag{24}$$

(Salamon [14, Theorem IV.2.11]).

ACKNOWLEDGEMENT

This work has been supported by the Forschungsschwerpunkt Dynamische Systeme.

REFERENCES

[1] Bernier, C., A. Manitius: On semigroups in $\mathbb{R}^n \times L^p$ corresponding to differential equations with delays, Can. J. Math. 30 (1978), 897-914.

[2] Burns, J.A., T.L. Herdman, H.W. Stech: Linear functional differential equations as semigroups in product spaces, Dept. of Mathematics, Virginia Polytechnic Institute and State University, Blacksburg, Virginia, 1981.

[3] Curtain, R.F., A.J. Pritchard: Infinite Dimensional Linear Systems Theory, LNCIS 8, Springer-Verlag, Berlin, 1978.

[4] Delfour, M.C.: The linear quadratic optimal control problem with delays in state and control variables: a state space approach, Centre de Recherche de Mathématiques Appliquées, Université de Montréal, CRMA-1012, March 1981.

[5] Delfour, M.C., E.B. Lee, A. Manitius: F-reduction of the operator Riccati equations for hereditary differential systems, Automatica 14 (1978), 385-395.

[6] Delfour, M.C., A. Manitius: The structural operator F and its role in the theory of retarded systems, Part 1: J. Math. Anal. Appl. 73 (1980), 466-490, Part 2: J. Math. Anal. Appl. 74 (1980), 359-381.

[7] Henry, D.: Linear autonomous functional differential equations of neutral type in the Sobolev space $W_2^{(1)}$, Technical Report, Dept. of Mathematics, University of Kentucky, Lexington, Kentucky, 1970.

[8] Ichikawa, A., A.J. Pritchard: Existence, uniqueness and stability of nonlinear evolution equations, J. Math. Anal. Appl. 68 (1979), 454-476.

[9] Manitius, A.: Completeness and F-completeness of eigenfunctions associated with retarded functional differential equations, J. Diff. Equations 35 (1980), 1-29.

[10] Manitius, A.: Necessary and sufficient conditions of approximate controllability for general linear retarded systems, SIAM J. Control Opt. 19 (1981), 516-632.

[11] Manitius, A.: F-controllability and observability of linear retarded systems, Applied Math. Opt. 9 (1982), 73-95.

[12] Salamon, D.: On controllability and observability of time delay systems, FS Dynamische Systeme, Universität Bremen, Report Nr. 38, 1981.

[13] Salamon, D.: On dynamic observation and state feedback for time delay systems in "Conference on Differential Equations and Delays", F.Kappel & W.Schappacher, eds., pp. 202-219, Pitman, London, 1982.

[14] Salamon, D.: On control and observation of neutral systems, Doctoral Dissertation, FS Dynamische Systeme, Universität Bremen, 1982.

[15] Salamon, D.: A duality principle for neutral functional differential equations, EQUADIFF 1982, K. Schmitt, ed., Springer-Verlag, Berlin, to appear.

[16] Vinter, R.B., R.H. Kwong: The infinite time quadratic control problem for linear systems with state and control delays: an evolution equation approach, SIAM J. Control Opt. 19 (1981), 139-153.

BOUNDARY OBSERVATION AND CONTROL OF A VIBRATING PLATE:

a preliminary report

T.I. Seidman

Department of Mathematics
University of Maryland Baltimore County
Catonsville, MD 21228, USA

1. INTRODUCTION

As suggested by the title, this is very much a report on work in progress. The further development of the ideas and methods presented here will be in collaboration with W. Krabs (Technische Hochschule Darmstadt). Indeed, the present work should be considered as the first steps in extending Krabs' work [6],[7], etc. on the one-dimensional case (vibrating beam) to higher dimensions. Nevertheless, the treatment here provides some new information (e.g., the $\mathcal{O}(e^{\beta/\delta})$ asymptotic estimate as the time interval 2δ shrinks) even for the one-dimensional case which has already been extensively investigated.

We will be considering vibrating systems whose motion is governed by the equation

$$\ddot{u} + \Delta^2 u = 0. \tag{1.1}$$

(While there are various objections to this as a model on physical grounds, it is widely employed and presents a problem which may certainly be felt to be of mathematical interest.) For our initial consideration it is essential that the spatial region $\Omega \in \mathbb{R}^m$ be a product region:

$$\Omega = (0,1) \times \hat{\Omega},\ \hat{\Omega} \text{ bounded in } \mathbb{R}^{m-1}.$$

(For a rectangular plate, of course, $m = 2$ and $\hat{\Omega}$ is an interval.) We begin with consideration of an observability problem for (1.1) and it will also be essential that the (homogeneous) boundary conditions are such that the operator $(\Delta^2;BC)$ is a square. Thus, for definiteness, we take the conditions

$$u_n = 0, \quad (\Delta u)_n = 0 \tag{1.2}$$

(where the subscript n denotes differentiation normal to the boundary $\partial\Omega$).

The classical technique of 'separation of variables' then gives a general solution of (1.1), (1.2) in series form:

$$u(t,x,y) = \sum_{k=0}^{\infty} \Big[\sum_{\pm,j=0}^{\infty} c^{\pm}_{j,k} e^{\pm i(j^2\pi^2+\nu_k)t} \cos j\pi x \Big] v_k(y) \tag{1.3}$$

where $(x,y) \in \Omega(0<x<1, y \in \hat{\Omega})$. Here $\{(\nu_k, v_k)\}$ is the sequence of eigenpairs of the $(m-1)$-dimensional Laplace operator on $\hat{\Omega}$:

$$-\Delta v_k = \nu_k v_k \quad \text{on } \hat{\Omega}, \quad (v_k)_n = 0 \quad \text{on } \partial\hat{\Omega}. \tag{1.4}$$

Assuming, as we do, that $\partial\hat{\Omega}$ is moderately smooth, it is standard that

$$0 = \nu_0 < \nu_1 \leq \nu_2 \leq \dots \quad (\nu_k \to \infty) \tag{1.5}$$

and that we may take $\{v_k\}$ to be <u>orthonormal</u> in $L^2(\hat{\Omega})$. For the present we take (1.3) to <u>define</u> the sense of u being a solution of (1.1), (1.2), leaving further questions of convergence and interpretation for later. We do note, however, that the orthonormality (fudging very slightly, especially for $j = 0$) of the set of functions $\{\cos j\pi x\, v_k(y)\}$ in $L^2(\Omega)$ gives

$$\|u(t,\cdot,\cdot)\|^2_{L^2(\Omega)} \leq 2\Sigma\Sigma|c^{\pm}_{j,k}|^2 = 2\|\vec{c}\|^2 \tag{1.6}$$

where we use $\vec{c}$ for the vector of coefficients, here taken in ℓ^2.

Suppose, then, it is known that u satisfies (1,1), (1.2) - hence is given by (1.3) with some coefficient vector $\vec{c}$ - and, without prior knowledge of the 'state' $[u,\dot{u}]$ at any initial time, we wish to determine that state from observation at the boundary. In particular, let us assume that we can observe

$$\hat{f}(t,y) := u(t,0,y) = \Sigma\Sigma c^{\pm}_{j,k} e^{\pm i(j^2\pi^2+\nu_k)t} v_k(y) \tag{1.7}$$

for a time interval of length 2δ (With no loss of generality it will be convenient to consider (1.1) for $-\delta < t < \delta$, translating t if necessary.) and for $y \in \hat{\Omega}$, i.e., we observe, for this time interval, the boundary trace of the solutions u on the 'base' $\{0\} \times \hat{\Omega}$ of the cylindrical boundary $\partial\Omega$. Our immediate aim is consideration of the state estimation map: $\hat{f} \to [u,\dot{u}]|_{t=0}$.

Using the product structure (spatial separation of variables) we can reduce the problem to consideration of a sequence of one-dimensional problems. Letting

$$f_k(t) := \langle \hat{f}, v_k \rangle_{\hat{\Omega}} := \int_{\hat{\Omega}} f(t,y) v_k(y)\,dy, \tag{1.8}$$

the orthonormality of $\{v_k\}$ gives

$$f_k(t) = \sum_{\pm,j=0}^{\infty} c^{\pm}_{j,k} e^{\pm i(j^2\pi^2+\nu_k)t} \tag{1.9}$$

and

$$||\hat{f}||^2_{L^2((-\delta,\delta)\times\hat{\Omega})} = \sum_{k=0}^{\infty} ||f_k||^2_{L^2(-\delta,\delta)} \quad . \tag{1.10}$$

In the next section we shall sketch the proof that there is a constant $M = M_\delta$ uniform in $\nu \in \{\nu_k\}$ such that

$$||\vec{c}||_{l^2} \leq M_\delta ||f||_{L^2(-\delta,\delta)} \tag{1.11}$$

for every function $f \in L^2(-\delta,\delta)$ of the form:

$$f(t) = \sum_{\pm,i=0}^{\infty} c_j^{\pm} e^{\pm i(j^2\pi^2+\nu)t}, \quad \vec{c} := (c_j^{\pm}). \tag{1.12}$$

Applying this to each f_k gives

$$||(u(t,..)||^2_{L^2(\Omega)} \leq 2\sum_k ||\vec{c}_k||^2_{l^2} \leq 2M_\delta^2 \sum_k ||f_k||^2_{L^2(-\delta,\delta)} \tag{1.13}$$

$$= 2M_\delta^2 ||\hat{f}||^2_{L^2((-\delta,\delta)\times\hat{\Omega})}$$

We defer consideration of $\dot{u}$ (to complete the state estimation) until later, but note that this already shows - given the harmonic analysis estimate (1.11)! - that the estimate of $u(t,..)$ in the sense of $L^2(\Omega)$ depends continuously on the observation of $\hat{f} := u|_{(-\delta,\delta)\times\{0\}\times\hat{\Omega}}$ in the sense of $L^2((-\delta,\delta)\times\hat{\Omega})$. This already motivates our concern for (1.11).

2. HARMONIC ANALYSIS

We are concerned here, motivated by the considerations of the first section, with showing the continuity of the coefficient map

$$\underset{\sim}{C}: f \mapsto \vec{c}: X \to l^2 \tag{2.1}$$

where $X = X_{\delta,\nu}(\delta > 0, \nu \geq 0)$ is the closed subspace of $L^2(-\delta,\delta)$ generated by the functions

$$e_j^{\pm} := \exp[ia_j^{\pm}t], \quad a_j^{\pm} := \pm(j^2\pi^2+\nu). \tag{2.2}$$

The complete details of the argument will be presented elsewhere [13] but it is perhaps instructive to the systems theoretician to see the line of argument, since similar trigonometric moment problems occur frequently in linear observability/ controllability problems of the present sort.

To consider the coefficient map $\underset{\sim}{C}$ we begin with the individual coefficient functionals

$$f \to c_k^{\pm} \quad (f \in X \text{ as in } (1.12)). \tag{2.3}$$

By Riesz' Theorem we have

$$c_k^{\pm} = \langle f, g_k^{\pm} \rangle_{L^2(-\delta,\delta)} = \int_{-\infty}^{\infty} f(t)\overline{g_k^{\pm}}(t)dt, \quad f \in X \tag{2.4}$$

where each $g_k^{\pm}$ is to be an L^2 function on $\mathbb{R}$ with support in $[-\delta,\delta]$. Taking the Fourier transforms (slightly modified), we set

$$G_k^{\pm}(\tau) := \int_{-\infty}^{\infty} \overline{g_k^{\pm}}(t)e^{i\tau t}dt \tag{2.5}$$

and note that

$$\langle e_j^{\pm}, g_k^{\pm} \rangle_{L^2(-\delta,\delta)} = \int \overline{g_k^{\pm}}(t)e^{ia_j^{\pm}t}dt = G_k^{\pm}(a_j^{\pm}). \tag{2.6}$$

We will construct a sequence $\{G_k^{\pm}\}$ for which

i) $G_k^{\pm}(a_j^{\pm}) = \{1 \text{ if } a_j^{\pm} = a_k^{\pm},\ 0 \text{ otherwise}\}$, (2.7)

ii) each $G_k^{\pm}$ is an entire analytic function with

$$|G_k^{\pm}(z)| = \mathcal{O}(e^{\delta'|z|}) \quad \text{with } \delta' < \delta,$$

iii) each $G_k^{\pm}$ is in $L^2(\mathbb{R})$ with

$$|\langle G_j^{\pm}, G_k^{\pm} \rangle| \leq \hat{M}_\delta^2\, e^{-\mu\sqrt{D}}, \qquad 2D := |a_j^{\pm} - a_k^{\pm}|.$$

From (2.7 ii, iii) it follows, by the Paley-Wiener Theorem, that there are, indeed, functions $g_k^{\pm} \in L(\mathbb{R})$ with support in $(-\delta,\delta)$ corresponding to $G_k^{\pm}$ as in (2.5). From (2.7 i), noting (2.6), one obtains (2.4). Finally, an elementary computation (which we do not reproduce here) shows that, for $f \in X$

$$||\vec{c}||^2 = ||\underset{\sim}{C}f||^2 \leq ||\Sigma c_k^{\pm}g_k^{\pm}||\cdot||f|| = C||\Sigma c_k^{\pm}G_k||\,||f||,$$

noting that (2.5) is essentially a unitary map: $g \mapsto G$ with a constant factor, and

$$||\Sigma c_k^{\pm}G_k^{\pm}||_{L^2(\mathbb{R})} \leq C_\mu \hat{M}_\delta ||\vec{c}||_{l^2}$$

where C_μ depends only on $(\mu > 0)$ in (2.7 iii) which may be fixed. This gives (1.11) with $M_\delta = C_\mu C\hat{M}_\delta$.

Note that we have suppressed expressing the possible dependence on $\nu \geq 0$ but that M_δ is uniform in ν to precisely the extent that $\hat{M}_\delta$ is. There is a slight anomaly in the case $\nu = 0$ since then the doublet $a_0^{\pm} = \pm\nu$ reduces to a singleton. We treat only the case $\nu \geq \nu_1 > 0$ and leave it to the reader to accept that $\nu = 0$ behaves in the same way. Thus, in particular, we have uniformity of the estimate for $\nu \in \{\nu_k\}$ as in (1.5) if we have $\hat{M}_\delta$ in (2.7 iii) uniform in $\nu \geq \nu_1 > 0$ and μ independent of ν. Besides merely showing existence of M_δ, we shall also be concerned with obtaining asymptotic estimates as $\delta \to 0^+$ and as $\delta \to \infty$ (very short and very long observation intervals).

We begin with the observation that the functions $\gamma_k^\pm$ defined by

$$\gamma_k^\pm(z) := \psi_k(z-y)\frac{\psi^*(-z-\nu)}{\psi^*(-a_k^+ - \nu)}, \quad \gamma_k^-(z) := \gamma_k^+(-z), \tag{2.8}$$

$$\psi^*(z) := \sqrt{z}\sin\sqrt{z},\ \psi_k(z) := \frac{(-1)^k\psi^*(z)}{z-k^2\pi^2}$$

and entire analytic functions with the correct zeroes and growth $\exp[\mathcal{O}(|z|^{1/2})]$ but are not in $L^2(\mathbb{R})$. Indeed, it requires some work, using properties of the sin functions to obtain the uniform estimate

$$|\gamma_k^\pm(a_k^\pm+s)| \leq \alpha(1+\sigma^2)e^\sigma, \qquad \sigma = |s|^{1/2}, \tag{2.9}$$

where α is a constant which depends on ν but is uniform for $\nu \geq \nu_1 > 0$. We will set

$$G_k^\pm(z) = \gamma_k^\pm(z)R_\delta(z-a_k^\pm) \tag{2.10}$$

where, depending on δ, R_δ is a function of suitable (complex) growth giving (2.7 ii) and decaying exponentially in σ to give (2.7 iii). The basic construction, with $R = R_\delta$ of the form

$$R(z) := \prod_{j>J}\phi(\theta^2 z/j^2), \qquad \phi(z) := \frac{\sin z}{z}, \tag{2.11}$$

can be traced through [8], [5] and appears in the system theoretic literature in, e.g., [9], [10], [12]. (This last reference includes, in the context of parabolic boundary control, an asymptotic estimate for short observation/control intervals closely related to that presented here.)

In considering $R = R_\delta$, we first note that (as $|\phi(z)| \leq e^{|z|}$ and from the growth asserted above for $\gamma_k^\pm$) one has (2.7 ii) provided

$$\theta^2\sum_{j>J} 1/j^2 < \delta, \tag{2.12}$$

which we take as a first basic constraint on the δ-dependent choice of (θ,J) in (2.11).

To estimate R on the reals, it is convenient to let $P_J(s) := \prod_{j>J}\phi(s/j^2)$ so

$R(z) = P_J(\theta^2 z)$ and consider $s \geq 0$, $\sigma = \sqrt{s}$, $K = [\sigma](K \leq \sigma < K+1)$. An elementary computation with Taylor series shows

$$0 < \phi(t) \leq e^{-t^2/6} \quad \text{for} \quad |t| \leq 6/\sqrt{5}$$

from which it follows that for $\rho := \sigma/J \leq \sqrt{(6/\sqrt{5})}$ one has

$$|P_J(s)| \leq \exp[-s^2 \sum_{j>J} 1/j^4] \leq \exp[\rho^{4/12} - (\rho^{3/18})\sigma]. \tag{2.13}$$

In particular, taking $J = K$ in (2.13) one obtains (note $\rho \approx 1$ for large s)

$$|P_K(s)| \leq Ce^{-\sigma/18} \text{ for } s \geq 0. \tag{2.14}$$

(Actually, one can show that one can take $C \approx 1$ for large s.) For large s, again, one writes

$$|P_J(s)| = [\prod_{J<j\leq K} |\phi(s/j^2)|]|P_K(s)| \leq (K!/J!)^2 s^{-(K-J)}|P_K(s)|$$

when $K \geq J$, noting that $|\phi(t)| \leq 1/t$. Using Stirling's approximation for $K!$ gives $(K!)^2 s^{-K} = 2\pi e^{-2\sigma}[\sigma + \mathcal{O}(1)]$ for large s so, using (2.14),

$$|P_J(s)| \leq (C[\sigma+B]/[J!]^2)\sigma^{2J} e^{-(37/18)\sigma} \text{ for } s \geq J \tag{2.15}$$

with constant C in (2.15) approximately 2π if one has to consider only large s, e.g., if J were large. This already shows that one can get $e^{-\mu\sigma}$ decay for arbitrarily large $\bar{\mu}$ by taking $\theta > (18/37)\bar{\mu}$ and then choosing J large enough to have (2.12).

For large δ (long observation/control intervals) we take $J = 0$ so (2.12) requires $\theta^2 < 6\delta/\pi^2$ ($\theta = \mathcal{O}(\sqrt{\delta})$) and, since $[\sigma+B]e^{-\varepsilon\sigma}$ is bounded for any $\varepsilon > 0$, one has

$$|R(s)| \leq C_\varepsilon e^{-\bar{\mu}\sigma} \text{ for } s \geq 0, \; \bar{\mu} = (37/8)(\sqrt{6\delta}/\pi) - 2\varepsilon \tag{2.16}$$

with C_ε independent of δ, σ and, certainly, $\bar{\mu} > 1$ for large δ.

For very small δ, on the other hand, we first determine λ, Γ by letting λ be the (unique) positive solution of

$$(\tfrac{9}{2}\lambda)^{1/3} = 2 \log [\frac{2}{37/18 - \lambda}] =: \Gamma \tag{2.17}$$

and then choose (any) $\theta < 1/\lambda$ corresponding to a choice of

$$\beta := \Gamma\theta^2 > \beta_0 := \Gamma/\lambda^2 = 1.16949859. \tag{2.18}$$

(This numerical value is obtained from the computed solution

$$\lambda = 1.22859784, \quad \Gamma = 1.76971776$$

of (2.17).) With (2.12) this determines J. Note that $\Sigma_{j>J} \sqrt{j}^2 \sim \sqrt{(J+\sqrt{2})}$ so fixing θ gives $J \sim \theta^2/\delta$ as $\delta \to 0^+$. For $s < J$ we use (2.13) with $\rho \in [0,1]$ to obtain

$$|P_J(s)|e^{\lambda\sigma} \leq e^{\sqrt{12}} e^{[\lambda-\rho^3/18]\sigma} = e^{\sqrt{12}} e^{[\lambda\rho-\rho^4/18]J} \leq e^{\sqrt{12}} e^{\Gamma J} \tag{2.19}$$

since (2.17) gives $\Gamma = \max_\rho \{\lambda\rho-\rho^4/18 : 0\leq\rho\leq 1\}$. For $s \geq J$ we use Stirling's approximation for J! in (2.15) to obtain

$$\begin{aligned} |P_J(s)|e^{\lambda\sigma} &\leq [C/(J!)^2] \sup\{\sigma^{2J+1} e^{-[37/18-\lambda]\sigma} : \sigma\geq 0\} \qquad (2.20)\\ &= C\left(\frac{2}{37/18-\lambda}\right)^{2J+1} [(1+\frac{1}{2J})^{2J+1} e^{-2J-1} J^{2J+1}/(J!)^2][1+O(\frac{1}{J})]\\ &\leq C\left(\frac{2}{37/18-\lambda}\right)^{2J} = Ce^{\Gamma J}. \end{aligned}$$

(C appears in (2.20) as a generic absolute constant.) Combining (2.19) with (2.20) thus gives $|P_J(s)| \leq Ce^{\Gamma J}e^{-\lambda\sigma}$ for all $s \geq 0$ with C uniform for large J so that, noting (2.18) and $J \sim \theta^2/\delta$,

$$|R(s)| \leq Ce^{\beta/\delta}e^{-\bar{\mu}\sigma} \text{ for } s \geq 0, \quad \bar{\mu} = \lambda\theta = \sqrt{\beta/\beta_0} > 1. \tag{2.21}$$

Now choose $\bar{\varepsilon} > 0$ (independent of δ for large or small δ) and use (2.9) with either (2.16) or (2.21), as appropriate, to estimate (2.10). One has, then, (with the above choices of (θ,J) for large or small θ)

$$|G_k^\pm(a_k^\pm+s)| \leq \alpha C^* e^{-\mu\sigma}, \ \sigma := |s|^{\sqrt{2}}, \ \mu := \bar{\mu} - 1 - \bar{\varepsilon} > 0 \tag{2.22}$$

with $\bar{\mu}$ as in (2.16) or (2.21) and C^* depending on the choices of ε, $\bar{\varepsilon}$ (for example, it includes bounding $(1+\sigma^2)e^{-\bar{\varepsilon}\sigma}$) but not on δ as $\delta \to \infty$ and with a factor $e^{\beta/\delta}$ as $\delta \to 0^+$. Using (2.22) one obtains

$$<|G_j^\pm,G_k^\pm|> \leq \alpha^2 C^{*2} \int_{-\infty}^{\infty} e^{-\mu(\sigma+\sigma')} ds$$

with $\sigma := |s - a_j^\pm|^{\sqrt{2}}$, $\sigma' := |s - a_k^\pm|^{\sqrt{2}}$ so

$$|<G_j^\pm,G_k^\pm>| \leq 4\sigma^2 C^{*2} e^{-\mu\sqrt{D}} \int_{\infty}^{\infty} e^{-\mu\sigma} ds. \tag{2.23}$$

As $\delta \to \infty$ we have from (2.16) that μ is, essentially, proportional to $\sqrt{\delta}$. Evaluating the integral on the right of (2.23) thus gives a factor of (const./δ) so (2.7 iii) holds as $\delta \to \infty$ with μ bounded away from 0 and $\hat{M}_\delta = O(1/\sqrt{\delta})$. As noted earlier, it follows that (1.11) holds as $\delta \to \infty$ with $M_\delta = O(1/\sqrt{\delta})$.

As $\delta \to 0^+$ we have $\mu > 0$ fixed in (2.23) so (2.7 iii) holds as $\delta \to 0^+$ with $\hat{M}_\delta = O(e^{\beta/\delta})$. Since $\beta > \beta_0$ was arbitrary, this gives $M_\delta = O(e^{2\beta/2\delta})$ for any choice giving $2\beta > 2\beta_0 = 2.3399718$.

We have already noted that there is no trouble as to the existence of M_δ for (1.11) for any (intermediate) value of δ. Indeed, it is clear from the definition that if M_δ is finite for any $\delta > 0$ then it is finite for every greater δ with M_δ nonincreasing. (A slightly more subtle argument shows M_δ strictly decreasing.) We have shown that (1.11) holds for every $\delta > 0$ with

$$M_\delta = \begin{cases} \mathcal{O}(e^{\beta/\delta}) & \text{as } \delta \to 0^+,\ \beta > \beta_0, \\ \mathcal{O}(1/\sqrt{\delta}) & \text{as } \delta \to \infty. \end{cases} \tag{2.24}$$

The estimates obtained giving (2.24) depend on ν only through α, as noted, and this is ok for $\nu = 0$ and is uniform in $\nu \geq \nu_1 > 0$.

3. BOUNDARY OBSERVATION AND CONTROL

Before applying the fundamental estimates (1.11), (2.24), we consider, briefly, the norms to be used. The functions on Ω which we consider are all presented in the form of (formal) expansions

$$w(x,y) = \sum_{j,k} \omega_{j,k} \cos jx\, v_k(y), \quad (x,y) \in (0,1) \times \hat{\Omega} \tag{3.1}$$

and we already have the norm

$$||w||_0 := [\sum_{j,k} |\omega_{j,k}|^2]^{1/2} \tag{3.2}$$

which is essentially the $L^2(\Omega)$ norm. We now introduce

$$||w||_s := [\sum_{j,k} (j^2\pi^2 + \nu_k)^s\, |\omega_{j,k}|^2]^{1/2}, \quad s \in \mathbb{R}. \tag{3.3}$$

Observe that applying $-\Delta$ to (3.1) gives

$$-\Delta w = \Sigma(j^2\pi^2 + \nu_k)\omega_{j,k} \cos jx\, v_k$$

so $||w||_2$ is just $||-\Delta w||_0$. Since, by standard elliptic theory, $-\Delta$ is an isomorphism from $H^2_*(\Omega) := \{w \in H^2(\Omega) : w_n = 0\}$ (i.e., "H^2 with the BC"), we are justified in referring to $||\cdot||_2$ as 'the' H^2 norm. Similarly, we view $||\cdot||_s$ given by (3.3) as the H^s norm for all $s \in \mathbb{R}$. Note that, starting with (3.2), we have $H^s_*(\Omega)$ given this way as the domain of $A^{s/2}$ where A is $-\Delta$ taken with domain $H^2_*(\Omega)$ as above. The precise comparison with the usual Sobolev spaces $H^s(\Omega)$ is given in [3], [4].

Returning now to the observation problem discussed in Section 1, we observe that for a solution u of (1.1), (1.2) given as in (1.3) one has

$$\dot{u} = \Sigma\Sigma[\pm i(j^2\pi^2 + \nu_k)c^{\pm}_{j,k}]e^{\pm i(j^2\pi^2 + \nu_k)t} \cos j\pi x\, v_k(y) \tag{3.4}$$

so that, corresponding to (1,6), one has

$$||\dot{u}||_{-2}^{2} \leq 2||\vec{c}||^{2}. \tag{3.5}$$

Combining this with (1.13) and the results of Section 2, we see that we have shown that:

The map: $\hat{f} \mapsto [u,\dot{u}]|_{\bar{t}}$, giving the state at time $\bar{t}$ from observation of the Dirichlet trace $\hat{f}$ of u (satisfying (1.1), (1.2)) on the base $\{0\} \times \hat{\Omega}$ of $\Omega = (0,1) \times \hat{\Omega}$ for a time interval of arbitrarily short length $T > 0$, is well-defined and continuous from the L^2 topology to the $L^2 \times H_*^{-2}$ topology; further, the norm of this map grows no worse than $\mathcal{O}(e^{2\beta/T})$ (any $2\beta > 2.3399718$) as $T \to 0^+$ and decreases as $\mathcal{O}(T^{-1/2})$ as $T \to \infty$.

In a similar fashion we see that if, instead of the Dirichlet trace $\hat{f} := u|_{\{0\}\times\hat{\Omega}}$, we would observe $\hat{f} := (\Delta u)|_{\{0\}\times\hat{\Omega}}$, then the map: $\hat{f} \to [u,\dot{u}]|_{\bar{t}}$ is continuous (with the same asymptotic estimates as $T \to 0^+,\infty$) with $\tilde{f}$ topologized in L^2 and $[u,\dot{u}]$ now topologized in $H_*^2 \times L^2$. This can be shown by applying the harmonic analysis estimate to $\langle \hat{f},v_k\rangle$ which has an expansion like (1.12) but with coefficients which include the same factor $(j^2\pi^2 + \nu_k)$ as for the definition of $||u||_2$. Alternatively, we can observe that for u satisfying (1.1), (1.2) one also has $\Delta u = w$ satisfying (1.1), (1.2). Applying the original result $(\tilde{f} = w|_{\{0\}\times\hat{\Omega}})$ one has $\hat{f} \to [w,\dot{w}] : L^2 \to L^2 \times H_*^{-2}$ but $w = \Delta u \in L^2$ corresponds to $u \in H_*^2$ while $\dot{w} = (\Delta u)^{\bullet} = \Delta\dot{u} \in H_*^{-2}$ corresponds to $\dot{u} \in L^2$.

Again, if one topologizes the original observation $\hat{f}$ in $H^1((-\delta,\delta) \to L^2(\hat{\Omega}))$ one goes continuously to $[u,\dot{u}]|_{\bar{t}}$ in $H_*^2 \times L^2$. As above, this can be seen in either of two ways after noting that $\hat{f} \in H^1(\ldots)$ is the same as having $\dot{\hat{f}} \in L^2$: apply the harmonic analysis estimate to the expansion of $\langle \dot{\hat{f}},v_k\rangle$ or apply the original theorem to observation of $w|_{\{0\}\times\hat{\Omega}}$ with $w = \dot{u}$ satisfying (1.1), (1.2) (which gives $[w,\dot{w}] = [\dot{u},-\Delta^2 u] \in L^2 \times H_*^{-2}$ so $[u,\dot{u}] \in H_*^2 \times L^2$).

Further such results can be obtained similarly and then by interpolation.

Next let us consider the relation between boundary observability and controllability (compare [2]). We begin with the identity

$$\int_\Omega [\dot{u}v - u\dot{v}]\Big|_0^T = \int_\Sigma [u(\Delta v)_n - u_n(\Delta v) + (\Delta u)v_n - (\Delta u)_n v] \tag{3.6}$$

holding for u, v satisfying (1.1). Specializing this to u, v such that also (1.2) holds for u and there is a boundary null-control ϕ such that v with

$$(\Delta v)_n = \phi \in L^2, \quad v_n = 0 \tag{3.7}$$

gives $[v,\dot{v}]|_{t=0} = [0,0]$, we have

$$\int_\Omega [\dot{u}v - u\dot{v}]|_T = \int_\Sigma \hat{f}\,\phi \tag{3.8}$$

where the right hand side is the inner product in $L^2((0,T)\times\{0\}\times\hat{\Omega})$ and the left hand side is interpreted as duality between $[u,\dot{u}]\in L^2\times H_*^{-2}$ and $[v,\dot{v}]\in H_*^2\times L^2$. (Compared with (3.3) there is some fudging of the norms here which we safely ignore.) This shows that the control map $\underset{\sim}{N}$: $[v,\dot{v}]|_T \mapsto \phi$ = [minimum norm nullcontrol in $L^2((0,T)\times\{0\}\times\hat{\Omega})$] is (<u>where defined</u>) exactly the adjoint of the state estimation map $\underset{\sim}{E}$: $\hat{f}\to[u,\dot{u}]|_T$. Since we have shown above that $\underset{\sim}{E}$ is continuous (with asymptotic estimates as $T\to 0^+$, ∞ for $||\underset{\sim}{E}||$), it follows that $\mathcal{D}(\underset{\sim}{N})$ is closed and $\underset{\sim}{N}$ is also continuous with $||\underset{\sim}{N}|| = ||\underset{\sim}{E}||$. To see that $\mathcal{D}(\underset{\sim}{N})$ is <u>all</u> of $H_*^2\times L^2$ we argue as follows:

Suppose we knew (see below) that $\underset{\sim}{E}$ was surjective. If $\mathcal{D}(\underset{\sim}{N})\neq H_*^2\times L^2$, then one could find non-trivial $[u_0,\dot{u}_0]\in L^2\times H_*^{-2}$ such that the left side of (3.8) vanishes for every $[v,\dot{v}]\in\mathcal{D}(\underset{\sim}{N})$. On the other hand $[v,\dot{v}]\in\mathcal{D}(\underset{\sim}{N})$ if and only if one can reach $[v,\dot{v}]$ at $t = T$ by solving (1.1) for v forward in time with 0 initial data and (3.7) with any $\phi\in L^2((0,T)\times\{0\}\times\hat{\Omega})$. Thus ϕ on the right of (3.8) ranges over all L^2 so if the left is always 0 one has $\hat{f} = 0$. By our result this gives $[u,\dot{u}]$ trivial. It is immediate that the range of $\underset{\sim}{E}$ is dense. [Using any finite combination of eigenfunctions as $[u,\dot{u}]$ and solving (1.1), (1.2) with this as data at $t = T$ certainly gives $\hat{f}\in L^2$]. On the other hand $u\in L^2$, $\dot{u}\in H^{-2}$ gives

$$\Sigma|c_{j.k}^+ + c_{j,k}^-|^2 < \infty,\quad \Sigma|c_{j,k}^+ - c_{j.k}^-|^2 < \infty$$

so $\Sigma|c_{j.k}^+|^2 < \infty$. Write

$$f_k(t) := e^{i\nu_k t} f_k^+(t) + e^{-i\nu_k t} f_k^-(t),$$

$$f_k^+(t) := \sum_j c_{j,k}^+ e^{ij^2\pi^2 t},\quad f_k^-(t) = \sum_j c_{j,k}^- e^{-ij^2\pi^2 t}.$$

For any interval whose length L is a multiple of $2/\pi$ the functions $\{e^{\pm ij^2\pi^2 t}\}$ are (to within a constant factor) orthonormal so, choosing such $L > T$,

$$||f_k||_{L^2(0,T)} \leq ||f_k^+||_{L^2(0,t)} + ||f_k^-||_{L^2(0,t)}$$
$$\leq \sqrt{2}\,(C\sum_j|c_{j,k}^+|^2 + C\sum_j|c_{j,k}^-|^2)$$

where $||\hat{f}||^2 \leq 2C||\vec{c}||^2$ showing continuity of $\underset{\sim}{E}^{-1}$. This shows that any $[v_0,\dot{v}_1]$ in $H_*^2\times L^2$ can be controlled to 0 by $\phi\in L^2(0,T)\times\{0\}\times\hat{\Omega})$. Since the equation is time reversible, one equally well can control from 0 to an arbitrary state and so, adding, from any state $[v,\dot{v}]|_{t=0}$ in $H_*^2\times L^2$ to any other at $t = T$ using an L^2 control ϕ in (3.7); further, one has the same $\mathcal{O}(e^{2\beta/T})$ and $\mathcal{O}(T^{-1/2})$ estimates as $T\to 0^+$ and $T\to\infty$.

It should be clear, of course, that to each of the possible variations suggested above for the observation problem there is a corresponding control problem.

4. COROLLARIES AND OPEN QUESTIONS

In this section we sketch, quite briefly, some related results which may be expected to follow from the above by 'standard methods' and some related questions which, on the other hand, would seem to require new ideas.

i) If the boundary conditions (1.2) were changed so that (still keeping $(\Delta^2;BC)$ a square) the conditions on the cylindrical boundary $(0,1) \times \partial\hat{\Omega}$ were different, then only the ν_k would change and the argument would go through exactly as above. If the conditions at the bases are changed, however, the sequence $\{j^2\pi^2\}$ may be modified. Using homogeneous Dirichlet conditions for u, Δu gives a set of problems which can be handled pretty much as above. For other (mixed) conditions one would need a generalization of the results of Section 2; the difficulty is getting an analogue of (2.9) without being able to use the convenient representation in terms of the sin function.

ii) Suppose one wished to work with other boundary conditions or with Ω not a product as above. One method (used effectively for the parabolic case - note the abstract Extension Theorem in [11]) is to contain Ω in a 'box', extend the data appropriately, control at (one face of) the boundary of the box and then restrict the resulting solution for $(0,T) \times$ box to $(0,T) \times \Omega$, taking the appropriate traces as controls. Some investigation is needed of (optimal) trace theorems to see that one could take these traces to get a control in L^2 - or whichever space is appropriate to the problem. Alternatively, the machinery of this paper permits one to construct smoother controls for smoother initial data and, with this, one can surely obtain the necessary traces. (A third alternative might be to note the arbitrariness of the choice of box, permitting - with the original smoothness of data - construction of a family of nullcontrols. The hope would then be that convolving these with a smoothing kernel would give a smooth control in any case. If this were to work, then a dual result might be well-posedness of state estimation from observation in a weaker space than otherwise needed provided one knew a priori that the data sought had support in some subset of the interior.) One difficulty with this procedure, however, is that one cannot, as in (3.7), choose to make one component of the boundary data vanish nor can one restrict the support of the control to a portion of $\partial\Omega$. This, of course, also involves more extensive observation for the dual state estimation problem.

iii) Suppose one wished to consider a nonlinear perturbation as replacing (1.1) by

$$\ddot{v} + \Delta^2 v = f(v) \tag{4.1}$$

with $f(0) = 0 = f'(0)$ so the linearization would again be (1.1). An argument along the lines of the Implicit Functions Theorem in Banach spaces should enable one to show control from one state to another for (4.1) provided both states are closed enough to 0 (with differentiable dependence). One would let

$$F(\phi,[v_0,\dot v_0],[v_1,\dot v_1]) := [v,\dot v]|_{t=T} - [v_1,\dot v_1]$$

where v was the solution of (4.1), (3.7) with initial date $[v_0,\dot v_0]$ at t = 0. Since one has invertibility of $\partial F/\partial\phi|_{0,0,0}$ - this is just the continuity of the control map for the linearization, given by (1.1) - one has local solvability for the control ϕ of the equation F(...) = 0. (Compare [1] for the wave equation.) More general nonlinearities - or even this, with larger data - would seem to require a new idea.

iv) The argument in Section 3 giving surjectivity of $\underset{\sim}{E}$ also shows the set M of $\hat f$'s obtainable as traces of solutions is a closed subspace of L^2. On the other hand, it is clearly <u>not all</u> of L^2 since the observability result for arbitrarily short time intervals shows $\hat f \in M$ is uniquely (indeed, continuously!) determined by its restriction to any time interval. This almost sounds like analyticity but the example

$$u(t,x,y) = \sum_{j=1}^{\infty} j^{-1} \cos j\pi x \cos j^2\pi^2 t$$

(i.e., $c^+_{j,k} = c^-_{j,k} = \sqrt{2}j$ for k = 0, j ≠ 0 and $c^{\pm}_{j,k} = 0$ otherwise) shows that $\hat f$ need not even be bounded. The identity (3.6) with (1.2) and (3.7) shows that whether or not ϕ is a nullcontrol depends only on its action as a linear function on M so optimal (minimum L^2 norm) controls are themselves in M. We thus obtain an 'optimality system' of the form

$$\begin{aligned}
&\ddot v + \Delta^2 v = 0, \quad v(0) = 0 = \dot v(0), \\
&v_n = 0, \quad (\Delta v)_n = u|_{(0,T)\times\{0\}\times\hat\Omega}, \\
&\ddot u + \Delta^2 u = 0, \\
&u_n = 0, \quad (\Delta u)_n = 0
\end{aligned} \qquad (4.2)$$

to which is adjoined the additional inhomogeneous condition: $v(T) = v_0$, $\dot v(T) = v_1$.

ACKNOWLEDGEMENTS

I should like to thank the organizers of this Workshop - Profs. F. Kappel, K. Kunisch and W. Schappacher - for the apportunity to have participated and to have presented this material. In addition, I am indebted to the generous hospitality of the Inst. für Math. of the Univ. of Graz for the period following the Workshop during which this was written. Finally, acknowledgement is due to the AFORS for financial support under Grant No. 820271.

REFERENCES

[1] Chewning, W.: Local controllability of a hyperbolic partial differential equation, Dynamical Systems I (Cesari, Hale, LaSalle, Eds.), Academic Press, N.Y., 1976, 303-306.

[2] Dolecki, S., D.L. Russell: A general theory of observation and control, SIAM J. Control & Opt. <u>15</u> (1977), 185-220.

[3] Fujiwara, D.: Complete characterization of the domains of fractional powers of some elliptic differential operators of 2-nd order, Proc. Japan Acad. 43 (1967), 82-86.

[4] Grisvard, P.: Caracterisation de quelques espaces d'interpolation, Arch. Rat. Mech. Anal. 25 (1967), 40-63.

[5] Ingham, A.E.: Some trigonometric inequalities with applications to the theory of series, Math. Z. 41 (1936), 367-379.

[6] Krabs, W.: On boundary controllability of one-dimensional vibrating systems, Math. Meth. in Appl. Sci. 1 (1979), 277-306.

[7] Krabs, W.: Optimal control of processes governed by partial differential equations, Part II: vibrations, Z. für Oper. Res. 26 (1982), 63-86.

[8] Redheffer, R.: Elementary remarks on completeness, Duke Math. J. 35 (1968), 103-116.

[9] Russell, D.L.: Nonharmonic Fourier series in control theory of distributed parameter systems, JMAA 18 (1967), 542-560.

[10] Russell, D.L.: A unified boundary controllability theory for hyperbolic and parabolic partial differential equations, Stud. Appl. Math. 52 (1973), 189-211.

[11] Seidman, T.I.: Exact boundary controllability for some evolution equations, SIAM J. Cont. & Opt. 16 (1978), 979-999.

[12] Seidman, T.I.: Two results on exact boundary control of parabolic equations, Appl. Math. Opt., to appear.

[13] Seidman, T.I.: The coefficient map for certain exponential sums, to appear.

BOUNDARY FEEDBACK STABILIZATION FOR A QUASI-LINEAR WAVE EQUATION

M. Slemrod *)

Dept. of Mathematical Sciences
Rensselaer Polytechnic Institute
Troy, NY 12180, USA

0. INTRODUCTION

The topic of feedback control stabilization of distributed parameter systems has come into renewed interest of late due to possible applications in the control of large space structures (see for example [1]). In such problems the control typically enters into the system through boundary conditions on some prescribed spatial domain. However in all such work to date it has been tacitly assumed that the structural dynamics are linear. Since any real structure must in fact have nonlinear dynamics it seems natural to investigate what the effects of these nonlinearities will be. In this paper I will pursue this topic. While similar results have recently been presented in a paper of Greenberg & Tsien [7] I hope that my control theoretic approach may make these ideas available to an audience that might otherwise be unfamiliar with their work.

1. REVIEW OF FINITE DIMENSIONAL THEORY

Consider the finite dimensional, nonlinear control system

$$\dot{x}(t) = f(x,u), \quad x(0) = x_0, \tag{1.1}$$

where $x \in \mathbf{R}^n, u \in \mathbf{R}^m$, $f: \mathbf{R}^n \times \mathbf{R}^m \to \mathbf{R}^n$ and is C^1, $f(0,0), = 0$. We are interested in finding a linear feedback controller

$$u = Kx, \quad K: \mathbf{R}^n \to \mathbf{R}^m, \quad K \text{ time invariant}, \tag{1.2}$$

so that the closed loop system

---

*) Sponsored by the Air Force Office of Scientific Research, Air Force Systems Command, USAF, under Contract/Grant No. AFORS-81-0172. The United States Government is authorized to reproduce and distribute reprints for government purposes not withstanding any copyright hereon.

$$\dot{x}(t) = f(x,Kx) \tag{1.3}$$

has the origin $x = 0$ exponentially asymptotically stable.

One natural approach (see e.g. [13]) is to linearize (1.1) about $x = 0$, $u = 0$ and write

$$\dot{x}(t) = f_x(0,0)x + f_u(0,0)u + g(x,u) \tag{1.4}$$

where $f_x(0,0)$, $f_u(0,0)$ denote the Jacobians of f taken with respect to x and u evaluated at $x = 0$, $u = 0$. If we define $A = f_x(0,0)$, $B = f_u(0,0)$ then (1.4) becomes

$$\dot{x}(t) = Ax + Bu + g(x,u). \tag{1.5}$$

We now use linear stabilization theory to assert the existence of a linear feedback control stabilizer for the nonlinear system (1.5). For example we may prove the following result.

Proposition 1.1. If A, B is a controllable pair i.e.

$$\text{rank } [B, AB, A^2B, \quad ,A^{n-1}B] = n,$$

then there exists a linear map, time invariant, $K: \mathbb{R}^n \to \mathbb{R}^m$, so that the closed loop system (1.3) has $x = 0$ exponentially asymptotically stable.

While the result is well known and the proof is more or less obvious I will provide one anyway. The reason being that the proof actually contains some subtle points that touch on the ability to extend Prop. 1.1 to distributed control systems.

Proof of Prop. 1.1. It is well known that if A, B is a controllable pair there exists a linear map $K: \mathbb{R}^n \to \mathbb{R}^m$ so that the linear map

$$C = A + BK \tag{1.6}$$

has all its eigenvalues strictly in the left half complex plane, i.e. Re[spectrum (C)] $\leq -\delta < 0$. Furthermore we know we can solve the Lyapunov matrix equation

$$PC^T + CP = -D \tag{1.7}$$

for P positive definite given any D positive definite. Hence the feedback control law $u = Kx$ means (1.3) can be written as

$$\dot{x}(t) = Cx + g(x,Kx) \tag{1.8}$$

where $g(x,Kx) = O(|x|^2)$. Furthermore the Lyapunov function

$$V(x) = x^T P x$$

is positive definite on $\mathbb{R}^n$ and possesses time derivative

$$\dot{V}(x) = -x^T D x + 2x^T P g(x, Kx). \tag{1.9}$$

Since $g(x,Kx) = O(|x|^2)$ we see $\dot{V}(x) \leq -$ const.$|x|^2$ if $|x|$ is sufficiently small for some const. > 0. This shows stability of $x = 0$. So we know that if $|x_0|$ is sufficiently small we stay in a region for which $\dot{V}(x) \leq -$ const. $|x|^2$. Examination of the ratio $\dot{V}(x)/V(x)$ now yields exponential asymptotic stability. □

Of course the same result could be obtained without recourse to Lyapunov functions. Namely use the variation of constants formula on (1.8) and then apply Gronwall's inequality (see e.g. [9]). In either case the underlying feature of the argument is the same: the linear decay of $|x(t)|^2$ in a neighborhood of $x = 0$ dominates the possibility of superlinear growth. This is the essence of (1.9).

Now that we know the basic theme it is a trivial business to extend this idea to some infinite dimensional control systems. I will spare the reader this excersize and point out the defects of the more obvious generalizations, at least with respect to problems motivated by nonlinear elastic systems. This is done in the next section.

2. BOUNDARY STABILIZATION IN ONE-DIMENSIONAL NONLINEAR ELASTICITY

Consider the one-dimensional elongation of an elastic body which is originally in a reference configuration shown in Figure 1a, i.e. one end is at X = 0, the other end at X = 1, where X denotes a generic point on the body.

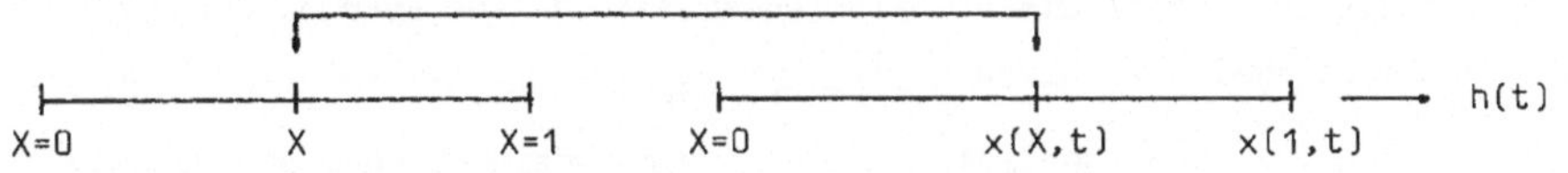

Fig. 1a Fig. 1b.

If we keep the left hand end point fixed and place a force h(t) on the right hand end our generic point X will move to a new position x(X,t). This is illustrates in Fig. 1b.

If $\rho(X)$ denotes the density of the material in the undeformed configuration (Fig. 1a) and τ denotes the Piola-Kirchoff stress, the balance of linear momentum asserts

$$\rho(X) x_{tt}(X,t) = \tau_X, \quad 0 < X < 1. \tag{2.1}$$

(For the uninitiated reader a discussion of these arguments may be found in any good book on nonlinear continuum mechanics, e.g. Gurtin [8], Truesdell & Wang [23]). To keep things simple we assume $\rho(X)$ is a constant, i.e. $\rho(X) = \rho_0 > 0$.

The boundary conditions are precisely as described above: the material point originally at X = 0 remains at X = 0, so

$$x(0,t) = 0; \tag{2.2}$$

the stress at the boundary point originally at X = 1 is prescribed to be h(t), so

$$\tau\big|_{X=1} = h(t). \tag{2.3}$$

Our system described by (2.1) - (2.3) is not yet closed since we still must specify τ. We do this by assuming our material is elastic, i.e.

$$\tau(X,t) = \hat{\sigma}(x_X(X,t)), \quad \hat{\sigma}' > 0.$$

Now if $\hat{\sigma}$ is a linear function of its argument we will be in the range of linear elasticity and (2.1) is just the one-dimensional wave equation. However no real material could ever be linearly elastic for all values of the deformation gradient x_X. At best linear theory is an approximation to nonlinear theory, i.e. we assume

$$\left.\begin{array}{l} \hat{\sigma}(x_X) = \sigma_0 x_X + \text{higher order terms in } x_X, \\ \sigma_0 \text{ a positive constant.} \end{array}\right\}$$

We then repeat the usual incantation that if x_X is small the higher order terms are negligible and the linear theory is a valid approximation. However anyone with a modicum of training in partial differential equations can see such an argument presumes an a priori smallness bound on $|x_X|_{L^\infty([0,1]\times[0,\infty))}$. Typically such bounds should occur as the result of analysis, not as a hypothesis. Second and equally disconcerting one often finds analyses given of elastic motions where linear theory is used and x_X can be large. For example imagine controlling our one dimensional elastic material with boundary control h(t). Such a control when applied in either an open or closed loop form to a linearly elastic approximation may lead to motions with x_X large and hence through us out of the range where linear theory is applicable. For these reasons it makes sense to study stabilization of nonlinear elastic systems directly without a priori neglecting the nonlinearities.

With the above motivation let us return to our problem. Define $\sigma = \hat{\sigma}/\rho_0$ so that (2.1) becomes

$$x_{tt}(X,t) = \sigma(x_X(X,t))_X, \quad 0 < X < 1. \tag{2.4}$$

The boundary condition at $X = 0$ is still the same

$$x(0,t) = 0, \tag{2.5}$$

and the boundary condition at $X = 1$ is now

$$\sigma(x_X(1,t)) = h_a(t) \tag{2.6}$$

where $h_a(t) = h(t)/\rho_o$. We prescribe initial conditions on displacement and velocity, namely

$$\begin{aligned} x(X,0) &= x^o(X), \\ x_t(X,0) &= x^1(X), \qquad 0 < X < 1. \end{aligned} \tag{2.7}$$

System (2.4) - (2.7) is a quasi-linear, hyperbolic, initial - boundary value problem. For (2.4), (2.7) and $-\infty < X < \infty$ (no boundary conditions) shocks (i.e. jumps in x_t, x_X) will generally occur in finite time no matter how smooth the initial data as long as σ is nonlinear [12]. One might conjecture that this is true for the uncontrolled boundary value problem (2.4), (2.5), (2.6), (2.7), $h_a \equiv 0$, as well. However Greenberg [6] has shown that this is not generally the case. He has shown that for certain choices of nonlinear σ the uncontrolled problem does possess global smooth solutions with the energy

$$E = \int_0^1 [\frac{x_t^2}{2} + \int_0^{x_X} (\sigma(\xi) - \sigma(0))d\xi]dX$$

conserved. This situation is reminiscent of the motion of the linearized uncontrolled problem

$$y_{tt} = \sigma_o\, y_{XX}, \qquad 0 < X < 1, \tag{2.8}$$

$$y(0,t) = 0, \tag{2.9}$$

$$y_X(1,t) = 0, \tag{2.10}$$

which admits the general solution

$$y(X,t) = \sum_{n=0}^{\infty} (A_n \sin \lambda_n t + B_n \cos \lambda_n t) \sin \lambda_n X \tag{2.11}$$

where $\lambda_n = \pi(n + 1/2)$, $n = 0,1,\ldots$. Here the motion is purely oscillatory and the absence of any damping mechanism makes decay to the rest state impossible.

The above situation motivates the need for implementing a stabilizing feedback control through $h_a(t)$ for (2.4) - (2.7). If we were to follow the ideas outlined in Section 1 we would write (2.4) - (2.6) as a nonlinear evolution equation on an infinite dimensional vector space, exponentially stabilize the linear system, and apply a Lyapunov or Gronwall argument to conclude exponential stability of the rest state for the nonlinear system. So formally one might set $x(X,t) = z_1(X,t)$, $x_t(X,t) = z_2(X,t)$ and write (2.4) as

$$\frac{d}{dt}\begin{pmatrix} z_1 \\ z_2 \end{pmatrix} = \underbrace{\begin{pmatrix} 0 & 1 \\ \sigma_0 \frac{d^2}{dx^2} & 0 \end{pmatrix}}_{A} \begin{pmatrix} z_1 \\ z_2 \end{pmatrix} + \underbrace{\begin{pmatrix} 0 & 0 \\ N & 0 \end{pmatrix}}_{J} \begin{pmatrix} z_1 \\ z_2 \end{pmatrix} \tag{2.12}$$

where $Nz_1 = \sigma(z_{1X})_X - \sigma_0 z_{1XX}$, $\sigma_0 = \sigma'(0)$. The boundary conditions could be incorporated into an appropriate definition of the domains of the operators appearing in (2.12). But this is not the point. The main issue is that the nonlinear operator J is from the functional analytic point of view as bad as A, i.e. they are both second order differential operators on z_1. This is no accident, it is just a fact of life in dealing with nonlinear elastic systems. No matter how one defines a state space Y (say a Banach space) for (2.12) one will not have both. A being the infinitesimal generator of a C_0-semigroup on Y and $J: Y \to Y$ continuous. This why I said that the obvious generalizations of Prop. 1.1 to infinite dimensions inevitably possess a defect, namely such generalizations tend to require some degree of continuity on J that system (2.12) just can't possess.

Does the above discussion mean our desire to find a stabilizing feedback control for (2.4) - (2.7) cannot be satisfied? On the contrary it does not. What it does mean is that we will have to work harder to get our result. The remarkable thing is that the underlying theme of the proof of Prop. 1.1 is still our main tool, i.e. exponential asymptotic decay for the linearized system should dominate the effect of nonlinearities if the initial data is sufficiently small. The next section is devoted to just this topic.

3. FEEDBACK STABILIZATION OF THE NONLINEAR SYSTEM

It has been known for the last few years that very mild damping mechanisms will actually prevent shock formation in quasi-linear wave equations provided the initial data is small and smooth. The first papers in this direction were those of Klainerman [10],[11] and Nishida [17], which has been followed by papers of Matsumura [16], Mac Camy [14],[15], Dafermos & Nohel [4] , [5] , Slemrod [22]. All these papers have the same underlying idea, i.e. a good decay estimate on the solutions of the linearized equation will carry over to the nonlinear case as well. Of course this is the very same idea used in the proof of Prop. 1.1. In this section we will carry over this approach to (2.4) - (2.7). Of course as mentioned earlier the idea used here has already been given by Greenberg & Tsien [7].

As a preliminary step we will rewrite (2.4) - (2.7) in terms of Riemann invariants [12],[3]. Define

$$\left.\begin{matrix} r(X,t) \\ \\ s(X,t) \end{matrix}\right\} = x_t \mp \int_0^{x_X} \sqrt{\sigma'(\xi)}\, d\xi,$$

$$\left.\begin{matrix} \lambda \\ \mu \end{matrix}\right\} = \pm \sqrt{\sigma'(x_X(X,t))} \,.$$

In this case (2.4) may be written as

$$r_t + \lambda r_X = 0$$

$$s_t + \mu s_X = 0 \,, \qquad 0 < X < 1. \tag{3.1}$$

Adding and subtracting r,s shows $x_t = \frac{1}{2}(r+s)$, $x_X = \phi(\frac{r-s}{2})$ for some smooth invertible function ϕ. Boundary conditions (2.5), (2.6) become

$$r = -s \quad \text{at} \quad X = 0, \tag{3.2}$$

and

$$\sigma(\phi(\frac{r-s}{2})) = h_a(t) \quad \text{at} \quad X = 1. \tag{3.3}$$

Since σ, ϕ are both globally invertible (3.3) can be written as

$$r-s = h_b(t) \quad \text{at} \quad X = 1 \tag{3.3'}$$

where $h_b(t) = 2\phi^{-1} \circ \sigma^{-1} \circ h_a(t)$.

We also need the notion of a characteristic curve. We say $x_1(t,\xi)$, $x_2(t,\eta)$ are characteristics if they satisfy the ordinary differential equations

$$\frac{dx_1}{dt}(t,\xi) = \lambda, \qquad x_1(0,\xi) = \xi,$$

$$\frac{dx_2}{dt}(t,\xi) = \mu, \qquad x_2(0,\eta) = \eta,$$

where x_1, x_2 are continuously differentiable in t. If we use the notation

$$' = \frac{\partial}{\partial t} + \lambda \frac{\partial}{\partial x} \,,$$

$$` = \frac{\partial}{\partial t} + \mu \frac{\partial}{\partial x} \,,$$

we see $'$ denotes (via the chain rule) differentiation with respect to t along the x_1 characteristic curve, $`$ denotes differention with respect to t along the x_2 characteristic curve. So (3.1) can be rewritten as

$$r' = 0, \; s` = 0, \quad 0 < X < 1. \tag{3.4}$$

Hence r is constant along x_1 characteristics and s is constant along x_2 characteristics. Of course if r and s remain constant they will never decay, hence we must introduce some dissipation at the boundary X = 1 via $h_b(t)$.

To find out one possible choice of $h_b(t)$ we resort to linearization. A little thought shows the linearized form of (2.4) - (2.7) written in Riemann invariant form is

$$\bar{r}' = 0, \quad \bar{s}` = 0, \quad 0 < X < 1, \tag{3.5}$$

$$\bar{r} = -\bar{s} \quad \text{at} \quad X = 0, \tag{3.6}$$

$$\bar{r} - \bar{s} = h_b(t) \quad \text{at} \quad X = 1, \tag{3.7}$$

where now

$$\left.\begin{matrix}\bar{r}\\ \bar{s}\end{matrix}\right\} = x_t \mp \sqrt{\sigma_o}\; x_X$$

and

$$\left.\begin{matrix}\lambda\\ \mu\end{matrix}\right\} = \pm\sqrt{\sigma_o}, \qquad \sigma_o = \sigma'(0).$$

In this case λ, μ are constants and the characteristics are actually straight lines.

Now let us return to our stabilization problem. How shall we choose $h_b(t)$? One method is to choose $h_b(t)$ so that the linearized system dissipated energy. To do this we formally differentiate the energy with respect to t, i.e.

$$\frac{d}{dt}\int_0^1[\bar{r}^2(X,t) + \bar{s}^2(X,t)]\,X = 2\int_0^1[\bar{r}\,\bar{r}_t + \bar{s}\,\bar{s}_t]dX = -\sqrt{\sigma_o}\,[\bar{r}^2 - \bar{s}^2]_{X=0}^{X=1} =$$

$$= -\sqrt{\sigma_o}\;(2\bar{s}(1,t)\;h_b(t) + h_b(t)^2).$$

So if we want the energy to dissipate we will need

$$2\bar{s}(1,t)\;h_b(t) + h_b(t)^2 \geq 0.$$

So if we set $h_b(t) = \beta\bar{s}(1,t)$ we will need

$$2\beta + \beta^2 > 0, \quad \text{i.e. } \beta > 0 \quad \text{or} \quad \beta < -2. \tag{3.8}$$

In this case we find

$$\frac{d}{dt}\int_0^1[\bar{r}^2(X,t) + \bar{s}^2(X,t)]dX = -\sqrt{\sigma_o}\;(2\beta + \beta^2)\;\bar{s}^2(1,t) \leq 0$$

so that the boundary condition at X = 1 actually dissipates energy. To prove the solutions of the linearized problem with $h_b(t) = \beta\bar{s}(1,t)$, $\beta > 0$ or $\beta < -2$, actually decay to zero one can use the method of characteristics (see Rauch & Taylor [19], Russell [20]), or seperation of variables (see Russell [21]). I won't repeat either of these arguments here. My point is that one choice of $h_b(t)$ motivated by linear analysis is to set

$$h_b(t) = \beta s(1,t) \tag{3.9}$$

for the nonlinear problem. Again we require $\beta > 0$ or $\beta < -2$. In this case the motion of the nonlinear system (2.4) - (2.7) is governed in Riemann invariant form by the system (3.1) , (3.2)

$$s = \left(\frac{1}{\beta+1}\right) r \quad \text{at } X = 1 \tag{3.10}$$

and initial condition

$$\begin{aligned} r(X,0) &= r^0(X) \\ & \qquad = x^1(X) \mp \int_0^{x_X^0} \sqrt{\sigma'(\xi)}\,d\xi, \qquad 0 < x < 1. \\ s(X,0) &= s^0(X) \end{aligned} \tag{3.11}$$

We will now use the method of characteristics to prove the following theorem.

<u>Theorem 3.1.</u> Consider system (3.1), (3.2), (3.10), (3.11) for $\beta < -2$ or $\beta > 0$. Assume r^0, s^0 are continuously differentiable on $[0,1]$ and satisfy the compatibility conditions $r^0(0) = -s^0(0)$, $r_X^0(0) = s_X^0(0)$, $r^0(1) = \left(\frac{1}{\beta+1}\right)s^0(1)$, $r_X^0(1) = -\left(\frac{1}{\beta+1}\right)s_X^0(1)$. If

$$M = \sup_{0\le X\le 1}\{|r^0(x)| + |s^0(X)| + |r_X^0(X)| + |s_X^0(X)|\}$$

is sufficiently small then the system has a unique classical solution r, s in $C^1([0,1]\times[0,\infty))$ and

$$\sup_{0\le X\le 1}\{|r(X,t)| + |s(X,t)| + |r_X(X,t)| + |s_X(X,t)|\} \le Ce^{-\alpha t} \tag{3.12}$$

for positive constants C,α which depend only on M.

<u>Proof.</u> The essense of the proof is illustrated by the following diagram in the X, t plane.

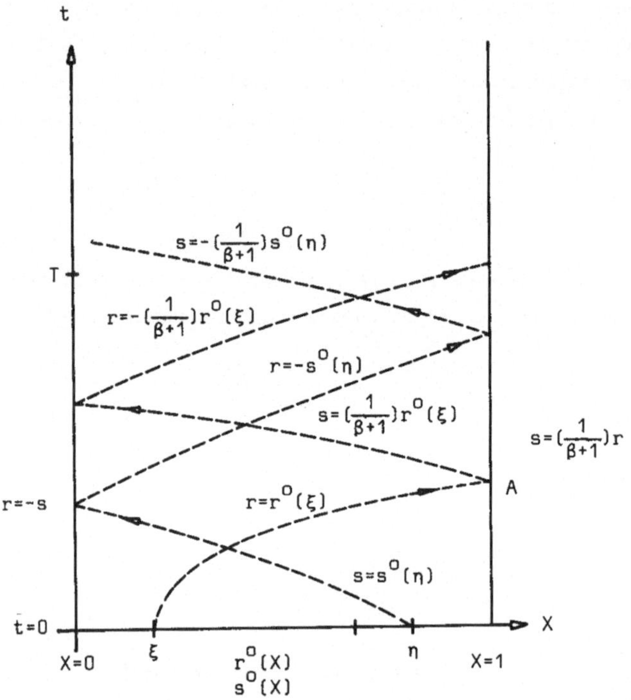

Fig. 2

Assume for the moment that a classical smooth solution exists to (3.1), (3.2), (3.10), (3.11). Then we know $r' = 0$, $s^{\cdot} = 0$ and r is constant along x_1 characteristics, s is constant along x_2 characteristics. Let us follow an x_1 characteristic emanating from a point $\xi \in [0,1]$. On this characteristic r is constant $(= r^o(\xi))$ and this determines a boundary value for r when the characteristic reaches X = 1 (the point A). The boundary condition $s = (\frac{1}{\beta+1})\, r$ at X = 1 determines the boundary value for s at X = 1. We then follow an x_2 characteristic out of A with the knowledge that s is constant along this characteristic $(= (\frac{1}{\beta+1})r^o(\xi))$. The rest is self explanatory from the picture. We see that r, s stay bounded and hence the slopes of the characteristics stay, respectively, positive and negative. So for all $T > T_o$, T_o sufficiently large (so that we can be sure of at least three reflections as shown in Fig. 2) we will have

$$|r(X,kT)| \le \left|\frac{1}{\beta+1}\right|^k \sup_{0\le X\le 1} |r^o(X)|,$$

$$|s(X,kT)| \le \left|\frac{1}{\beta+1}\right|^k \sup_{0\le X\le 1} |s^o(X)|.$$

We define $\alpha = \frac{1}{T} \log |\beta+1|$ and hence

$$\sup_{0\le X\le 1} (|r(X,kt)| + |s(X,kt)|) \le e^{-\alpha kT} \sup_{0\le X\le 1} (|r^0(X)| + |s^0(X)|)$$

and since T is any number sufficiently large we have

$$\sup_{0\le X\le 1} (|r(X,t)| + |s(X,t)|) \le e^{-\alpha t} \sup_{0\le X\le 1} (|r^0(X)| + |s^0(X)|)$$

for t sufficiently large. Thus we know

$$\sup_{0\le X\le 1} (|r(X,t)| + |s(X,t)|) \le Ce^{-\alpha t} \sup_{0\le X\le 1} (|r^0(X)| + |s^0(X)|)$$

for $t > 0$ where C, α depend on M.

Our next goal is to obtain <u>a priori</u> estimates on r_X, s_X. To do this we differentiate (3.1), (3,2) with respect to X. Consider (3.1) differentiated with respect to X, i.e. we have

$$r_{tX} + \lambda_r r_X^2 + \lambda_s r_X s_X + \lambda r_{XX} = 0$$

or

$$(r_X)' + \lambda_r r_X^2 + (\lambda_s s_X) r_X = 0. \tag{3.13}$$

Unfortunately (3.13) contains an unpleasant s_X term but recall

$$s' = s_t + \lambda s_X ,$$

$$s^{\backprime} = s_t - \lambda s_X ,$$

so $s' - s^{\backprime} = 2\lambda s_X$. However $s^{\backprime} = 0$ by (3.4) so $s' = 2\lambda s_X$ and (3.13) becomes

$$(r_X)' + \lambda_r r_X^2 + (\frac{\lambda_s s'}{2\lambda}) r_X = 0. \tag{3.14}$$

Now write

$$(\frac{1}{2} \log \lambda)' = \frac{1}{2} (\frac{\lambda'}{\lambda}) = \frac{\lambda_r r' + \lambda_s s'}{2\lambda} .$$

Since $r' = 0$ we see

$$(\frac{1}{2} \log \lambda)' = \frac{\lambda_s s'}{2\lambda}$$

and hence (3.14) can be written as

$$(r_X)' + \lambda_r r_X^2 + (\frac{1}{2} \log \lambda)' r_X = 0. \tag{3.15}$$

So $\lambda^{1/2}$ becomes an integrating factor for (3.15) and we see multiplication of (3.15) by $\lambda^{1/2}$ yields

$$(\lambda^{1/2} r_X)' + \lambda^{1/2} \lambda_r r_X^2 = 0. \tag{3.16}$$

A similar computation for s_X yields

$$(\lambda^{1/2} s_X)^{\backprime} + \lambda^{1/2} \mu_s s_X^2 = 0. \tag{3.17}$$

Now that we have evolution equations for r_X, s_X along the x_1, x_2 characteristics, respectively, we will now find boundary conditions for r_X, s_X. First differentiate (3.2) with respect to t and use (3.1). We find

$$r_X = s_X \quad \text{at} \quad X = 0. \tag{3.18}$$

Similarly we differentiate (3.10) with respect to t and use (3.1) to find

$$s_X = -\left(\frac{1}{\beta+1}\right) r_X \quad \text{at } X = 1. \tag{3.19}$$

Define $\lambda^{1/2} r_X = m$, $\lambda^{1/2} s_X = n$ so that (3.16) - (3.19) can be written as

$$m' + \lambda^{-1/2} \lambda_r m^2 = 0, \tag{3.20}$$

$$0 < X < 1,$$

$$n^{\backprime} + \lambda^{-1/2} \mu_s n^2 = 0, \tag{3.21}$$

$$m = n \quad \text{at } X = 0, \tag{3.22}$$

$$n = -\left(\frac{1}{\beta+1}\right)m \quad \text{at } X = 1. \tag{3.23}$$

Notice that the form of (3.20) - (3.23) is almost the same as (3.1), (3.2), (3.10) with r replaced by m and s replaced by n. The difference of course is the additional quadratic terms in (3.20), (3.21). Nevertheless we can still get estimates on m and n in a manner similar to what was used for r and s.

To get our estimates first set

$$p(X,t) = \lambda^{-1/2} \lambda_r m, \quad q(X,t) = \lambda^{-1/2} \mu_s n,$$

so that (3.20), (3.21) become

$$m' + pm = 0, \tag{3.24}$$

$$n^{\backprime} + qn = 0, \tag{3.25}$$

and hence along an x_1 characteristic passing through a point x_1^o at time t_o, $x_1(t, x_1^o, t_o)$, we have

$$m(x_1(t,x_1^o,t_o),t) = e^{-\int_{t_o}^{t} p(x_1(\tau,x_1^o,t_o),\tau)d\tau} m(x_1^o,t_o) \quad . \tag{3.26}$$

Similarly along an x_2 characteristic passing through a point x_2^o at time t_o, $x_2(t,x_2^o,t_o)$, we have

$$n(x_2(t,x_o^2,t_o),t) = e^{-\int_{t_o}^{t} q(x_2(\tau,x_2^o,t_o),\tau)d\tau}\, n(x_2^o,t_o). \tag{3.27}$$

So for smooth solution of (3.1), (3.2), (3.10), (3.11) we can compute the values of m, n at any point by using (3.26), (3.27) and (3.22), (3.23). This is illustrated in Fig. 3.

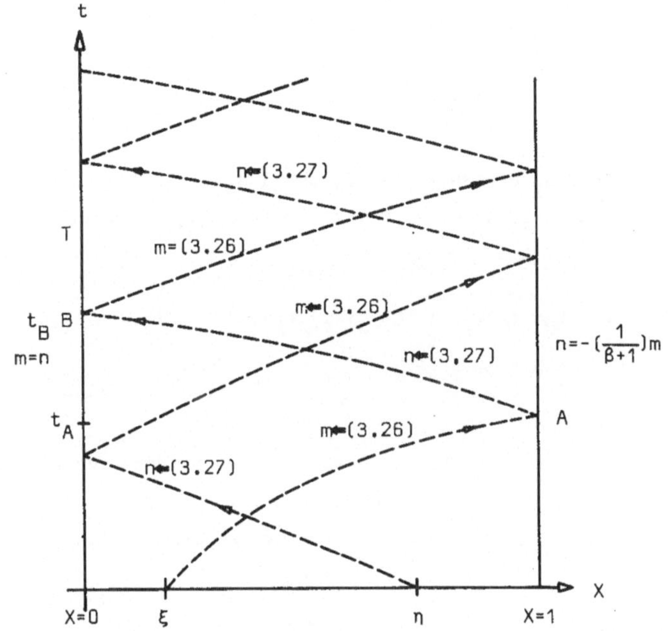

Fig. 3

For example the value of m at boundary point A is given by

$$m_A = e^{-\int_0^{t_A} p(x_1(\tau,\xi,0),\tau)d\tau}\, m^o(\xi)d\xi \, , \tag{3.28}$$

where

$$m^o = \lambda^{1/2}(r^o - s^o)r_X^o \, ,$$

and hence from the boundary condition at X = 1 the value of n at A is

$$n_A = -\left(\frac{1}{\beta+1}\right) e^{-\int_0^{t_A} p(x_1(\tau,\xi,0),\tau)d\tau}\, m^o(\xi). \tag{3.29}$$

To find the value of n at B we use (3.27) which gives

$$n_B = e^{-\int_{t_A}^{t_B} q(x_2(\tau;1,t_A),\tau)d\tau}\, n_A$$

and hence from (3.29) we see

$$n_B = -\left(\frac{1}{\beta+1}\right) \exp\left(- \int_{t_A}^{t_B} q(x_2(\tau,1,t_A),\tau)d\tau - \int_0^{t_A} p(x_1(\tau,\xi,0),\tau)d\tau\right)m^o(\xi). \quad (3.30)$$

From the boundary condition at X = 0 we know

$$m_B = n_B. \quad (3.31)$$

So for $T > t_B$, $0 < X < 1$, we find

$$m(X,T) = -\left(\frac{1}{\beta+1}\right) \exp\left(- \int_{t_B}^{T} p(x_1(\tau,0,t_B),\tau)d\tau - \int_{t_A}^{t_B} q(x_2(\tau,1,t_A),\tau)d\tau - \int_0^{t_A} p(x_1(\tau,\xi,0),\tau)d\tau\right)m^o(\xi). \quad (3.32)$$

Now let's assume r, s, r_X, s_X are sufficiently small at $t = 0$. In fact so small that p, q which depend continuously on r, s, r_X, s_X (note $p = q = 0$ when $r_X = s_X = 0$) satisfy

$$\exp\left(- \int_{t_B}^{T} p(x_1(\tau,0,t_B),\tau)d\tau - \int_{t_A}^{t_B} q(x_2(\tau,1,t_A),\tau)d\tau - \int_0^{t_A} p(x_1(\tau,\xi,0),\tau)d\tau\right) < |\beta + 1| . \quad (3.33)$$

(Recall $|\beta + 1| > 1$ by our restriction on β so this can be done for p, q small). The fact that there is a unique classical solution statisfying (3.32) can be shown using a standard local existence argument based on the contraction mapping principle. From (3.32), (3.33) we find

$$|m(X,T)| \le K|m^o(\xi)| , \qquad 0 < K < 1. \quad (3.34)$$

Hence proceeding as in our earlier estimate for r, s we find

$$\sup_{0\le X\le 1} (|m(X,t)| + |n(X,t)|) \le Ce^{-\alpha t}$$

where C, $\alpha > 0$ depend only on M.

Having established a priori bounds on m, n, r, s we can go back using the continuity of λ to conclude similar a priori decay estimates on r_X, s_X. But now a standard continuation argument (which I omit) will show these a priori estimates in fact show that the local solution obtained by the contraction mapping theorem is indeed a global, unique solution. □

I hope the reader will notice the main feature of the proof hidden in (3.33). Namely the ability to obtain (3.33) is precisely based on the fact that p, q can be kept small, at least for a long enough time, so that dissipative boundary conditions can take effect. So even though the argument of Prop. 1.1. doesn't go through to the nonlinear elastic problem in an obvious way the essential idea is still intact. Namely the linear decay (here at the boundaries) will dominate the potential for super linear growth (as reflected in non-zero p, q) if the initial data is sufficiently small and smooth.

We now state the following corollary which applies to our original system for $x(X,t)$.

Corollary 3.2. Consider system (2.4), (2.5), (2.6), (2.7) where $h_a(t)$ is given by the feedback law $h_a(t) = \sigma(\phi(\frac{\beta}{2}\, s(1,t)),\ \beta < -2$ or $\beta > 0$. Assume $x^o(X)$, $x^1(X)$ are in $C^2[0,1], C^1[0,1]$ respectively and satisfy compatibility conditions $x^o(0) = 0$, $x^1(0) = 0$, $x^o_{XX}(0) = 0$, $r^o(1) = (\frac{1}{\beta+1})s^o(1)$, $r^o_X(1) = -(\frac{1}{\beta+1})s^o_X(1)$. If

$$N = \sup_{0\le X\le 1}\{|x^o(X)| + |x^1(X)| + |x^o_X(X)| + |x^1_X(X)| + |x^o_{XX}(X)|\}$$

is sufficiently small then the system has a unique, classical smooth solution $x(t,X)$ in $C^2([0,1]\times[0,\infty))$ and

$$\sup_{0\le X\le 1}(|x(X,t)| + |x_t(X,t)| + |x_X(X,t)| + |x_{tX}(X,t)| + |x_{XX}(t,X)|) \le Ce^{-\alpha t}$$

the positive constants C,α depend only on N.

Proof of Corollary. Under the above assumptions Theorem 3.1 implies that the systems (2.4), (2.6), (2.7), $x_t = 0$ at $X = 0$, and this choice of $h_a(t)$, will have a unique (up to a constant) classical, smooth solution $\bar{x}(X,t)$ satisfying

$$\sup_{0\le X\le 1}(|\bar{x}_t(X,t)| + |\bar{x}_X(X,t)| + |\bar{x}_{tX}(t,X)| + |\bar{x}_{XX}(t,X)|) \le Ce^{-\alpha t}, \tag{3.35}$$

where C, α depend only on N. Pick any such solution. It must satisfy the boundary condition $\bar{x}(0,t) = c_o$ for a constant c_o. So if we set $x(X,t) = \bar{x}(X,t) - c_o$, $x(X,t)$ is uniquely defined and satisfies $x(0,t) = 0$. Furthermore

$$x(0,X) = \bar{x}(0,X) - c_o = \int_0^X x^o_X(X)dX = x^o(X)$$

so the initial condition (2.7a) is satisfied. (Of course (2.7b) is also satisfied). Finally since

$$x(t,X) = \int_0^X x_X(t,X)dX = \int_0^X \bar{x}_X(t,X)dX,$$

(3.35) yields the decay estimate advertised in the statement of the corollary. □

One final remark is perhaps in order. The methods used in the analysis were one space dimensional in nature and do not generalize to higher dimensions. For higher dimensional problems some variant of Matsamura's energy method [16],[4],[5],[22], may be in order. However such an approach will require good estimates on the solutions to the linearized problem which is itself a difficult issue except in special spatial domains (see Russell [21], Quinn and Russell [18], Chen [2]).

REFERENCES

[1] Balas, M.J.: Trends in Large Space Structure Control Theory, Fondest Hopes, Wildest Dreams, IEEE Transactions on Automatic Control, AC-27 (1982), 522-535.

[2] Chen, G.: Energy Decay Methods and Control Theory for the Wave Equation in a Bounded Domain, Ph.D. Thesis, Dept. of Mathematics, Univ. of Wisconsin, Madison (1977).

[3] Courant, R., K.O. Friedrichs: Supersonic Flow and Shock Waves, John Wiley, Intersciences, New York, 1948.

[4] Dafermos, C.M., J. Nohel: Energy Methods for Nonlinear Hyperbolic Volterra Integro-differential Equations, Comm. in Partial Differential Equations 4 (1979), 219-278.

[5] Dafermos, C.M., J. Hohel: A Nonlinear Hyperbolic Volterrra Equation in Viscoelasticity, American J. Mathematics, to appear.

[6] Greenberg, J.M.: Smooth and Time Periodic Solutions to the Quasi-linear Wave Equation, Archive for Rational Mechanics and Analysis, Vol. 60 (1975), 29-50.

[7] Greenberg, J.M., L.T. Tsien: The Effect of Boundary Damping for the Quasi-linear Wave Equation, submitted to J. Differential Equations.

[8] Gurtin, M.E.: An Introduction to Continuum Mechanics, Academic Press, New York, 1981.

[9] Hale, J.K.: Ordinary Differential Equations, Krieger Publish. Co., Huntington, N.Y., 1980.

[10] Klainerman, S.: Global Existence for Nonlinear Wave Equations, Comm. Pure and Applied Mathematics, Vol. XXXIII (1980), 43-101.

[11] Klainerman, S.: Classical Solutions to Nonlinear Wave Equations and Nonlinear Scattering, Proceedings I.S.I.M.M. Conference: Trends in Applications of Pure Mathematics in Mechanics, R.J. Knops, ed., Pitman, London, 1980.

[12] Lax, P.D.: Hyperbolic Systems of Conversation Laws, Conference Board of Mathematicians, Regional Conference Series in Applied Mathematics, SIAM Publications, Philadelphia, 1973.

[13] Lee, E.B., L. Markus: Foundations of Optimal Control Theroy, Wiley, New York, 1967.

[14] Mac Camy, R.C.: A Model for One-Dimensional, Nonlinear Viscoelasticity, Quarterly of Applied Mathematics 35 (1977), 21-33.

[15] Mac Camy, R.C.: An Integro-differential Equation with Applications in Heat Flow, Quarterly of Applied Mathematics 35 (1977), 1-20.

[16] Matsumura, A.: Global Existence and Aysmptotics of the Solutions of the Second-Order Quasilinear Hyperbolic Equation with First-Order Dissipation, Publications Research Institute for Mathematical Sciences, Kyoto University, Series A 13 (1977), 349-379.

[17] Nishida, T.: Global Smooth Solutions for the Second Order Quasilinear Wave Equation with First Order Dissipation, unpublished note (1975).

[18] Quinn, J.P., D.L. Russell: Asymptotic Stability and Energy Decay Rates for Solutions of Hyperbolic Equations with Boundary Damping, Proc. Royal Society of Edinburgh 77A(1977), 97-127.

[19] Rauch, J., M. Taylor: Exponential Decay of Solutions of Hyperbolic Equations in Bounded Domains, Indiana Univ. Math. J. 24 (1974), 79-86.

[20] Russell, D.L.: Controllability and Stabilizability Theory for Linear PDE's: Recent Progress and Open Questions, SIAM Review 20 (1978), 639-739.

[21] Russell, D.L.: Nonharmonic Fourier Series in the Control of Distributed Parameter Systems, J. Math. Analysis and Applications 18 (1967), 542-560.

[22] Slemrod, M.: Global Existence, Uniqueness, and Asymptotic Stability of Classical Smooth Solutions in One-dimensional Nonlinear Thermoelasticity, Archive for Rational Mechanics and Analysis 76 (1981), 97-134.

[23] Wang, C.C., C. Truesdell: Introduction to Rational Elasticity, Noordhoff, Leyden, (1973).

BOUNDARY FEEDBACK STABILIZATION PROBLEMS FOR HYPERBOLIC EQUATIONS *)

I. Lasiecka and R. Triggiani

Mathematics Department
University of Florida
Gainesville, Fl. 32611, USA

In line with the Workshop's expressed request, we intend to present here a brief summary of recent results in the area of boundary feedback stabilization for hyperbolic equations, along with some new results of more recent origin. We shall mainly consider three stabilization problems and draw our material mostly from our papers [2] - [5].

Let Ω be a bounded open domain in R^{ν} with boundary Γ assumed to be on $(\nu-1)$-dimensional variety with Ω locally on one side of Γ. For the most part, in particular for the second and third problem considered below, Γ may have finitely many conical points with Ω convex. In the first two problems, the feedback is an "interior observation" of the "position", which acts in the Dirichlet B.C., while in the third problem the feedback is a "boundary observation" of the velocity, which acts in the Neumann B.C.

Problem 1 (Almost periodic stabilization; no damping) [2]
We begin with a differential operator $A(\xi,\partial)$ which, along with homogeneous Dirichlet B.C., is realized as an operator A: $L_2(\Omega) \supset \mathcal{D}(A) \to L_2(\Omega)$. It is assumed throughout that A generates a strongly continuous (s.c) cosine operator $C(t)$ on $L_2(\Omega)$. We then consider the hyperbolic equation

$$\begin{aligned} &x_{tt}(t,\xi) = -A(\xi,\partial)x(t,\xi) && t > 0,\ \xi \in \Omega \\ &x(0,\xi) = x_0(\xi),\ x_t(0,\xi) = x_1(\xi) && \xi \in \Omega \\ &x(t,\xi) = f(t,\zeta) && t > 0,\ \zeta \in \Gamma . \end{aligned} \tag{P1.1}$$

Here, we study the case where $f(t,\zeta)$ is realized as a bounded, finite rank operator, acting only on the position x in the interior (no damping) of the form

$$f(t,\zeta) = \sum_{j=1}^{J} \langle x(t,.),w_j(.)\rangle g_j(\zeta) \tag{P1.2}$$

*) Paper presented at the Workshop by the second named author.

where here and here after <,> denotes the $L_2(\Omega)$-inner product. We first need examine the well posedness of the "closed loop system" (P.1) - (P1.2).

Theorem 1.1. [2],[3]. Let $w_j \in \mathcal{D}(cI-A)^{1/4+\rho}$, $\rho > 0$ for some c for which the factional powers are well defined. Then, the feedback closed loop solutions $x(t,x_o,x_1)$ of (P1.1) - (P1.2) can be expressed simply as

$$x(t,x_o,x_1) = C_F(t)x_o + S_F(t)x_1, \quad x_o \in L_2(\Omega), \quad x_1 \in H^{-1}(\Omega), \ t \in \mathbb{R}$$

where $C_F(t)$ defines a s.c. (feedback) cosine operator on $L_2(\Omega)$ and $S_F(t)$ is the corresponding sine operator. Actually, C_F extends/restricts as a s.c. cosine operator on each fixed interpolation space between $[\mathcal{D}(cI-A)^{3/4+\rho}]'$ and $\mathcal{D}(A^{1/4-\rho})$□.

With the wellposedness question settled, we now turn to a problem which may be viewed as being part of the general area of stabilization. We then assume that the operator A be selfadjoint and unstable, in the sense that its eigenvalues $\{-\lambda_k\}$ satisfy

$$\ldots < -\lambda_K < 0 < -\lambda_{K-1} < \ldots < -\lambda_2 < -\lambda_1 \qquad \text{(P1.3)}$$

and are all simple (multiplicity one). Let $\{\Phi_k\}$ denote the corresponding orthonormal basis of eigenvectors in $L_2(\Omega)$. Thus, the free system $(f(t,\zeta) \equiv 0)$ has the eigen-solutions for $1 \le k \le K-1$ that blow up exponentially in time. We then pose the problem: can we select general classes of vectors $w_j \in L_2(\Omega)$, $g_j \in L_2(\Omega)$ for $j = 1,2,\ldots J_{minimum}$ which will restore the typical oscillatory behavior of all solutions of the closed loop system (P1.1) - (P1.2)? An answer in spectral terms of the feedback generator A_F corresponding to $C_F(t)$ of Theorem 1.1 is given by

Theorem 1.2. [2]. Let $\nu = \dim \Omega \ge 2$ and let Ω either have C^∞-boundary Γ or else be a parallelopiped. Let A be selfadjoint with simple eigenvalues satisfying (P1.3). Let the vectors $w_j \in L_2(\Omega)$ satisfy the following algebraic conditions at the unstable eigenvalues

$$\text{rank } W = l_w$$

with

$$K-1 \le l_T + l_w - 1$$

where l_T is the number of linearly independent Neumann traces $[\partial\Phi_k/\partial\eta]_\Gamma$, $k = 1,\ldots,K-1$, and $W = [W_1,W_2,\ldots,W_{K-1}]$ with

$$W_k = [\langle w_1,\Phi_k\rangle, \langle w_2,\Phi_k\rangle, \ldots \langle w_J,\Phi_k\rangle] .$$

Finally, let the vectors w_j satisfy the growth condition:

$$0 \neq \langle w_j, \Phi_k \rangle \leq \text{const}/k^{1+2/\nu(V4+\rho)} \qquad k = 1,2,\ldots; \quad j = 1,\ldots J,$$

(which implies $w_j \in \mathcal{D}(cI-A)^{V4+\rho)}$ consistently with Theorem 1.1).

Then, if such w_j are suitably small, there exist boundary vectors $g_j \in L_2(\Gamma)$ (whose minimum number can also be specified) such that the feedback cosine generator A_F guaranteed by Theorem 1.1 has all (real) negative eigenvalues, denoted by

$$\{-c_i^2\}_{i=1}^{K-1} \quad \text{and} \quad \{-\alpha_r^2\}_{r=K}^{\infty}$$

with corresponding (normalized) eigenvectors

$$\{e_{F,i}\}_{i=1}^{K-1} \quad \text{and} \quad \{e_{F,r}\}_{r=K}^{\infty}$$

which form a Schauder basis (non orthogonal as the w_j and g_j are not all zero). Thus, if $x \in L_2(\Omega)$, then

$$x = \sum_{i=1}^{K-1} \eta_i(x) e_{F,i} + \sum_{r=K}^{\infty} \eta_r(x) e_{F,r}$$

$$A_F x = \sum_{i=1}^{K-1} - c_i^2 \eta_i(x) e_{F,i} + \sum_{r=k}^{\infty} - \alpha_r^2 \eta_r(x) e_{F,r}, \quad x \in \mathcal{D}(A_F)$$

where the bounded linear functionals η_k and the eigenvectors $e_{F,k}$ are biorthogonal. Also

$$C_F(t)x = \sum_{i=1}^{K-1} \eta_i(x) \cos c_i t \; e_{F,i} + \sum_{r=K}^{\infty} \eta_r(x) \cos \alpha_r t \; e_{F,r} \; \square .$$

The spectral interpretation of the above result is that the suitable vectors w_j and g_j claimed in Theorem 1.2 have the effect of (i) replacing the unstable (positive) original eigenvalues $\{-\lambda_i\}$, $1 \leq i \leq K-1$ of A with the stable ones $\{-c_i^2\}$, and, in so doing, (ii) the original stable (negative) eigenvalues $\{-\lambda_k\}$, $K \leq k < \infty$ of A are perturbed into new ones, which however are still negative (stable) and are given by $\{-\alpha_k^2\}$, $K \leq k$. The proof shows also that c_i^2's, the α_k^2's and the λ_k's, $k \geq K$, must all be distinct (to avoid "resonance phenomena").

The proof of Theorem 1.2 is based upon a functional analytic input-solution model to describe (P1.1), which will be introduced in various forms in the third problem.

In the next problem damping will be introduced.

Problem 2. (Damping observed in the interior and acting in the Dirichlet B.C) [4]. We now analyze the open loop system

$$x_{tt}(t,\xi) = \Delta x(t,\xi) \qquad t > 0,\ \xi \in \Omega$$

$$x(0,\xi) = x_0(\xi),\ x_t(0,\xi) = x_1(\xi) \qquad \xi \in \Omega \tag{P2.1}$$

$$x(t,\zeta) = f(t,\zeta) \qquad t > 0,\ \zeta \in \Gamma$$

which becomes "closed loop system" under the feedback

$$f(t,\zeta) = \langle x_t(t,.),w(.)\rangle g(\zeta) \tag{P2.2}$$

$\langle,\rangle = L_2(\Omega)$-inner product, $w \in L_2(\Omega)$, $g \in L_2(\Gamma)$. Thus, the unstable differential operator of Problem 1 (canonically, the Laplacian Δ translated to the right) is replaced here by Δ. The free system ($f(t,\zeta) \equiv 0$), written as a first order system, defines a unitary s.c. group on the space $E = H_0^1(\Omega) \times L_2(\Omega)$, $H_0^1(\Omega) = \mathcal{D}(A^{1/2})$ where A is the positive self-adjoint operator obtained from Δ by imposing homogeneous Dirichlet B.C. Our main objective is then "to stabilize" the (unitary group of the) free system, by means of the boundary feedback (P2.2). This means, this time, that we seek general classes of vectors $w \in L_2(\Omega)$, $g \in L_2(\Gamma)$ such that the resulting boundary feedback closed loop system (P2.1) - (P2.2) - written as a first order system - (be well-posed and) have all solutions which decay to zero as $t \to \infty$ in the strongest possible (Sobolev) norm.

On the negative side, one can prove that: stabilization in the uniform operator topology can never occur for any choice of the vectors w, g, whenever the feedback closed loop system defines a s.c. semigroup: this is the case, e.g. if $w \in \mathcal{D}(A^{3/4+\rho})$ $\rho > 0$ and $g \in L_2(\Gamma)$. This can be proved by extending the original argument of Russell [7] from a bounded compact perturbation to a perturbation which is unbounded but of finite rank (hence unclosable).

Nevertheless, "strong stabilization" is indeed possible for classes of vectors w, g. Hereafter in Problem 2, $\{\lambda_i\}$ are the positive eigenvalues of A and $\{\Phi_i\}$ are the corresponding (normalized) eigenvectors. Moreover, D is the "Dirichlet map", i.e. $v = Dh$, where $\Delta v = 0$ in Ω and $v|_\Gamma = h$ on Γ.

<u>Theorem 2.1.</u> (Strong stabilization) [4] . Assume that the vectors $w \in L_2(\Omega)$ and $g \in L_2(\Gamma)$ satisfy the following conditions

(i) $\langle w,\Phi_i\rangle \langle Dg,\Phi_i\rangle < 0$, $i = 1,2,\ldots$

equivalent to $\langle w,\Phi_i\rangle (g,\frac{\partial\Phi_i}{\partial\eta}|_\Gamma)_\Gamma < 0$, $i = 1,2,\ldots;$

(ii) $0 < c \le \left|\frac{\langle w,\Phi_i\rangle\lambda_i^{1/2+2\rho}}{\langle Dg,\Phi_i\rangle}\right| \le c$, $i = 1,2,\ldots$.

Then, the boundary feedback closed loop system (P2.1) - (P2.2) defines a s.c. (feedback) group $G_F(t)$ on the interpolation spaces

$$I_\theta \equiv \mathcal{D}(A^\alpha) \times \mathcal{D}(A^\beta),\ \alpha = 1/4 - \rho - \theta/2;\quad \beta = -1/4 - \rho - \theta/2,\quad 0 \le \theta \le 1,$$

where the convention $\mathcal{D}(A^{-s}) = [\mathcal{D}(A^s)]'$, $s > 0$ is used.

Moreover, $G_F(t)$ decays strongly to zero as $t \to \infty$ on each I_θ. The corresponding feedback generator A_F is still with compact resolvent with eigenvalues all in $\mathbb{C}^- = \{\lambda: \operatorname{Re}\lambda < 0\}$. □

It is worthwhile noting that in Theorem 2.1 the claimed stable feedback group need not be dissipative, particularly on the spaces of interest, i.e. the spaces I_θ with $0 \le \theta/2 \le 1/4 - \rho$ between $L_2(\Omega) \times H^{-1}(\Omega)$ (for $\theta/2 = 1/4 - \rho$) and $\mathcal{D}(A^{1/4-\rho}) \times [\mathcal{D}(A^{1/4+\rho})]'$. This facts is the contrast with much of the known literature on distributed or boundary feedback stabilization for hyperbolic equations, which correspond to dissipativity on the spaces of interest, see e.g. [2], [6],[8]. In our proof, the hypotheses (i) and (ii) guarantee that the image of the feedback generator A_F under a suitable similarity transformation becomes dissipative on the largest space $I_{\theta=1}$ of all with weakest topology. Restriction to smaller spaces with stronger topology need not preserve dissipativity. The special case $w = -k^2 A^{-1/2-2\rho} Dg$ corresponds to dissipativity of A_F on the largest space $I_{\theta=1}$.

Problem 3. (Damping observed on the boundary and acting in the Neumann B.C) [5]. Here we consider instead

$$\begin{aligned} &x_{tt}(t,\xi) = \Delta x(t,\xi) && t > 0,\ \xi \in \Omega \\ &x(0,\xi) = x_0(\xi),\ x_t(0,\xi) = x_1(\xi) && \xi \in \Omega \qquad\qquad \text{(P3.1)} \\ &\frac{\partial x}{\partial \eta}\Big|_\Gamma = f(t,\zeta) && t > 0,\ \zeta \in \Omega \end{aligned}$$

which becomes a boundary feedback closed loop system under the following choice of the feedback

$$f(t,\zeta) = (x_t|_\Gamma, w)_\Gamma g. \qquad \text{(P3.2)}$$

Now, not only $g \in L_2(\Gamma)$ as before, but $w \in L_2(\Gamma)$ as well. Let $L_0^2(\Omega) = L^2(\Omega)/N(A)$ be the $L^2(\Omega)$ space quotient the null space $N(A)$ of the operator A given by $-\Delta$ plus homogeneous Neumann B.C. Let $\{\lambda_n\}$ and $\{\Phi_n\}$ be the (positive) eigenvalues and corresponding (normalized) eigenvectors of the positive selfadjoint operator A on $L_0^2(\Omega)$. As before in Problem 2, the free system $(f(t,\zeta) \equiv 0)$, written as a first order system, defines a s.c. unitary group on the space $E = \mathcal{D}(A^{1/2}) \times L_0^2(\Omega)$, and our objective is "to stabilize" it. Than is, we seek classes of vectors w, $g \in L_2(\Gamma)$ such that the corresponding closed loop system (P3.1) - (P3.2) written as a first order system (be well posed and) have all its solutions which decay to zero as $t \to \infty$ in the strongest possible norm. As in Problem 2, stabilization in the uniform operator topology can be ruled out for all w, $g \in L_2(\Omega)$; and for essentially the same reason. Instead, we have a positive result regarding strong stabilization.

Theorem 3.1. (Strong stabilization) [5]. Assume that the vectors w, $g \in L_2(\Gamma)$ satisfy the two hypotheses:

(i) $(g,\Phi_n|_\Gamma)_\Gamma(\Phi_n|_\Gamma,w)_\Gamma < 0, \quad n = 1,2,\dots,$

(ii) $0 < c \leq \left| \frac{(g,\Phi_n|_\Gamma)_\Gamma}{(\Phi_n|_\Gamma,w)_\Gamma} \right| \leq C, \quad n = 1,2,\dots.$

Then, the corresponding feedback closed loop system (P3.1) - (P3.2), written as a first order system, defines a. s.c. semigroup on E which is strongly stable on E as $t \to \infty$ □.

Remarks. a) The case $g = -k^2w$ corresponds to the dissipative feedback system on E. However, the result includes also non-necessarily dissipative feedback semigroups on the desired space E, as is the case of Problem 2. Contrast with [1],[6],[8].
b) The proof of Theorem 3.1 (Problem 3) has some conceptual elements in common with that of Theorem 2.1 (Problem 2). Yet, the two proofs have also technical and conceptual differences; for instance, in Problem 3 the "perturbation due to the trace term on the boundary is not relatively bounded" any more!

Proof of Theorem 3.1. (sketch) To study (P3.1) - (P3.2), we employ the integral model on $L^2_0(\Omega)$, see [5] for details:

$$x(t) = C(t)x_0 + S(t)x_1 + \int_0^t A^{1/4+\rho}S(t-\tau)A^{3/4-\rho}Ng(\dot{x}(\tau)|_\Gamma,w)d\tau$$

where C(t) is the cosine operator generated by A, with sine operator S(t), and N is the Neumann map (solution of elliptic problem on $L^2_0(\Omega)$); or the corresponding second order differential equation

$$\ddot{x} = -\tilde{A}x + \tilde{A}^{1/4+\rho}A^{3/4-\rho}Ng(\dot{x}|_\Gamma,w)$$

where $\ddot{x} \in [\mathcal{D}(A^{1/4+\rho})]'$, $A^{3/4-\rho}Ng \in L^2_0(\Omega)$ and $\tilde{A}$ is the isomorphic extension of A from $\mathcal{D}(A^{3/4-\rho})$ onto $[\mathcal{D}(A^{1/4+\rho})]$; with associated first order system in $y = [y_1,y_2]$, $y_1 = x$ and $y_2 = \dot{y}_1$:

$$\dot{y} = (\tilde{A} + \tilde{P})y \; ; \quad \tilde{A}_F = \tilde{A} + \tilde{P}$$

on $Y = Y_1 \times Y_2$, $Y_1 \equiv \mathcal{D}(A^{1/4-\rho})$; $Y_2 \equiv [\mathcal{D}(A^{1/4+\rho})]'$;

$$\tilde{A} = \begin{vmatrix} 0 & J \\ -\tilde{A} & 0 \end{vmatrix} \; ; \quad \tilde{P} = \begin{vmatrix} 0 & 0 \\ 0 & \tilde{P} \end{vmatrix} \; ;$$

$$\tilde{P}y_2 = \tilde{A}^{1/4+\rho}A^{3/4-\rho}Ng(y_2|_\Gamma, w)$$

Step 1. Under assumption $(g,\Phi_n|_\Gamma)_\Gamma(\Phi_n|_\Gamma,w)_\Gamma \leq 0$, $n = 1,2,3\ldots$ it follows that $\sigma(\tilde{A}_F)$ is contained in $\{\lambda\colon \mathrm{Re}\,\lambda < 0\} \cup \{\pm i\mu_n\}$, and $\lambda I-\tilde{A}_F$ is onto E for $\mathrm{Re}\,\lambda > 0$.

Here $\mu_n^2 = \lambda_n$ = eigenvalues of A. This follows from

$$R(\lambda,\tilde{A}_F) = [I - R(\lambda,\tilde{A})\tilde{P}]^{-1}R(\lambda,\tilde{A})$$

where

$$R(\lambda,\tilde{A})\tilde{P}y = \begin{vmatrix} 0 & R(\lambda^2,-\tilde{A})\tilde{P} \\ 0 & \lambda R(\lambda^2,-\tilde{A})\tilde{P} \end{vmatrix} y \quad .$$

Step 2. Under hypotheses (i) - (ii) of Theorem 3.1, the generator $\tilde{A}_F$ is transformed by a similarity map into a generator $\tilde{\mathfrak{a}}_F$ which is dissipative in the E-topology; moreover $\sigma(\tilde{A}_F) = \sigma(\tilde{\mathfrak{a}}_F) \subset \{\lambda\colon \mathrm{Re}\,\lambda < 0\}$.

Introduce the multiplication operator K

$$Kx = \sum_{n=1}^{\infty} \sqrt{|g_n/w_n|}\, \langle x,\Phi_n\rangle\, \Phi_n; \quad g_n = (g,\Phi_n|_\Gamma), \quad w_n = (\Phi_n|_\Gamma,w)$$

bounded on $L_0^2(\Omega)$ with bounded inverse

$$K^{-1}z = \sum_{n=1}^{\infty} \sqrt{|w_n/g_n|}\, \langle z,\Phi_n\rangle\, \Phi_n$$

and the operators $\mathcal{K}$ and $\mathcal{K}^{-1}$ on $L_0^2(\Omega) \times L_0^2(\Omega)$:

$$\mathcal{K} = \begin{vmatrix} K & 0 \\ 0 & K \end{vmatrix} , \qquad \mathcal{K}^{-1} = \begin{vmatrix} K^{-1} & 0 \\ 0 & K^{-1} \end{vmatrix} \quad .$$

Then

$$\mathcal{K}^{-1}\mathcal{A}\mathcal{K} = \mathcal{A}$$

and it turns out that

$$\tilde{K}^{-1}\tilde{P}Ky^2 = -\langle y_2,\beta\rangle \quad \beta \in Y_2$$

where β is a suitable (explicitly known) vector in Y_2 and $\tilde{\mathfrak{a}}$ denotes extension. Thus

$$\langle\tilde{K}^{-1}\tilde{P}Ky_2,y_2\rangle = -|\langle y_2,\beta\rangle|^2, \quad y_2 \in \mathcal{D}(A^{1/4+\rho})$$

and $\tilde{K}^{-1}\tilde{P}K$ as an operator from its domain $\mathcal{D}(A^{1/4+\rho}) \to Y_2$ is dissipative with respect to the $L_0^2(\Omega)$ - topology (duality pairing on $\mathcal{D}(A^{1/4+\rho}) \times Y_2$). Likewise, the operator

$$\tilde{K}^{-1}\tilde{P}K = \begin{vmatrix} 0 & 0 \\ & \\ 0 & \tilde{K}^{-1}\tilde{P}K \end{vmatrix}$$

as an operator from its domain $Y_1 \times \mathcal{D}(A^{1/4+\rho}) \to Y$ is dissipative in the E-topology. Thus, if we set $\tilde{\mathfrak{a}}_F \equiv \tilde{K}^{-1}\tilde{A}_F\tilde{K}$ we see that

$$\tilde{\mathfrak{a}}_F = \tilde{A} + \tilde{K}^{-1}\tilde{P}\tilde{K}$$

and thus $\tilde{\mathfrak{a}}_F$ with domain $\mathcal{D}(A^{3/4-\rho}) \times \mathcal{D}(A^{1/4+\rho}) \to \mathcal{D}(A^{1/4+\rho}) \times Y_2 \subset Y$ is dissipative in the E-topology. Thus, we similarity map leaves $\tilde{A}$ invariant and transforms the perturbation $\tilde{P}$ into an E-dissipative perturbation.

<u>Step 3.</u> Application of Nagy-Foias-Fogel theory of decomposition to the E-dissipative generator $\tilde{\mathfrak{a}}_F$, combined with Stone's theorem and the result of step 1.

Details are to be found in [5] .

REFERENCES

[1] Chen, G.: A note on the boundary stabilization of the move equation, SIAM J. Control & Opt. 19 (1981), 106-113.

[2] Lasiecka, I., R. Triggiani: "Hyperbolic equations with Dirichlet boundary feedback via position vector: regularity and almost periodic stabilization", Part I, II, III; Applied Math. & Optimiz. 8, 1-37(1981); 8, 103-130(1982); 8, 199-221 (1982).

[3] Lasiecka, I., R. Triggiani: "Feedback semigroups and cosine operators for boundary feedback parabolic and hyperbolic equations", J.Diff. Eqs. Dec. 1982.

[4] Lasiecka, I., R. Triggiani: "Dirichlet boundary stabilization of the move equation with damping feedback", J. Math. Anal. & Appl., to appear.

[5] Lasiecka, I., R. Triggiani: "Nondissipative boundary stabilization of hyperbolic equations with boundary observation", to appear.

[6] Quinn, J.P., D.L. Russell: Asymptotic stability and energy decay rates for solutions of hyperbolic equations with boundary damping, Proc. Royal Soc. Edinburgh 77A (1977), 97-127.

[7] Russell, D.L.: Decay rates for weakly damped systems in Hilbert space obtained via control-theoretic methods, J.Diff. Eqs. 19 (1975), 344-370.

[8] Slemrod, M.: Stabilization of boundary control systems, J. Diff. Eqs. 22 (1976), 402-415.

Lecture Notes in Control and Information Sciences

Edited by A. V. Balakrishnan and M. Thoma

Vol. 22: Optimization Techniques
Proceedings of the 9th IFIP Conference on Optimization Techniques,
Warsaw, September 4–8, 1979
Part 1
Edited by K. Iracki, K. Malanowski, S. Walukiewicz
XVI, 569 pages. 1980

Vol. 23: Optimization Techniques
Proceedings of the 9th IFIP Conference on Optimization Techniques,
Warsaw, September 4-8, 1979
Part 2
Edited by K. Iracki, K. Malanowski, S. Walukiewicz
XV, 621 pages. 1980

Vol. 24: Methods and Applications in Adaptive Control
Proceedings of an International Symposium
Bochum, 1980
Edited by H. Unbehauen
VI, 309 pages. 1980

Vol. 25: Stochastic Differential Systems – Filtering and Control
Proceedings of the IFIP-WG7/1 Working Conference
Vilnius, Lithuania, USSR, Aug. 28 – Sept. 2, 1978
Edited by B. Grigelionis
X, 362 pages. 1980

Vol. 26: D. L. Iglehart, G. S. Shedler
Regenerative Simulation of Response Times in Networks of Queues
XII, 204 pages. 1980

Vol. 27: D. H. Jacobson, D. H. Martin, M. Pachter, T. Geveci
Extensions of Linear-Quadratic Control Theory
XI, 288 pages. 1980

Vol. 28: Analysis and Optimization of Systems
Proceedings of the Fourth International Conference on Analysis and Optimization of Systems
Versailles, December 16–19, 1980
Edited by A. Bensoussan and J. L. Lions
XIV, 999 pages. 1980

Vol. 29: M. Vidyasagar,
Input-Output Analysis of Large-Scale Interconnected Systems – Decomposition, Well-Posedness and Stability
VI, 221 pages. 1981

Vol. 30: Optimization and Optimal Control
Proceedings of a Conference Held at Oberwolfach, March 16–22, 1980
Edited by A. Auslender, W. Oettli, and J. Stoer
VIII, 254 pages. 1981

Vol. 31: Berc Rustem
Projection Methods in Constrained Optimisation and Applications to Optimal Policy Decisions
XV, 315 pages. 1981

Vol. 32: Tsuyoshi Matsuo,
Realization Theory of Continuous-Time Dynamical Systems
VI, 329 pages, 1981

Vol. 33: Peter Dransfield
Hydraulic Control Systems – Design and Analysis of Their Dynamics
VII, 227 pages, 1981

Vol. 34: H.W. Knobloch
Higher Order Necessary Conditions in Optimal Control Theory
V, 173 pages, 1981

Vol. 35: Global Modelling
Proceedings of the IFIP-WG 7/1 Working Conference Dubrovnik, Yugoslavia, Sept. 1–5, 1980
Edited by S. Krčevinac
VIII, 232 pages, 1981

Vol. 36: Stochastic Differential Systems
Proceedings of the 3rd IFIP-WG 7/1 Working Conference
Visegrád, Hungary, Sept. 15–20, 1980
Edited by M. Arató, D. Vermes, A.V. Balakrishnan
VI, 238 pages, 1981

Vol. 37: Rüdiger Schmidt
Advances in Nonlinear Parameter Optimization
VI, 159 pages, 1982

Vol. 38: System Modeling and Optimization
Proceedings of the 10th IFIP Conference
New York City, USA, Aug. 31 – Sept. 4, 1981
Edited by R.F. Drenick and F. Kozin
XI, 894 pages. 1982

Vol. 39: Feedback Control of Linear and Nonlinear Systems
Proceedings of the Joint Workshop on Feedback and Synthesis of Linear and Nonlinear Systems
Bielefeld/Rom
XIII, 284 pages. 1982

Vol. 40: Y.S. Hung, A.G.J. MacFarlane
Multivariable Feedback: A Quasi-Classical Approach
X, 182 pages. 1982

Vol. 41: M. Gössel
Nonlinear Time-Discrete Systems – A General Approach by Nonlinear Superposition
VIII, 112 pages. 1982

Vol. 42: Advances in Filtering and Optimal Stochastic Control
Proceedings of the IFIP-WG 7/1 Working Conference
Cocoyoc, Mexico, February 1–6, 1982
VIII, 391 pages. 1982

Lecture Notes in Control and Information Sciences

Edited by A. V. Balakrishnan and M. Thoma

Vol. 43: Stochastic Differential Systems
Proceedings of the 2nd Bad Honnef Conference
of the SFB 72 of the DFG at the University of Bonn
June 28 – July 2, 1982
Edited by M. Kohlmann and N. Christopeit
XII, 377 pages. 1982.

Vol. 44: Analysis and Optimization of Systems
Proceedings of the Fifth International
Conference on Analysis and Optimization of Systems
Versailles. December 14–17, 1982
Edited by A. Bensoussan and J. L. Lions
XV, 987 pages, 1982

Vol. 45: M. Arató
Linear Stochastic Systems
with Constant Coefficients
A Statistical Approach
IX, 309 pages. 1982

Vol. 46: Time-Scale Modeling of Dynamic Networks
with Applications to Power Systems
Edited by J. H. Chow
X, 218 pages. 1982

Vol. 47: P. A. Ioannou, P. V. Kokotovic
Adaptive Systems with Reduced Models
V, 162 pages. 1983

Vol. 48: Yaakov Yavin
Feedback Strategies for Partially
Observable Stochastic Systems
VI, 233 pages, 1983

Vol. 49: Theory and Application of Random Fields
Proceedings of the IFIP-WG 7/1
Working Conference
held under the joint auspices of the
Indian Statistical Institute
Bangalore, India, January 1982
Edited by G. Kallianpur
VI. 290 pages. 1983

Vol. 50: M. Papageorgiou
Applications of Automatic Control Concepts
to Traffic Flow Modeling and Control
IX, 186 pages. 1983

Vol. 51: Z. Nahorski, H.F. Ravn, R.V.V. Vidal
Optimization of Discrete Time Systems
The Upper Boundary Approach
V, 137 pages 1983

Vol. 52: A. L. Dontchev
Perturbations, Approximations and Sensitivity Analysis
of Optimal Control Systems
IV, 158 pages. 1983

Vol. 53: Liu Chen Hui
General Decoupling Theory of Multivariable
Process Control Systems
XI, 474 pages. 1983

Vol. 54: Control Theory for Distributed
Parameter Systems and Applications
Edited by F. Kappel, K. Kunisch,
W. Schappacher
VII, 245 pages. 1983.